# HANDBOOK OF ELECTRONIC MATERIALS
Volume 8

# HANDBOOK OF ELECTRONIC MATERIALS

Compiled by:
ELECTRONIC PROPERTIES INFORMATION CENTER
*Hughes Aircraft Company*
*Culver City, California*

Sponsored by:
*U.S. DEFENSE SUPPLY AGENCY*
*Defense Electronics Supply Center*
*Dayton, Ohio*

*Volume 1:*
OPTICAL MATERIALS PROPERTIES, 1971

*Volume 2:*
III-V SEMICONDUCTING COMPOUNDS, 1971

*Volume 3:*
SILICON NITRIDE FOR MICROELECTRONIC APPLICATIONS, PART I:
PREPARATION AND PROPERTIES, 1971

*Volume 4:*
NIOBIUM ALLOYS AND COMPOUNDS, 1972

*Volume 5:*
GROUP IV SEMICONDUCTING COMPOUNDS, 1971

*Volume 6:*
SILICON NITRIDE FOR MICROELECTRONIC APPLICATIONS, PART II:
APPLICATIONS AND DEVICES, 1972

*Volume 7:*
III-V TERNARY SEMICONDUCTING COMPOUNDS—DATA TABLES, 1972

*Volume 8:*
LINEAR ELECTROOPTIC MODULATOR MATERIALS

*Volume 9:*
ELECTRONIC PROPERTIES OF COMPOSITE MATERIALS

HANDBOOK OF ELECTRONIC MATERIALS
Volume 8

# Linear Electrooptic Modular Materials

J.T. Milek and M. Neuberger
*Electronic Properties Information Center*
*Hughes Aircraft Company, Culver City, California*

IFI/PLENUM · NEW YORK-WASHINGTON-LONDON · 1972

This document has been approved for public release and sale; its distribution is unlimited. Sponsored by U.S. Defense Supply Agency, Defense Electronics Supply Center, Dayton, Ohio. Under Contract No. DSA 900-72-C-1182

*Library of Congress Catalog Card Number 76-147312*
ISBN 0-306-67108-5


227 West 17th Street, New York, N.Y. 10011

United Kingdom edition published by Plenum Press, London
A Division of Plenum Publishing Company, Ltd.
Davis House (4th Floor), 8 Scrubs Lane, Harlesden, NW10 6SE, London, England



Printed in the United States of America

# FOREWORD

This survey of 13 electrooptic materials includes both a review and compilation of all materials properties relevant to their use in linear (Pockels) electrooptic modulator applications. Information on actual electrooptic modulator design as well as applications for these materials, and data on materials exhibiting a quadratic (Kerr) electrooptic effect, are not included.

With these restrictions in mind, every attempt was made to be as comprehensive as possible by utilizing all available sources of literature: books, periodicals, reports, and vendor literature. The files of the Electronic Properties Information Center and full resources of the Hughes Aircraft Company Library were searched for pertinent data, and approximately 1000 articles were reviewed for this publication.

A brief Introduction to the survey is followed by a description of the Principles of Electrooptic Modulation, emphasizing the importance of crystal symmetry on the electrooptic properties of materials, and including the relationships between the electrooptic, piezooptic, elastooptic and piezoelectric effects in crystals.

The survey consists of 13 independent sections, each section covering the properties of one material: crystallographic, optical, electrooptic, photoelastic, piezoelectric, dielectric and thermal. References appearing in the text are listed at the conclusion of each section. Tables and Figures are numbered separately for each section.

TABLE OF CONTENTS

# INTRODUCTION

The modulation of light and electrooptics have had a long history. Linear electrooptic phenomena were discovered by Roentgen in quartz and thoroughly investigated in several crystals before the turn of the century by Pockels. However, prior to the advent of the laser in 1960, electrooptic materials were considered relatively insignificant and somewhat of a laboooratory curiosity. Since the laser and its growing applications in science and technology, electrooptic materials have played an important role in the development of laser communication systems, optical displays, photographic recording and computer memory devices.

There are two basic types of light modulators - temporal and spatial. Temporal modulators vary electro-magnetic characteristics such as phase, frequency, amplitude, etc. Spatial modulators vary the physical qualities of the light beam such as direction, intensity, beam width, etc. Of the temporal electrooptic modulators, two types of retardation can occur - linear retardation, called the Pockels effect, and quadratic, called the Kerr effect.

The Kerr effect, usually observed in liquids such as nitrobenzene, is due to the alignment of the molecules of the liquid in the direction of the field. Its characteristics generally make it undesirable for modulation use except for pulse-type modulators. A typical application of Kerr cells is as Q-switches for lasers.

The phenomenological principles of the Pockels effect are described in the following section. The fundamental nature of the effect is not well understood, and the search for suitable electrooptic materials has been guided mainly by practical considerations. The following materials properties and parameters (all considered in this survey) are important in electrooptic materials selection and modulator design:

- Electrooptic Coefficient (r). Relates the induced birefringence to the applied electric field.
- Refractive Index (n). Determines the speed of light in the crystal, the reflection losses at its surface, and the "figure of merit" ($n^3r$) for electrooptic modulation.
- Transmission. Determines the spectral region available for modulation.

- Relative Dielectric Constant ($\varepsilon$). Determines the capacitance of the crystal and the speed of electric fields in the crystal.

- Loss Tangent ( $\tan \delta$). Determines the electrical loss (modulation power) in the crystal.

- Electrical Resistivity ($\rho$). Related to space charge effects and heating of the crystal.

- Crystal Growth Techniques. Determines the laboratory and commercial availability of large crystals of good optical quality.

- Crystal Symmetry. Determines the non-zero electrooptic, piezoelectric and photoelastic coefficients and the applicability of the crystal to specific device configurations.

- Hardness. Determines the resistance of the crystal to strain during cutting and polishing.

- Solubility. Determines the applicability of the crystal to a moist ambient environment.

- Photoelastic and Piezoelectric Properties. Determines the frequency of undesirable piezoelectric resonances and relates the clamped to the unclamped electrooptic coefficients (see Principles of Electrooptic Modulation Section).

- Thermal Conductivity. Determines the deleterious birefringence due to thermal gradients in the crystal heated by electrical or optical sources.

Kaminow and Turner (see Appendix) have emphasized that each particular modulator application determines which material parameters are of greater importance, but often in contradicting fashion. For example, it is often found that r is proportional to $\varepsilon$, so that a large $\varepsilon$ is desirable. On the other hand, when $\varepsilon$ is large, the reactive power required to drive a lumped modulator is large. Additionally, large values of $\varepsilon$ generally imply a large disparity in n and $\sqrt{\varepsilon}$, so that broadband velocity matching becomes unefficient for a colinear traveling-wave type of modulator.

This survey has revealed that major interest has centered around the following electrooptic materials:

* Ammonium dihydrogen arsenate (ADA)
* Ammonium dihydrogen phosphate (ADP)
 Antimony-sulfur-iodine (SbSI)
* Barium sodium niobate ($Ba_2NaNb_5O_{15}$)
 Barium titanate ($BaTiO_3$)
* Bismuth germanium oxide ($Bi_{12}GeO_{20}$)
 Cadmium sulfide (CdS)
 Cadmium telluride (CdTe)
* Calcium pyroniobate ($Ca_2Nb_2O_7$)
* Copper chloride (CuCl)
 Gallium arsenide (GaAs)
 Gallium phosphide (GaP)
 Germanium (Ge)
 Hexamethylenetetramine (HMTA)
 Lead niobate ($PbNbO_3$)
 Lead tantalate ($PbTaO_3$)
* Lithium niobate ($LiNbO_3$)
* Lithium tantalate ($LiTaO_3$)
* Potassium dideuterium phosphate (KDDP)
* Potassium dihydrogen arsenate (KDA)
* Potassium dihydrogen phosphate (KDP)
 Potassium niobate ($KNbO_3$)
 Potassium tantalate ($KTaO_3$)
* Potassium tantalate niobate (KTN)
* Proustite ($Ag_3AsS_3$)
 Quartz
 Rubidium dihydrogen arsenate ($RbH_2AsO_4$)
 Rubidium dihydrogen phosphate ($RbH_2PO_4$)
 Selenium (Se)
 Sodium niobate ($NaNbO_3$)
 Strontium titanate ($SrTiO_3$)
 Zinc selenide (ZnSe)
 Zinc sulfide (ZnS)
 Zinc telluride (ZnTe)

Thirteen of these materials (*) are covered in a comprehensive manner in this survey. Electrooptic properties of several semiconductor materials not covered in the survey are given in Table 1 below. References to informative reviews of other electrooptic materials, as well as surveys of the principles of electrooptic modulator design and various modulator applications, are detailed in the Appendix.

TABLE 1. ELECTROOPTIC PROPERTIES OF SEMICONDUCTORS (Kaminow and Turner).

| | Symmetry | $r_{ij}$ ($10^{-12}$ m/V) | λ (microns) | $n_i$ | λ (microns) | $\varepsilon_i$ |
|---|---|---|---|---|---|---|
| ZnO | 6mm | (S) $r_{33} = 2.6$<br>(S) $r_{13} = 1.4$<br>$r_{33}/r_{13} < 0$ | .63<br>.63 | $n_3 = 2.123$<br>$n_2 = n_1 = 2.106$<br>$n_3 = 2.015$<br>$n_3 = n_1 = 1.999$ | .45<br>.45<br>.60<br>.60 | $\varepsilon = 8.15$ |
| ZnS | $\bar{4}3m$<br><br><br>6mm | (T) $r_{41} = 1.2$<br>2.0<br>2.1<br>(S) $r_{33} = 1.85$<br>(S) $r_{13} = .92$<br>$r_{33}/r_{13} < 0$ | .40<br>.546<br>.65<br>.63<br>.63 | $n_o = 2.471$<br>2.364<br>2.315<br>$n_3 = 2.709$<br>$n_2 = n_1 = 2.705$<br>$n_3 = 2.368$<br>$n_2 = n_1 = 2.363$ | .45<br>.60<br>.8<br>.36<br>.36<br>.60<br>.60 | (T) 16<br>(S) 12.5<br>8.3 |
| ZnSe | $\bar{4}3m$ | (T) $r_{41} = 2.0$ | .546 | $n_o = 2.66$ | .546 | 9.1<br>8.1 |
| ZnTe | $\bar{4}3m$ | (T) $r_{41} = 4.55$<br>3.95<br>(S) $r_{41} = 4.3$ | .59<br>.69<br>.63 | $n_o = 3.1$<br>2.91 | .57<br>.70 | 10.1 |
| CuCl | $\bar{4}3m$ | (T) $r_{41} = 5.6$ | .546 | $n_o = 1.99$ | .546 | (T) 10<br>(S) 8.3 |
| CuBr | $\bar{4}3m$ | (T) $r_{41} = .85$ | | $n_o = 2.16$<br>2.09 | .535<br>.656 | |
| GaP | $\bar{4}3m$ | (S) $r_{41} = .5$<br>(S) $r_{41} = 1.06$ | <br>.63 | $n_o = 3.4595$<br>3.315 | .54<br>.60 | 10<br>12 |
| GaAs | $\bar{4}3m$ | (T) $r_{41} = .27$ to 1.2<br><br>(S-T) $r_{41} = 1.3$ to 1.5<br>(S) $r_{41} = 1.2$<br>(T) $r_{41} = 1.6$ | 1 to 1.8<br><br>1 to 1.8<br>.9 to 1.08<br>3.39 & 10.6 | $n_o = 3.60$<br>3.50<br>3.42<br>3.30 | .90<br>1.02<br>1.25<br>5.0 | (T) 12.5<br>(S) 10.9<br>(S) 11.7 |
| CdS | 6mm | (T) $r_{51} = 3.7$<br>(T) $r_c = 4$<br>(S) $r_{33} = 2.4$<br>(S) $r_{13} = 1.1$<br>$r_{33}/r_{13} < 0$ | .589<br>.589<br>.63<br>.63 | $n_3 = 2.726$<br>$n_2 = n_1 = 2.743$<br>$n_2 = n_1 = 2.493$ | .515<br>.515<br>.60 | (T) $\varepsilon_1 = 10.6$<br>(T) $\varepsilon_3 = 7.8$<br>(S) $\varepsilon_1 = 8.0$<br>(S) $\varepsilon_3 = 7.7$ |

(T) = constant stress, (S) = constant strain.

# PRINCIPLES OF ELECTROOPTIC MODULATION

The electrooptic effect in crystals is essentially a change in the crystalline birefringence which occurs when a crystal is placed in an electric field. Application of a properly-directed electric field within the material gives rise to a perturbation of its refractive properties, characterized by two orthogonal directions, called the "fast" and "slow" axes. An optical beam, initially plane-polarized at 45° to these axes and directed normal to their plane, will split into two orthogonal components, travelling along the same path but at different velocities, determined by their respective indices of refraction ($n_{fast}$ and $n_{slow}$). The phase difference between these components as they leave the crystal will depend on the difference between the refractive indices $n_f$ and $n_s$ as well as on the length of the light path L through the material:

$$\Gamma = 2\pi L\Delta n/\lambda = 2\pi L(n_s - n_f)/\lambda,$$

where the phase retardation or optical phase shift $\Gamma$ is expressed in radians, $\lambda$ is the wavelength of light in vacuum and $\Delta n$ is the difference between the "fast" and "slow" refractive indices.

If now the emerging beam is passed through a linear polarizer whose preferred direction is perpendicular to that of the incident beam polarization, as shown in Figure 1, some of the elliptically-polarized beam will be transmitted. Its intensity I, relative to incident intensity $I_o$, is given by

$$I = I_o \sin^2(\Gamma/2),$$

assuming no losses in the material.

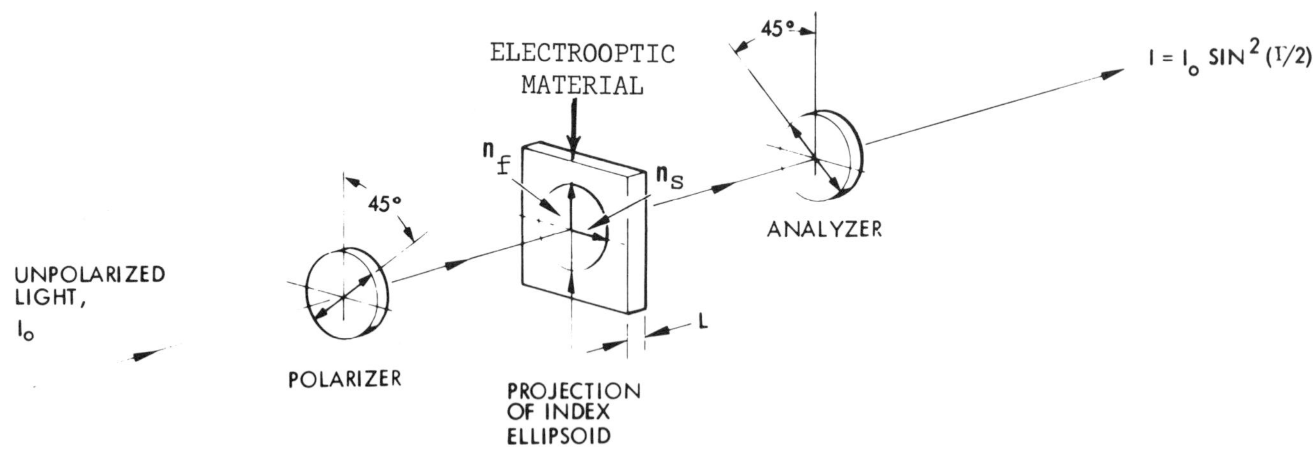


Fig. 1. Birefringent properties of an electro-optic modulator material in the presence of an electric field.

The indices of refraction $n_f$ and $n_s$ and, thus, the phase retardation Γ can be described in terms of the crystal's index ellipsoid or indicatrix, as shown in Figure 1. The dependence of Γ on the electric field can be described in terms of a change in orientation and dimensions of the index ellipsoid.

The index ellipsoid has the following important property: a plane passed through the center of the ellipsoid at right angles to the direction of propagation of an electromagnetic wave will cut the ellipsoid in an ellipse whose semi-axes are the two refractive indices associated with the two possible components of the wave, plane polarized along these semi-axes. The equation of the index ellipsoid is given by

$$x^2/n_x^2 + y^2/n_y^2 + z^2/n_z^2 = 1,$$

where $n_x$, $n_y$ and $n_z$ are the principle refractive indices. The equation of the index ellipsoid in a uniaxial crystal is given by (Figure 2)

$$(x^2 + y^2)/n_o^2 + z^2/n_e^2 = 1,$$

where $n_o$ is the ordinary and $n_e$ the extraordinary refractive index.

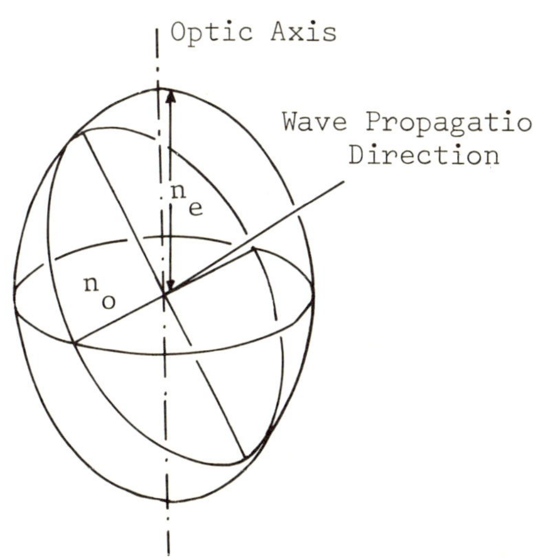


Fig. 2. The index ellipsoid for a (positive) uniaxial crystal.

The equation of the index ellipsoid, in the presence of an electric field and referred to the crystallographic axes, becomes

$$a_{11}x^2 + a_{22}y^2 + a_{33}z^2 + 2a_{23}yz + 2a_{31}zx + 2a_{12}xy = 1.$$

The constants $a_{ij}$, known as the polarization constants, can be written as

$$a_{11}-(1/n_x)^2 = r_{11}E_x + r_{12}E_y + r_{13}E_z$$

$$a_{22}-(1/n_y)^2 = r_{21}E_x + r_{22}E_y + r_{23}E_z$$

$$a_{33}-(1/n_z)^2 = r_{31}E_x + r_{32}E_y + r_{33}E_z$$

$$a_{23} = r_{41}E_x + r_{42}E_y + r_{43}E_x$$

$$a_{31} = r_{51}E_x + r_{52}E_y + r_{53}E_z$$

$$a_{12} = r_{61}E_x + r_{62}E_y + r_{63}E_z,$$

where the $r_{ij}$ are the electrooptic coefficients and $E_x$, $E_y$, $E_z$ are the components of the field along the three crystallographic axes.

The symmetry of the crystal determines which electrooptic coefficients are non-zero. The electrooptic matrix for the crystal classes considered in this survey are given in Table 1.

As an example, consider the uniaxial crystals of the type $XH_2PO_4$ exhibiting $\bar{4}2m$ ($D_{2d}$) symmetry. As shown in Table 1, the electrooptic properties of these crystals are described by only two independent coefficients: $r_{41} = r_{41}$ and $r_{63}$. The polarization constants reduce to

$$a_{11} = a_{22} = (1/n_o)^2$$

$$a_{33} = (1/n_e)^2$$

$$a_{23} = r_{41}E_x$$

$$a_{31} = r_{41}E_y$$

$$a_{12} = r_{63}E_z,$$

resulting in the following index ellipsoid:

$$z^2/n_e^2 + (x^2 + y^2)/n_o^2 + 2r_{41}\,(yzE_x + zxE_y) + 2r_{63}\,xyE_z = 1.$$

TABLE 1. ELECTROOPTIC MATRICES

Cubic $\bar{4}3m$ $(T_d)$ and 23 (T)

$$\begin{pmatrix} 0 & 0 & 0 \\ 0 & 0 & 0 \\ 0 & 0 & 0 \\ r_{41} & 0 & 0 \\ 0 & r_{41} & 0 \\ 0 & 0 & r_{41} \end{pmatrix}$$

Trigonal 3m $(C_{3v})$

$$\begin{pmatrix} 0 & -r_{22} & r_{13} \\ 0 & r_{22} & r_{13} \\ 0 & 0 & r_{33} \\ 0 & r_{42} & 0 \\ r_{42} & 0 & 0 \\ -r_{22} & 0 & 0 \end{pmatrix}$$

Tetragonal $\bar{4}2m$ $(D_{2d})$

$$\begin{pmatrix} 0 & 0 & 0 \\ 0 & 0 & 0 \\ 0 & 0 & 0 \\ r_{41} & 0 & 0 \\ 0 & r_{41} & 0 \\ 0 & 0 & r_{63} \end{pmatrix}$$

Tetragonal 4mm $(C_{4v})$

$$\begin{pmatrix} 0 & 0 & r_{13} \\ 0 & 0 & r_{13} \\ 0 & 0 & r_{33} \\ 0 & r_{42} & 0 \\ r_{42} & 0 & 0 \\ 0 & 0 & 0 \end{pmatrix}$$

Orthorhombic mm2 $(C_{2v})$

$$\begin{pmatrix} 0 & 0 & r_{13} \\ 0 & 0 & r_{23} \\ 0 & 0 & r_{33} \\ 0 & r_{42} & 0 \\ r_{51} & 0 & 0 \\ 0 & 0 & 0 \end{pmatrix}$$

Monoclinic 2 $(C_2)$

$$\begin{pmatrix} 0 & r_{12} & 0 \\ 0 & r_{22} & 0 \\ 0 & r_{32} & 0 \\ r_{41} & 0 & r_{43} \\ 0 & r_{52} & 0 \\ r_{61} & 0 & r_{63} \end{pmatrix}$$

Now consider the electrooptic modulator configuration in which an electric field is applied parallel to the crystal optic axis ($E_z \neq 0$; $E_x = E_y = 0$). The index ellipsoid becomes

$$z^2/n_e^2 + (x^2 + y^2)/n_o^2 + 2r_{63}\,E_z\,xy = 1.$$

This equation describes an ellipse with axes x' and y' making a 45 degree angle with the x and y crystallographic axes (Figure 3). That is, at zero voltage, the projection of the index ellipsoid on the xy plane is a circle with radius equal to $n_o$; as the voltage is increased, the circle elongates in the direction parallel to x'.

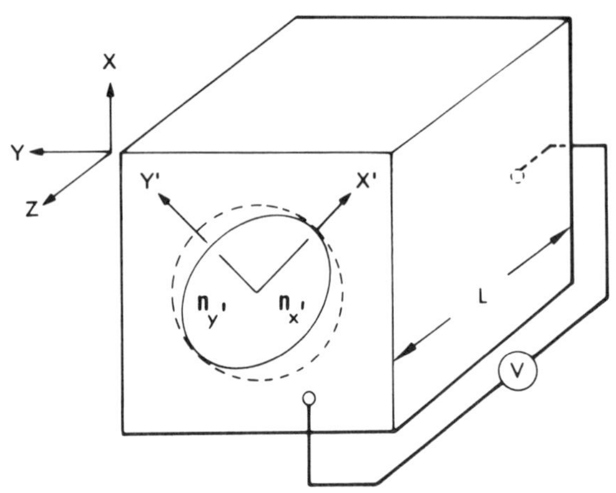


Fig. 3. Change in shape induced in x and y axes of index ellipsoid when an electric field is applied parallel to the crystal optic axes.

A beam of linearly polarized light propagating parallel to the optic axis (z-axis) and with its plane of polarization parallel to either the crystallographic x or y axes will experience an optical phase shift given by

$$\begin{aligned}\Gamma &= 2\pi L\Delta n/\lambda = 2\pi L(n_s - n_f)/\lambda = 2\pi L(n_{x'} - n_{y'})/\lambda \\ &= 2\pi L\,[(n_o + \tfrac{1}{2}\,n_o^3 r_{63}E) - (n_o - \tfrac{1}{2}\,n_o^3 r_{63}E)]/\lambda \\ &= 2\pi\,n_o^3 r_{63}(E\,.\,L)/\lambda \\ &= 2\pi\,n_o^3 r_{63}\,V/\lambda,\end{aligned}$$

where $V = E \cdot L$ is the applied voltage. Thus, in this configuration, the induced optical phase shift $\Gamma$ is independent of the interaction length L. The voltage required to produce a half wave of optical phase shift ($\Gamma = \pi$) is known as the half-wave voltage, $V_{\frac{1}{2}}$:

$$V_{\frac{1}{2}} = \lambda/2n_o^3\,r_{63}.$$

As another example, consider a crystal with 3m symmetry with the electric field in the z direction. The index ellipsoid reduces to

$$(n_o^{-2} + r_{13} E)\, x^2 + (n_o^{-2} + r_{13} E)\, y^2 + (n_e^{-2} + r_{33} E)\, z^2 = 1.$$

This equation describes an ellipse with axes corresponding to the crystallographic axes; the change in the birefringence due to the field is simply

$$n_s = n_{y'} = n_{x'} = n_o - \frac{1}{2} n_o^3 r_{13} E$$
$$n_f = n_{z'} = n_e - \frac{1}{2} n_e^3 r_{33} E$$

A beam of linearly polarized light propagating parallel to the y axis and with its plane of polarization making an angle of 45-degrees with the x and z axes will experience an optical phase shift of

$$\Gamma = 2\pi L \Delta n/\lambda = 2\pi L\,(n_{x'} - n_{z'})/\lambda$$
$$= \underbrace{2\pi L(n_e - n_o)/\lambda}_{\text{natural birefringence}} + \underbrace{\pi L\,(n_e^3 r_{33} - n_o^3 r_{13}) V/\lambda d}_{\text{induced birefringence}},$$

where V = E.d is the applied voltage, d is the thickness of the crystal in the field direction and, as before, L is the optical path length in the crystal.

The full effect of natural birefringence encountered here causes an undesirable optical bias which must be compensated for two reasons. First because the natural birefringence is directly a function of the temperature of the device (were it not for this strong temperature dependence, this bias could simply be bucked out by a fixed dc voltage), and second the emerging rays must be colinear on the same axis to properly recombine. Manufacturers have been able to compensate for this effect by utilizing a pair of crystals of identical length, maintained at a common uniform temperature, and placed along the path of the light beam such that the natural birefringence of one crystal accurately concels out that produced by the other. Thus, the problem of natural birefringence is soluble, but only after applying the expensive material and manufacturing tolerances required.

## PRINCIPLES OF ELECTROOPTIC MODULATION

In discussing the electrooptic effect, it is important to specify the mechanical constraints on the crystal; for example, the crystal might be free so that the stress is zero, or clamped, so that the strain is zero. This is important, because, if the crystal is free, a static electric field will cause a strain by the converse piezoelectric effect and this in turn produces a change in the refractive indices through the elastooptic effect. The primary or "true" electrooptic coefficient $r_{ij}^{S}$ may be measured using a clamped crystal (zero or constant strain) while the electrooptic coefficient $r_{ij}^{T}$ observed using a free crystal (zero or constant stress) is due to both the primary and secondary effects.

The secondary effect may be suppressed by clamping the crystal. Mechanical constraints may be used for clamping or, more often, the modulating frequency of the electric field is set much higher than the elastic resonance frequency of the crystal body. That is, at frequencies well above the fundamental elastic resonance and beginning typically at about 1 MHz if the overtone resonances are suppressed by the crystal's mount, mechanical deformation is excluded by the inertia of the crystal and the clamped effect is obtained.

The general relationship between the clamped ($r_{ij}^{S}$) and unclamped ($r_{ij}^{T}$) electrooptic coefficients is given in Table 2 and the specific relationships for the crystal classes considered in this survey are detailed in Table 3.

TABLE 2. GENERAL RELATIONSHIP BETWEEN CLAMPED AND UNCLAMPED ELECTROOPTIC COEFFICIENTS.

$$r_{ij}^{T} = r_{ij}^{S} + \sum_{k} p_{ik}\, d_{jk} \qquad i = 1 \text{ to } 6$$

$$p_{ik} = \sum_{\ell} \pi_{i\ell}\, c_{\ell k} \qquad j = 1 \text{ to } 3$$

$$k = 1 \text{ to } 6$$

$$\pi_{ik} = \sum_{\ell} p_{i\ell}\, s_{\ell k} \qquad \ell = 1 \text{ to } 6$$

| Symbol | Definition | Units, rationalized mks | conversion factor cgs-mks |
|---|---|---|---|
| $r_{ij}^{S}$ | "true", clamped or high frequency electrooptic coefficient | m/V | $\frac{1}{3} \cdot 10^{-4}$ m/V per cm/statvolt |
| $r_{ij}^{T}$ | unclamped, free or static (dc) electrooptic coefficient | | |
| $p_{ij}$ | elastooptic (strain-optic) coefficient (photostatic) | dimensionless | |
| $d_{ij}$ | piezoelectric strain constant | C/N (=m/V) | $\frac{1}{3} \cdot 10^{-4}$C/N per statcoulomb/dyne |
| $\pi_{ij}$ | piezooptic coefficient | $m^2/N$ | 10 $m^2/N$ per $cm^2$/dyne |
| $s_{ij}$ | elastic compliance constant | | |
| $c_{ij}$ | elastic stiffness constant | $N/m^2$ | $10^{-1}$ $N/m^2$ per dyne/$cm^2$ |

TABLE 3. SPECIFIC RELATIONSHIPS BETWEEN CLAMPED AND UNCLAMPED ELECTROOPTIC COEFFICIENTS.

| Symmetry | $r_{ij}^T - r_{ij}^S = \sum_k p_{ik} d_{jk}$ | |
|---|---|---|
| | ij | $\sum_k p_{ik} d_{jk}$ |
| Cubic, $\bar{4}3m$ ($T_d$) and 23 (T) | 41 | $p_{44}d_{14}$ |
| Trigonal, 3m ($C_{3v}$) | 22<br>13<br>33<br>42 | $(p_{11} - p_{12})\, d_{22} - p_{14}\, d_{15}$<br>$(p_{11} + p_{12})\, d_{31} + p_{13}\, d_{33}$<br>$2\, p_{31}\, d_{31} + p_{33}\, d_{33}$<br>$-2\, p_{41}\, d_{22} + p_{44}\, d_{15}$ |
| Tetragonal, $\bar{4}2m$ ($D_{2d}$) | 41<br>63 | $p_{44}d_{14}$<br>$p_{66}d_{36}$ |
| Tetragonal, 4mm ($C_{4v}$) | 42<br>13<br>33 | $p_{44}d_{15}$<br>$(p_{11} + p_{12})\, d_{31} + p_{13}\, d_{33}$<br>$2\, p_{31}\, d_{31} + p_{33}\, d_{33}$ |
| Orthorhombic, mm2 ($C_{2v}$) | 51<br>42<br>13<br>23<br>33 | $p_{55}d_{15}$<br>$p_{44}d_{24}$<br>$p_{11}\, d_{31} + p_{12}\, d_{32} + p_{13}\, d_{33}$<br>$p_{21}\, d_{31} + p_{22}\, d_{32} + p_{23}\, d_{33}$<br>$p_{31}\, d_{31} + p_{32}\, d_{32} + p_{33}\, d_{33}$ |
| Monoclinic, 2 ($C_2$) | 41<br>61<br>12<br>22<br>32<br>52<br>43<br>63 | $p_{44}\, d_{14} + p_{46}\, d_{16}$<br>$p_{64}\, d_{14} + p_{66}\, d_{16}$<br>$p_{11}\, d_{21} + p_{12}\, d_{22} + p_{13}\, d_{23} + p_{15}\, d_{25}$<br>$p_{21}\, d_{21} + p_{22}\, d_{22} + p_{23}\, d_{23} + p_{25}\, d_{25}$<br>$p_{31}\, d_{21} + p_{32}\, d_{22} + p_{33}\, d_{23} + p_{35}\, d_{25}$<br>$p_{51}\, d_{21} + p_{52}\, d_{22} + p_{53}\, d_{23} + p_{55}\, d_{25}$<br>$p_{44}d_{34} + p_{46}\, d_{36}$<br>$p_{64}d_{34} + p_{66}\, d_{36}$ |

# AMMONIUM DIHYDROGEN ARSENATE (ADA)

## Introduction

Ammonium dihydrogen arsenate (ADA) has been considered a potential electrooptic material and has been explored by a number of investigators for modulation purposes. The amount of available data and information is quite limited when compared to other electrooptic materials. Yap and Bicknell reported ADA to have high bandwidth performance characteristics.

## Chemical and Physical Properties

| | | |
|---|---|---|
| Chemical Formula | $NH_4H_2AsO_4$ | |
| Molecular Weight | 159.07 | |
| Density | 2.311 g/cm$^3$ | [Lynch,] |
| Solubility in Water | 33.74 g/100 ml cold $H_2O$ | [Pulvari et al.] |
| | 122.4 g/100 ml 90°C $H_2O$ | |

## Crystallography

Ammonium dihydrogen arsenate belongs to the non-centrosymmetrical point group $\bar{4}2m$ ($D_{2d}$) and crystallizes in the tetragonal system in the paraelectric phase at room temperature. It undergoes an antiferroelectric phase transition at 216.1 ± 0.5 °K [Stephenson and Adams]. Later work by Loiacono with a differential thermal analysis method, yielded values for the Curie temperature of 219.2 °K on the heating cycle and 218.6 °K on the cooling.

Cook has reviewed the literature on the lattice constants of ADA, as shown in Table 1.

TABLE 1. LATTICE CONSTANTS OF ADA

| $a_o$ (Å) | $c_o$ (Å) | T (°C) | Reference |
|---|---|---|---|
| 7.6944 ± 0.0004<br>7.6790 ± 0.0012 | 7.720 ± 0.002<br>7.715 ± 0.002 | 25<br>-57 | Cook |
| 7.694 ± 0.008 | 7.718 ± 0.008 | | Haussühl |
| 7.6996 ± 0.0009 | 7.7158 ± 0.0025 | 20 | Deshpande and Khan |
| 7.699 ± 0.005 | 7.729 ± 0.005 | | Delain |

## AMMONIUM DIHYDROGEN ARSENATE (ADA)

Optical Properties [Sliker]

TABLE 2. REFRACTIVE INDEX OF ADA

| $n_o$ | $n_e$ | λ (Å) |
|---|---|---|
| 1.5721 | 1.5186 | 6563 |
| 1.5766 | 1.5217 | 5893 |
| 1.5859 | 1.5296 | 4861 |

Region of >50% transmission: 0.260 to >0.75 microns (2 mm thickness)

Electrooptic Properties

Recently, Adhav has determined the following room temperature values for the half-wave voltage $V_{1/2}$ and the electrooptic constant $r_{63}$ for ADA at 5500 Å:

$$V_{1/2} = 7.20 \text{ kV}$$
$$r_{63} = 9.2 \times 10^{-12} \text{ m/V}.$$

These data differ considerably from the often-quoted value of $V_{1/2}$ = 13 kV reported in the 1957 AIP Handbook [e.g., Sliker].

Adhav's data indicate that $r_{63}$ varies with wavelength in a linear fashion from $9 \times 10^{-12}$ m/V at 4000 Å to $10 \times 10^{-12}$ m/V at 7000 Å. The variation of half-wave voltage with wavelength is shown in Figure 1.

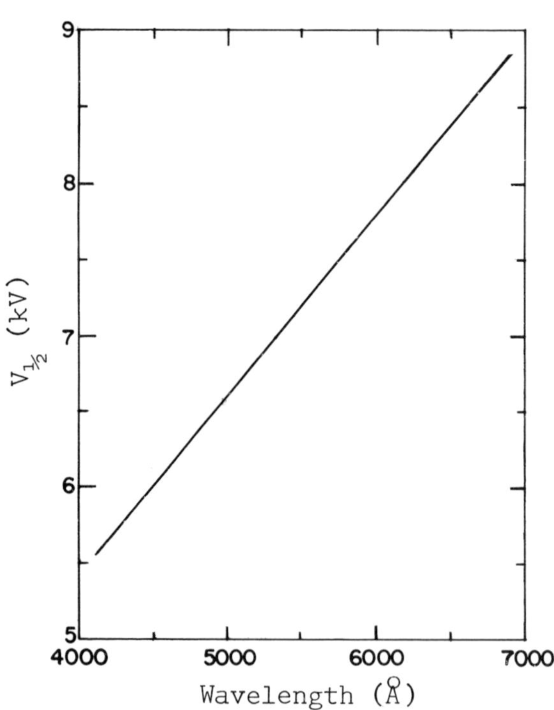


Fig. 1. Half-wave voltage of ADA as a function of wavelength [Adhav].

## Piezooptic Coefficients

$\pi^{E}_{66} = -8.3 \times 10^{-12}\ m^2/N$ [Landolt-Börnstein ]

## Elastic Constants

TABLE 3. ELASTIC COMPLIANCE CONSTANTS OF ADA

| | Elastic Constant at 25°C ($10^{-12}\ m^2/N$) | Temperature Coefficient of Elastic Constant (ppm/°C) | Reference |
|---|---|---|---|
| $s_{11}$ | 19.3 | 680 | Adhav |
| $s_{12}$ | 6.5 | 1610 | |
| $s_{13}$ | -14.1 | -290 | |
| $s_{33}$ | 48.5 | 40 | |
| $s_{44}$ | 146.0 | 220 | Adhav |
| | 136 | | Sliker |
| $s_{66}$ | 156.5 | 720 | Adhav |
| | 156 | | Sliker |

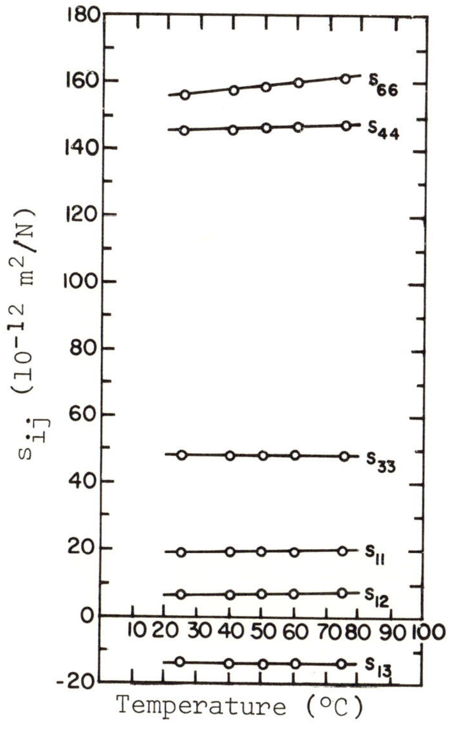


Fig. 2. Elastic compliance constants of ADA as a function of temperature [Adhav]

## Piezoelectric Properties

The following piezoelectric constants of ADA at 25°C have been reported by Mason:

$$d_{14} = +41 \times 10^{-12}\ \mathrm{C/N}$$

$$d_{36} = +31 \times 10^{-12}\ \mathrm{C/N}$$

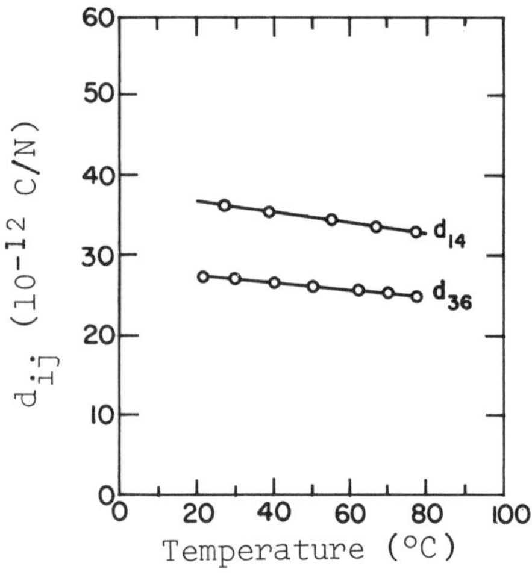


Fig. 3. Piezoelectric constants of ADA as a function of temperature [Adhav]

The following electromechanical coupling coefficients have been reported by Sliker:

$$k_{14} = 0.136$$

$$k_{36} = 0.24$$

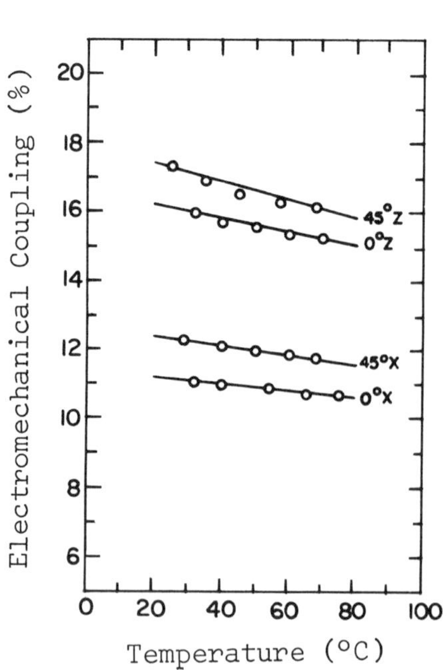


Fig. 4. Electromechanical coupling coefficients of ADA as a function of temperature for shear modes of vibration of 0°Z and 0°X plates and extensional modes of 45°Z and 45°X plates [Adhav]

## Dielectric Properties - Dielectric Constant

The relative dielectric constants of ADA for electric fields along the c-axis ($\varepsilon_c = \varepsilon_{33}/\varepsilon_o$) and for fields perpendicular to the c-axis ($\varepsilon_a = \varepsilon_{11}/\varepsilon_o$) are reported as follows:

$\varepsilon_c^T = 14$ [Mason]

$\varepsilon_a^T = 75$

$\varepsilon_c^S = 13$ [Sliker]

$\varepsilon_a^S = 74$

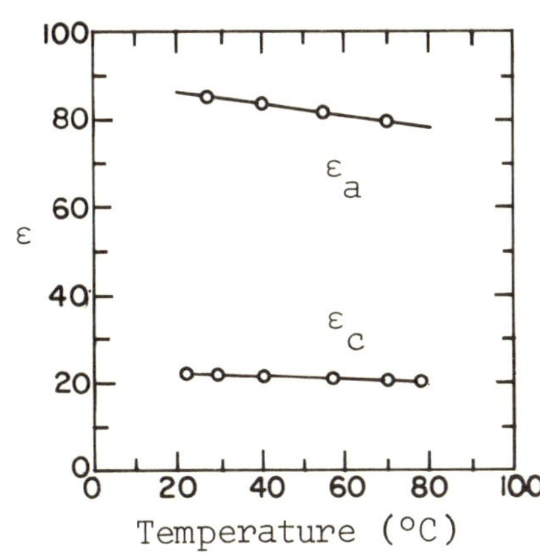


Fig. 5. Dielectric constant of ADA as a function of temperature at 1 kHz [Adhav]

## Dielectric Properties - Loss Tangent

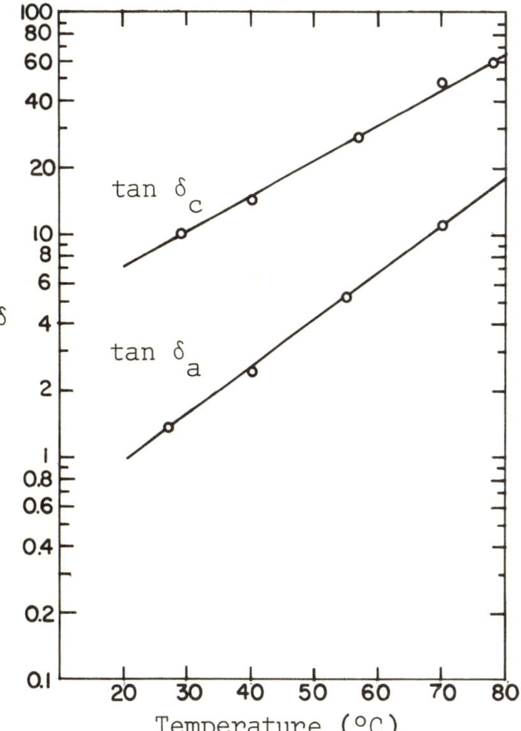


Fig. 6. Loss tangent of ADA as a function of temperature at 1 kHz [Adhav]

## Dielectric Properties - Electrical Resistivity

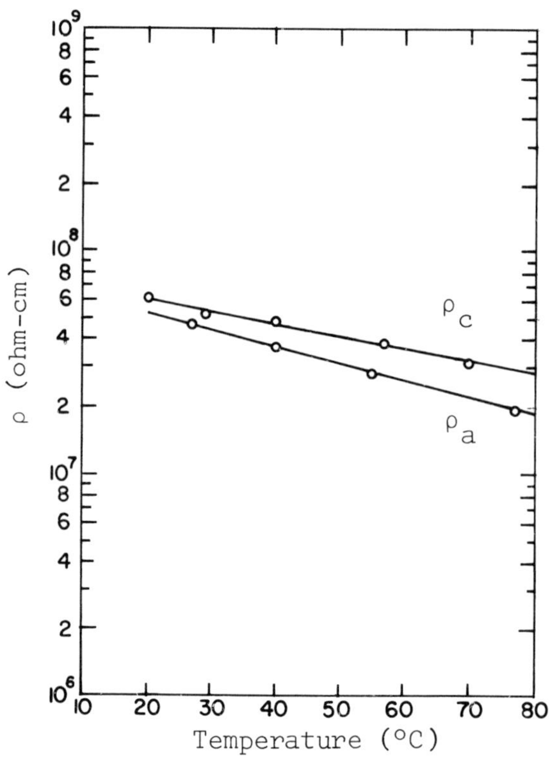


Fig. 7. Electrical resistivity of ADA as a function of temperature [Adhav].



## Thermal Properties

| | | |
|---|---|---|
| Thermal Conductivity: | 0.0043 W/cm°K | [Yap and Bicknell] |
| Estimated maximum safe operating temperature: | 80°C | [Sliker] |

Cook has reviewed the literature on the thermal expansion of ADA, as shown in Table 4.

TABLE 4. THERMAL COEFFICIENT OF EXPANSION OF ADA

| Temperature Range (°C) | Expansion Coefficient ($10^{-6}$/°C) | | Reference |
|---|---|---|---|
| | along a-axis | along c-axis | |
| -50 to +50 | 27.4 | 5.8 | Cook |
| 20 | 28 | 2 | Haussühl |
| 20 | 18.6 | 1.0 | Deshpande and Kahn |

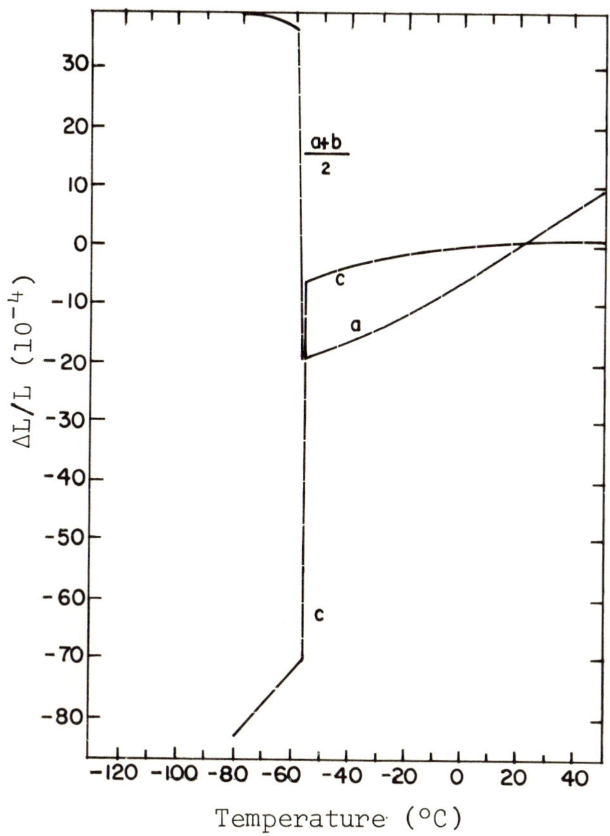


Fig. 8. Thermal expansion of ADA as a function of temperature. The curves are slightly offset at the Curie point ($T_c$ = -57°C) to increase the readability of the graph [Cook].

## AMMONIUM DIHYDROGEN ARSENATE (ADA)

ADHAV, R.S. Elastic, Piezoelectric, and Dielectric Properties of Ammonium Dihydrogen Arsenate (ADA). I. ACOUSTICAL SOC. OF AMERICA, J., v. 43, no. 4, Apr. 1969. p. 835-838.

ADHAV, R.S. Linear Electro-Optic Effects in Tetragonal Phosphates and Arsenates. OPTICAL SOC. OF AMERICA, J., v. 59, no. 4, Apr. 1969. p. 414-418.

BUSCH, G. New Ferroelectrics (In Ger.). HELV. PHYS. ACTA, v. 11, 1938. p. 269-298.

COOK, W.R. Jr. Thermal Expansion of Crystals with $KH_2PO_4$ Structure. J. OF APPLIED PHYS., v. 38, no. 4, Mar. 15, 1967. p. 1637-1642.

DELAIN, C. Structure of $AsO_4H_2(NH_4)$ (In Fr.). ACAD DES SCI., C.R., v. 247, no. 18, Nov. 1958. p. 1451-1452.

DESHPANDE, V.T. and A.A. KHAN. Current Sci. (India), v. 34, 1965. p. 633.

HAUSSÜHL, S. Elastic and Thermoelastic Properties of $KH_2PO_4$, $KH_2AsO_4$, $NH_4H_2PO_4$, and $RbH_2PO_4$. Z. FUER KRYSTALLOGRAPHIE, v. 120, 1964. p. 401-414.

LANDOLT-BÖRNSTEIN. Numerical Data and Functional Relationships in Science and Technology; New Series, v. 1, Elastic, Piezoelectric, Piezooptic and Electro-optic Constants of Crystals. Group III: Crystal and Solid State Phys., ed. by: HELLWEGE, K.-H. and A.M. HELLWEGE. Berlin, Germany, Springer-Verlag, 1966.

LOIACONO, G.M. A.D.T.A. Study of the Ferroelectric Transition in Potassium Dihydrogen Phosphate Type Crystals. MAT. RES. BULL., v. 5, no. 9, Sept. 1970. p. 775-782.

POST OFFICE RES. STATION (London). Piezoelectric Crystals. Data Tables. Res. Report No. 13389 (Rev.), by: LYNCH, A.C. Aug. 1957.

MASON, W.P. Elasto-Electric Constants of $KH_2PO_4$ Type Crystal. Table III. PHYSICAL ACOUSTICS, v. 1, Pt. A. Academic Press, N.Y., 1964. p. 181.

CATHOLIC UNIV. OF AMERICA. Ferroelectricity in Crystals. Rept. No. ASD-TDR-62-636. By: PULVARI, C.F. Contract No. AF 33(616)-6233, Nov. 1962. AD 292 983.

CLEVITE CORP. Reference Data on Linear Electro-Optic Effects. Engineering Memorandum 64-10. By: SLIKER, T.R. May 15, 1964. 9pp.

STEPHENSON, C.C. and H.E. ADAMS. The Heat Capacity of Ammonium Dihydrogen Arsenate from 15 to 300°K. The Anomaly at the Curie Point. AMERICAN CHEM. SOC., J., v. 66, 1944. p. 1409-1412.

WIENER, A.E. et al. Antiferroelectric Transitions in Ammonium Dihydrogen Phosphate and Ammonium Dihydrogen Arsenate Studied by Infrared Absorption. J. OF CHEM. PHYS., v. 52, no. 6, Mar. 15, 1970. p. 2891-2900.

SYLVANIA ELECTRONIC SYSTEMS. Solid State Techniques for Modulating Optical Waves. Final Report, by: YAP, B.K. and W.E. BICKNELL. Tech. Report ECOM-01283-F, Oct. 1966. AD 803 162.

# AMMONIUM DIHYDROGEN PHOSPHATE (ADP)

## Introduction

Ammonium dihydrogen phosphate (ADP) is a transparent, piezoelectric crystal containing no water of crystallization. Single crystals of this material were originally developed for use in underwater sound projectors and hydrophones. Transducer problems during World War II fostered a joint study program on this material by Bell Telephone Laboratories, Brush Development Company and the Naval Research Laboratories. An adequate and relatively simple technology was developed in this joint effect to grow ADP crystals and to fabricate them into sonar transducers. ADP is comparable to ordinary table salt in workability, brittleness and solubility. Blumenthal was one of the first investigators to demonstrate the successful microwave modulation of light by using the electrooptic effect in ADP.

## Chemical and Physical Properties

| | | |
|---|---|---|
| Chemical Formula | $NH_4H_2PO_4$ | |
| Molecular Weight | 115.04 | |
| Density | 1.803 g/cm$^3$ | Mason |
| Solubility in water, at 0°C | 22.7 g/100g water | Ballard et al. |
| at 100°C | 173.2 g/100g water | |

ADP is very hygroscopic and is insoluble in alcohol and acetone. Blumenthal notes that because of the brittle and hygroscopic nature of ADP, great care must be exerted in cutting, polishing, and subsequent handling. A constant optical check must be made during cutting and polishing to insure that the crystal is on axis.

## Crystallography

The crystal structure of ADP is discussed in detail by Känzig. The room temperature phase of ADP is isomorphous with potassium dihydrogen phosphate (KDP). Above 148°K, ADP belongs to the tetragonal crystal class $\bar{4}2m$; it lacks a center of inversion and exhibits a linear electrooptic effect. ADP undergoes a transition at 148°K to an antiferroelectric state, the antiferroelectric axes being parallel to the a-axis of the room temperature tetragonal (paraelectric) phase.

## AMMONIUM DIHYDROGEN PHOSPHATE (ADP)

Crystal Structure

| | | |
|---|---|---|
| Above $T_c$ | Tetragonal | $\bar{4}2m$ or $D_{2d}$ |
| Below $T_c$ | Orthorhombic | 222 or $D_2$ |

Data on the lattice constants of ADP have been reviewed by Cook:

| | |
|---|---|
| 20°C | $a_o = 7.4991 \pm 0.0004$ Å |
| | $c_o = 7.5493 \pm 0.0012$ Å |
| -126°C | $a_o = 7.4710 \pm 0.0008$ Å |
| | $c_o = 7.5374 \pm 0.0018$ Å |

### Optical Properties - Refractive Index

Zernike has made perhaps the most comprehensive refractive index measurements on ADP. His data are summarized in Table 1 and Figure 1. The temperature dependence of the refractive indices of ADP is detailed in Tables 2 and 3 and Figures 2 and 3.

TABLE 1. REFRACTIVE INDICES OF ADP AT 24.8°C [Zernike].

| Wavelength (μ) | Index in air | | Absolute index corrected to vacuum | |
|---|---|---|---|---|
| | $n_o$ | $n_e$ | $n_o$ | $n_e$ |
| 0.2000 | 1.648335 | 1.587012 | 1.649083 | 1.587632 |
| 0.3000 | 1.563478 | 1.512318 | 1.563951 | 1.512787 |
| 0.4000 | 1.540308 | 1.492136 | 1.540785 | 1.492571 |
| 0.5000 | 1.529792 | 1.483315 | 1.530276 | 1.483737 |
| 0.6000 | 1.523539 | 1.478412 | 1.524024 | 1.478828 |
| 0.7000 | 1.519047 | 1.475202 | 1.519528 | 1.475614 |
| 0.8000 | 1.515340 | 1.472818 | 1.515813 | 1.473227 |
| 0.9000 | 1.511969 | 1.470859 | 1.512433 | 1.471268 |
| 1.0000 | 1.508705 | 1.469123 | 1.509156 | 1.469530 |
| 1.1000 | 1.505418 | 1.467494 | 1.505853 | 1.467901 |
| 1.2000 | 1.502029 | 1.465904 | 1.502447 | 1.466311 |
| 1.3000 | 1.498490 | 1.464311 | 1.498888 | 1.464718 |
| 1.4000 | 1.494766 | 1.462687 | 1.495142 | 1.463094 |
| 1.5000 | 1.490834 | 1.461012 | 1.491187 | 1.461419 |
| 1.6000 | 1.486676 | 1.459272 | 1.487004 | 1.459679 |
| 1.7000 | 1.482279 | 1.457458 | 1.482580 | 1.457865 |
| 1.8000 | 1.477631 | 1.455562 | 1.477903 | 1.455970 |
| 1.9000 | 1.472724 | 1.453578 | 1.472965 | 1.453986 |
| 2.0000 | 1.467548 | 1.451501 | 1.467756 | 1.451910 |

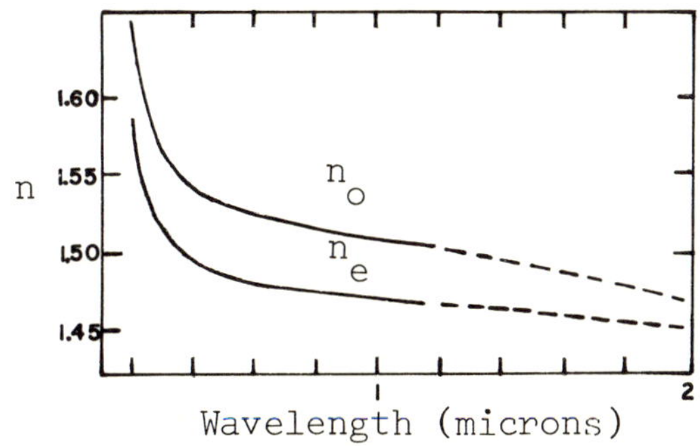


Fig. 1. Refractive indices of ADP as a function of wavelength at 24.8°C. Curves are dashed in those spectral regions where no actual measurements were made [Zernike].

TABLE 2. REFRACTIVE INDICES OF ADP AT THREE TEMPERATURES [Vishnevskii and Stefanskii].

| T(°C) | λ = 5770 Å | | λ = 5460 Å | |
|---|---|---|---|---|
| | $n_o$ | $n_e$ | $n_o$ | $n_e$ |
| -196 | 1.5334 | 1.4805 | 1.5350 | 1.4821 |
| -50 | 1.5273 | 1.4798 | 1.5290 | 1.4813 |
| +20 | 1.5249 | 1.4793 | 1.5265 | 1.4808 |

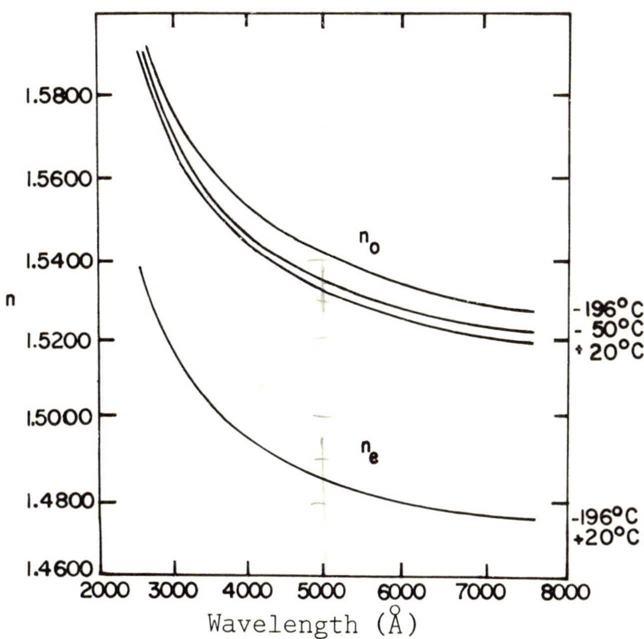


Fig. 2. Refractive indices of ADP as a function of wavelength at various temperatures [Vishnevskii and Stefanskii].

## AMMONIUM DIHYDROGEN PHOSPHATE (ADP)

TABLE 3. CHANGE IN THE REFRACTIVE INDICES OF ADP WITH TEMPERATURE [Phillips].

| Wavelength (Å) | Index at 298°K | | Increase at 201°K from 298°K | | Increase at 150°K from 298°K | |
|---|---|---|---|---|---|---|
| | $n_o$ | $n_e$ | $n_o$ | $n_e$ | $n_o$ | $n_e$ |
| 6907 | 1.5192 | 1.4753 | | | 0.0059 | 0.0002 |
| 6438 | | | 0.0039 | 0.0001 | | |
| 6234 | 1.5223 | 1.4775 | | | 0.0060 | 0.0003 |
| 5779 | 1.5246 | 1.4792 | 0.0039 | 0.0001 | 0.0060 | 0.0003 |
| 5461 | 1.5265 | 1.4808 | 0.0040 | 0.0001 | 0.0062 | 0.0003 |
| 4916 | 1.5303 | 1.4838 | 0.0040 | 0.0001 | 0.0061 | 0.0003 |
| 4358 | 1.5357 | 1.4882 | 0.0041 | 0.0001 | 0.0063 | 0.0003 |
| 4078 | 1.5392 | 1.4912 | | | | |
| 4047 | 1.5396 | 1.4915 | 0.0043 | 0.0001 | 0.0067 | 0.0003 |
| 3653 | 1.5457 | 1.4970 | | | | |

Temperature tunable, 90° phase matchable, parametric fluorescence has been observed in this material by Dowley. Fluorescence was pumped by 2573 Å CW radiation and was observed over the visible spectrum by temperature tuning through -12°C to 40°C. The temperature variation of the refractive endex was calculated and was found to be considerably larger than the Phillips values:

$d(n_o-n_e)/dT = 4.0 \times 10^{-5}/°C$; $n_o$ at 6943 Å, $n_e$ at 3452 Å [Phillips]

$d(n_o-n_e)/dT = 5.65 \times 10^{-5}/°C$; $n_o$, $n_e$ at 2573 Å [Dowley]

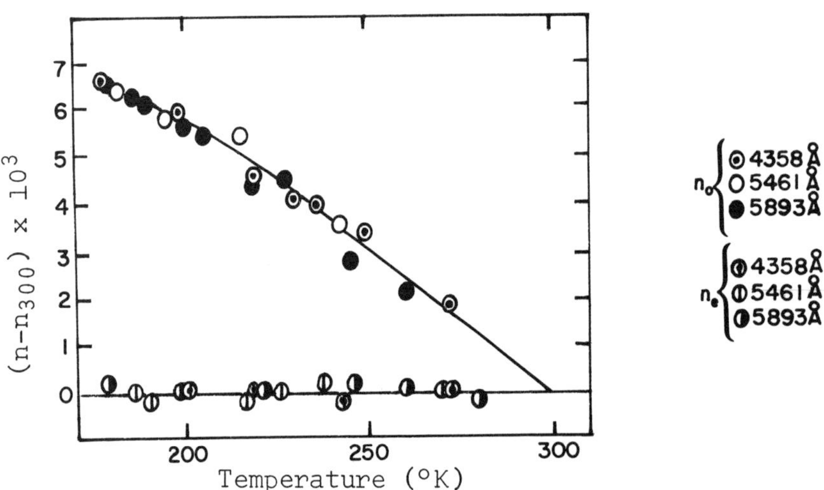


Fig. 3. Change in the refractive indices of ADP as a function of temperature [Yamazaki and Ogawa].

## Optical Properties - Birefringence

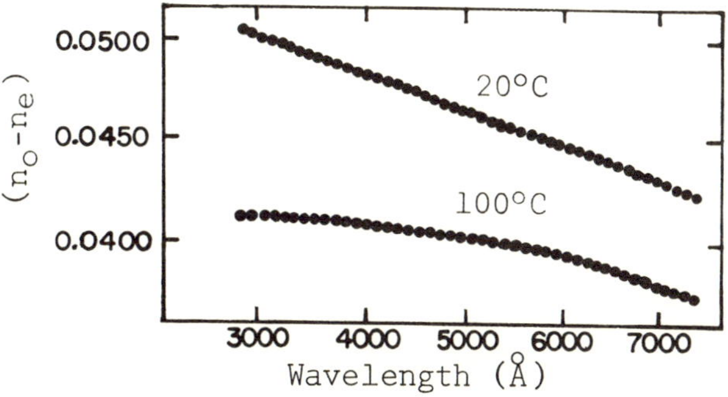


Fig. 4. Birefringence in ADP as a function of wavelength [Vishnevskii et al.].

## Optical Properties - Transmission

The transmission characteristics of a 0.25 in. thick ADP crystal are shown in Figure 5. It is apparent that the wavelength region of greater than 50% transmission is from 0.2 to 1.2 micron. Ballard et al. report transmission data for a 7.8 mm thick crystal that agree with the data shown in Figure 5. These authors note that the reflection loss of ADP for two surfaces at 0.7 micron is 10.5%. Morlot presents absorption data for single crystals at 70 to 500 μ at low temperatures. Absorption is highly anisotropic, especially at the Curie temperature.

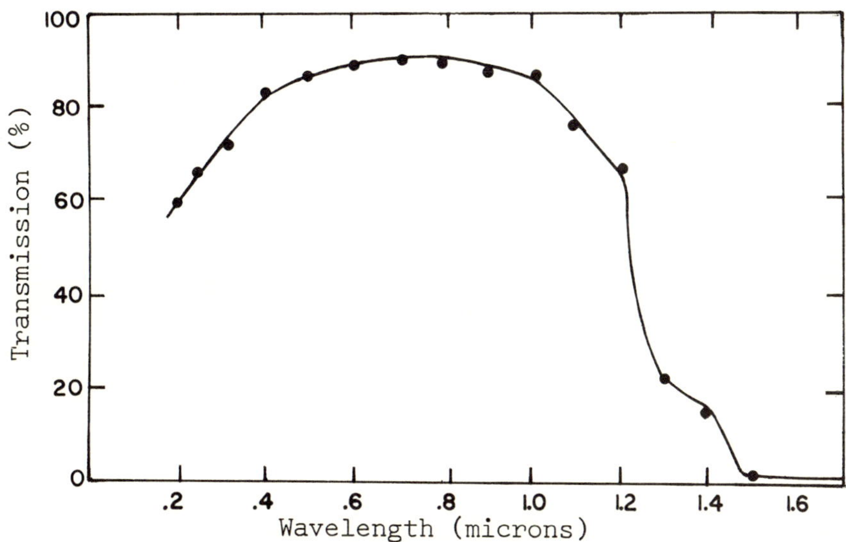


Fig. 5. Transmission as a function of wavelength for ADP crystal 0.25 in. thick [Burnett].

## AMMONIUM DIHYDROGEN PHOSPHATE ( ADP)

### Optical Properties - Nonlinear Optical Behavior

The nonlinear coefficients for second harmonic generation in ADP, relative to $d_{36}^{2\omega}$ in $KH_2PO_4$ (KDP) are:

| | [Bjorkholm] | [Miller et al.] | |
|---|---|---|---|
| | 1.15 microns | 1.06 microns | 0.643 microns |
| $d_{36}^{2\omega}/d_{36}^{2\omega}$ (KDP) | 1.03 ± 0.16 | 0.99 ± 0.06 | 0.93 ± 0.06 |
| $d_{14}^{2\omega}/d_{36}^{2\omega}$ (KDP) | | 0.98 ± 0.05 | 0.89 ± 0.0 |

An absolute measurement of $d_{36}^{2\omega}$ at 6328 Å by McMahon and Franklin gave $d_{36}^{2\omega} = 2.0 \pm 0.5 \times 10^{-9}$ esu. More recent absolute measurements of this coefficient at 6328 Å are:

$d_{36}^{2\omega} = 1.36 \pm 0.16 \times 10^{-9}$ esu [Francois]

$d_{36}^{2\omega} = 1.38 \pm 0.22 \times 10^{-9}$ esu [Bjorkholm and Siegman]

Van der Ziel and Bloembergen have indicated that the nonlinear coefficients of ADP are independent of temperature.

The following optical rectification coefficients have been reported for ADP at 6943 Å:

$\frac{1}{2}(X^{o}_{xyz} + X_{xzy}) = +6.70 \pm 0.90 \times 10^{-8}$ esu [Ward]

$\frac{1}{2}(X^{o}_{zxy} + X_{zyx}) = +1.97 \pm 0.28 \times 10^{-8}$ esu [Ward and New]

### Electrooptic Properties

Ammonium dihydrogen phosphate belongs to the crystal class $\bar{4}2m$ and exhibits only two independent linear electrooptic coefficients: $r_{63}$ and $r_{41} = r_{52}$. Reported experimental values of these coefficients are listed in Table 4.

TABLE 4. ELECTROOPTIC COEFFICIENTS OF ADP AT ROOM TEMPERATURE

| Electrooptic Coefficient ($10^{-12}$ m/V) | | | λ (Å) | Reference | |
|---|---|---|---|---|---|
| $r_{41}^T$ | $r_{63}^T$ | $r_{63}^S$ | | | |
| 20.8 ± 0.3 | 8.47 ± 0.17 | 5.5 | 5560 | 1950 | Carpenter |
| | 8.4 ± 0.2 | | 5000 | 1961 | Namba |
| 24.5 ± 0.4 | | | 5460 | 1964 | Ott and Sliker |
| 23.41 | | | 6328 | 1966 | Ley |
| | | 4.05 ± 0.45 | 6328 | 1966 | Silverstein and Sucher |
| | 8.55 ± 0.15 | | 6328 | 1967 | Koetser |
| | 8.4 | | 5500 | 1969 | Adhav |
| 24.7 ± 0.3 | | | 6328 | 1969 | Trevelyan |
| | 7.0 | 4.3 | 5350 | 1970 | Vasilevskaya and Sonin |

The dc electrooptic coefficient $r_{63}^T$ corresponds to a "free" crystal, whereas the coefficient $r_{63}^S$, measured at high modulating frequencies, corresponds to a "clamped" crystal. The frequency response of the electrooptic effect in ADP, in terms of $r_{63}$, is shown in Figure 6, indicating that $r_{63}^S/r_{63}^T = 0.65$ at 5560 Å. It appears that both $r_{63}^T$ and $r_{63}^S$ are dependent only slightly upon wavelength in the region 4000 to 7000 Å [Adhav, Carpenter, Vasilevskaya, Blokh and Lutsiv-Shumskii].
Vasilevskaya reports $r_{63}^S/r_{63}^T$ values ranging from 0.65 to 0.57 in the wavelength region from 4000 to 7000 Å, respectively, while Popova and Rips report lower values of 0.54 to 0.37 from 4500 to 5400 Å.

The dependence of $r_{41}^T$ on wavelength is indicated in Figure 7.

The temperature dependence of the electrooptic coefficients is shown in Figures 8 and 8a.

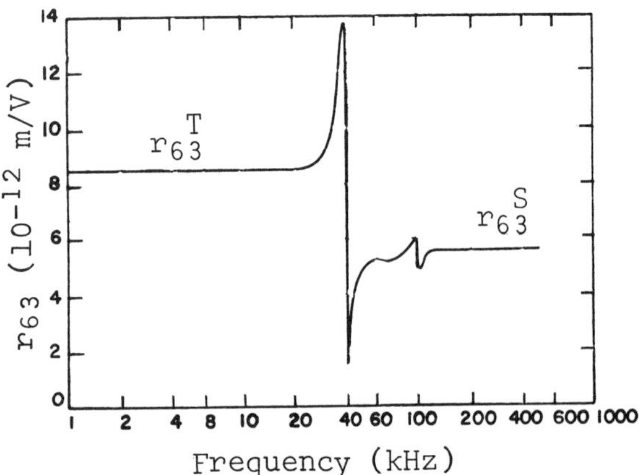


Fig. 6. Electrooptic coefficient $r_{63}$ of ADP as a function of modulating frequency at 5560 Å [Carpenter].

Fig. 6a. Dispersion of $r_{63}$ as a function of wavelength.

[Blokh].

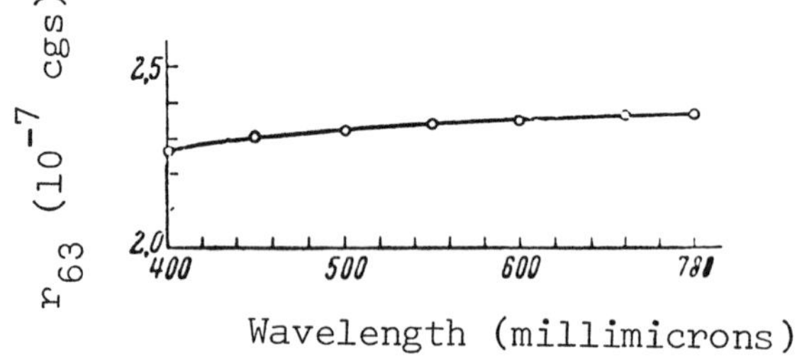


Fig. 7. Electrooptic coefficient $r_{41}^T$ of ADP as a function of wavelength [Ley].

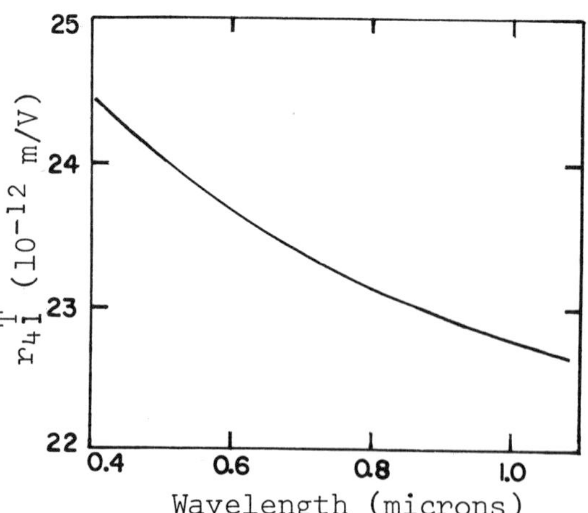

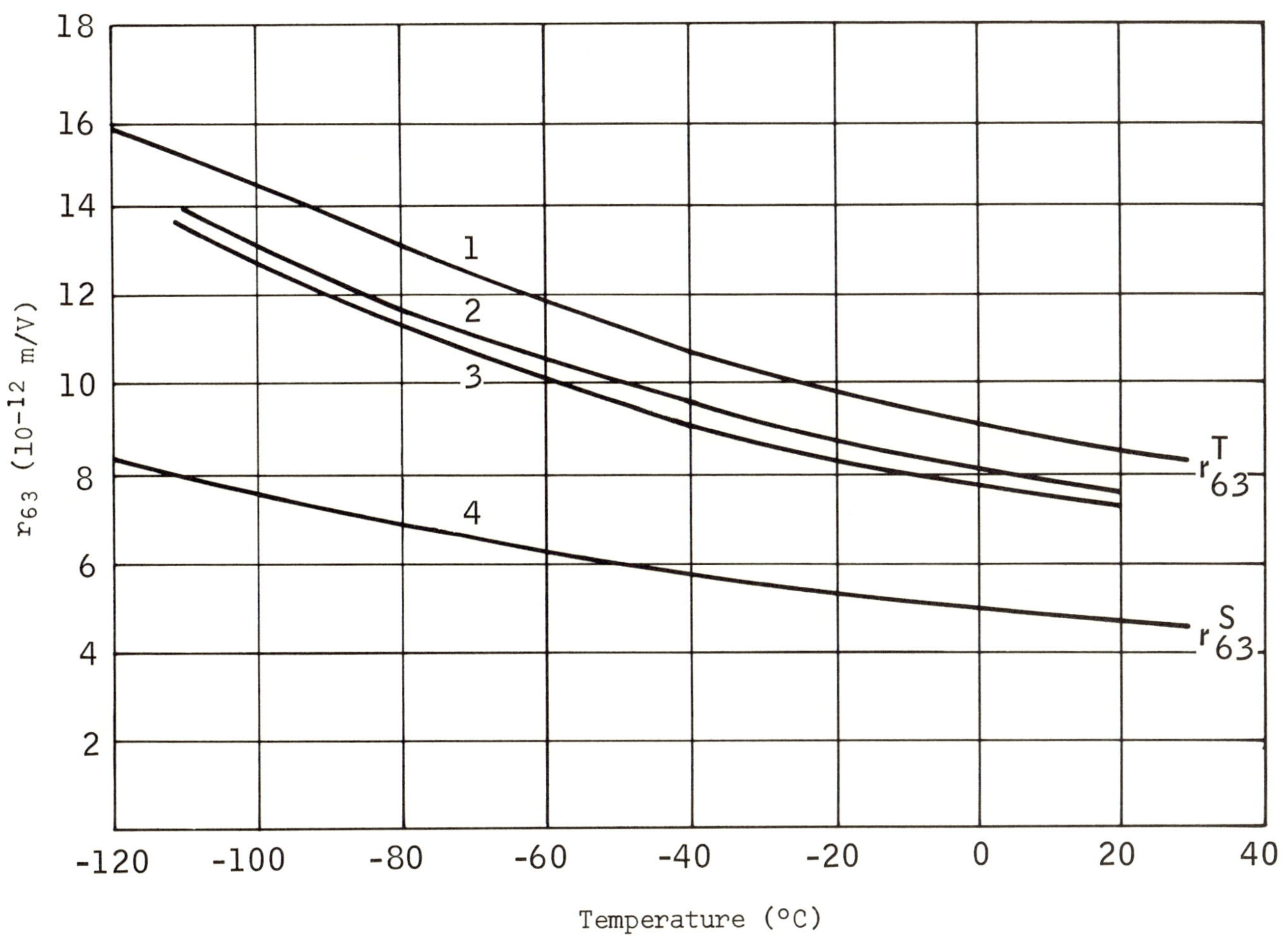


Fig. 8. Electrooptic coefficient $r_{63}$ of ADP as a function of temperature.

| Curve | Coefficient | λ(Å) | Reference |
|---|---|---|---|
| 1 | $r^T_{63}$ | 5460 | Landolt-Börnstein |
| 2 | $r^T_{63}$ | 6800 | Blokh and |
| 3 | $r^T_{63}$ | 4200 | Lutsiv-Shumskii |
| 4 | $r^S_{63}$ | 5460 | Landolt-Börnstein |

Fig. 8a. Electrooptic coefficients of ADP as a function of temperature. Since below the Curie point (-128°C) the crystallographic x and y axes were not identified by x-ray diffraction, the subscripts of the coefficients $r_{41}$ and $r_{52}$ have a conditional meaning [Apkaryants and Sonin].

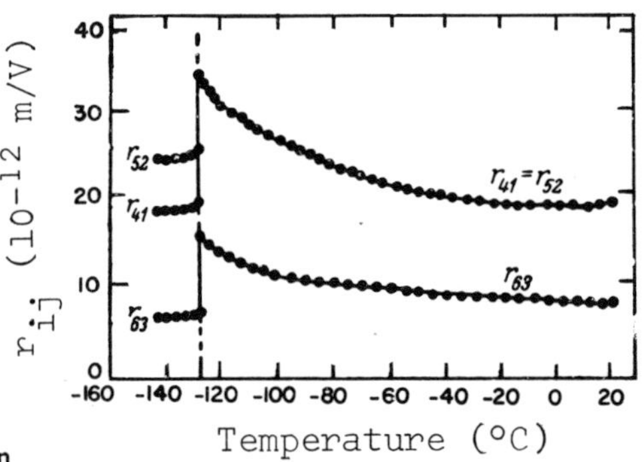




The electrooptic coefficient $r_{63}$ is appropriate in the longitudinal configuration in which the optical beam and the electrical field are parallel to the optic axis. The $r_{41}$ electrooptic coefficient which, as can be seen from Table 4, is approximately equal to $3r_{63}$, is the coefficient appropriate in the transverse mode of operation in which the electric field is applied normal to the direction of propagation of the light beam. In general, ADP is the most useful isomorph of KDP in applications that depend upon $r_{41}$.

Using ADP in the longitudinal configuration, the half-wave voltage is given by

$$V_{\frac{1}{2}} = \lambda/2n_o^{\,3}\, r_{63},$$

where $\lambda$ is the wavelength and $n_o$ the ordinary refractive index. Using a value of $r_{63} = 8.5 \times 10^{-12}$ m/V and the refractive index data given in Table 1, the dependence of $V_{\frac{1}{2}}$ on wavelength which results is shown in Figure 9 for the longitudinal configuration.

Koetser in his studies on ADP found that the retardation produced by the crystal was a linear function of the applied voltage up to breakdown at 200 kV/cm, showing no variation of the electro-optic coefficient with field strength.

In the transverse configuration a y-cut ADP crystal is oriented such that the electric field is directed parallel to the crystallographic y-axis; the light beam traverses the material in the x-z plane along a direction oriented at 45° relative to the x- and z-axes. The half-wave voltage in this configuration is given by

$$V_{\frac{1}{2}} = \left(\frac{1}{n_o^{\,2}} + \frac{1}{n_e^{\,2}}\right)^{3/2} \left(\frac{\lambda}{r_{41}}\right)\left(\frac{1}{2\sqrt{2}}\right)\left(\frac{d}{\ell}\right)$$

where $n_o$ and $n_e$ are the ordinary and extraordinary refractive indices, d the crystal thickness and $\ell$ the crystal length. A plot of $V_{\frac{1}{2}}$ versus the ratio $(d/\ell)$ is shown in Figure 10, representing the toleranced values of $r_{41}$ given in Table 4.

All isomorphs of KDP are capable of operation in this configuration; however, ADP enjoys the unique advantage not only of having a large value of $r_{41}$, but also of exhibiting an extremely small piezoelectric effect for y-directed fields.

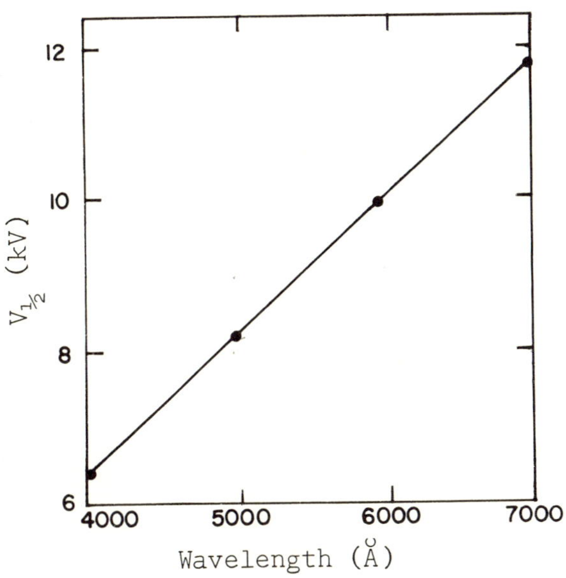


Fig. 9. Half-wave voltage of ADP in the longitudinal configuration as a function of wavelength.

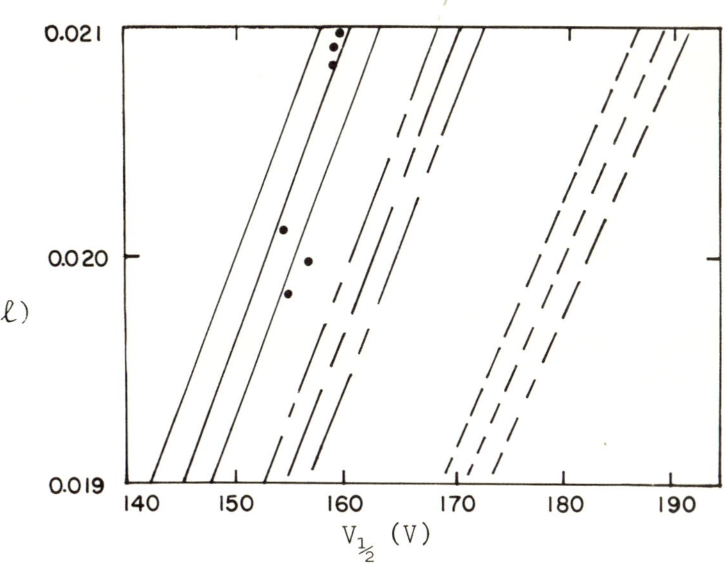


Fig. 10. Crystal ratio $(d/\ell)$ as a function of half-wave voltage [Trevelyan].

$r_{41}$

• $24.7 \pm 0.3 \times 10^{-12}$ m/V Trevelyan

———— $24.5 \pm 0.4 \times 10^{-12}$ m/V Ott and Sliker

— — — $23.06 \pm 0.25 \times 10^{-12}$ m/V Ley

– – – – $20.8 \pm 0.3 \times 10^{-12}$ m/V [Carpenter]

## AMMONIUM DIHYDROGEN PHOSPHATE (ADP)

The quadratic electrooptic effect in ADP has been investigated by Perifilova and Sonin.

### Photoelastic Properties - Elastooptic Coefficients

The following values of elastooptic coefficient ratios of ADP have been reported:

| $p_{12}/p_{11}$ | $p_{13}/p_{33}$ | $p_{31}/p_{11}$ | |
|---|---|---|---|
| 1.46 | 1.20 | -1.75 | Ziauddin and Narasimhamurty |
| 1.17 | 1.18 | -1.74 | Achyuthan and Breazeale |

Ziauddin and Narasimhamurty have derived a complete set of elastooptic (strain-optic) coefficients ($p_{ij}$) using the ratio $p_{31}/p_{11}$, the piezooptic coefficients ($\pi_{ij}$) reported by Carpenter and the elastic constants reported by Bechmann. Similarly, Achyuthan and Breazeale have derived a complete set of coefficients using their own piezooptic data and the elastic constants reported by Price and Huntington. These results are given in Table 5 along with absolute magnitude data for the elasto-optic coefficients as reported by Dixon.

The temperature dependence of the elastooptic coefficient $p_{66}$ measured under constant field (E), constant displacement (D) and constant polarization (P) is indicated in Table 6.

TABLE 5. ELASTOOPTIC COEFFICIENTS OF ADP AT ROOM TEMPERATURE

| $p_{11}$ | $p_{12}$ | $p_{44}$ | $p_{31}$ | $p_{13}$ | $p_{33}$ | $p_{66}^E$ | $\lambda(\mu)$ | Reference |
|---|---|---|---|---|---|---|---|---|
| -0.11 ± 0.01 | -0.16 ± 0.01 | -0.056 ± 0.002 | +0.18 ± 0.01 | -0.84 ± 0.05 | -0.70 ± 0.05 | -0.099 ± 0.008 | | Ziauddin and Narasimhamurty |
| -0.11 ± 0.02 | -0.15 ± 0.02 | | +0.20 ± 0.03 | -0.93 ± 0.41 | -0.71 ± 0.32 | | 0.546 | Achyuthan and Breazeale |
| \|0.302\| | \|0.246\| | | \|0.195\| | \|0.236\| | \|0.263\| | \|0.075\| | 0.63 | Dixon |
| | | | | | | -0.0744 | 0.546 | Landolt-Börnstein citing Carpenter |
| | | | | | | -0.1107 | 0.560 | West and Makas |

TABLE 6. ELASTOOPTIC COEFFICIENT $p_{66}$ OF ADP AS A FUNCTION OF TEMPERATURE AT 5460 Å [Landolt-Börnstein, citing Carpenter].

| $p_{66}^E$ | $p_{66}^D$ | $p_{66}^P$ | T(°C) |
|---|---|---|---|
| -0.085 | -0.060 | -0.058 | -120 |
| -0.089 | -0.065 | -0.064 | -110 |
| -0.092 | -0.070 | -0.069 | -100 |
| -0.094 | -0.073 | -0.072 | -90 |
| -0.093 | -0.074 | -0.073 | -80 |
| -0.093 | -0.074 | -0.073 | -70 |
| -0.091 | -0.074 | -0.073 | -60 |
| -0.089 | -0.073 | -0.072 | -50 |
| -0.088 | -0.072 | -0.071 | -40 |
| -0.086 | -0.071 | -0.070 | -30 |
| -0.084 | -0.070 | -0.069 | -20 |
| -0.081 | -0.068 | -0.067 | -10 |
| -0.080 | -0.067 | -0.067 | 0 |
| -0.079 | -0.066 | -0.066 | 10 |
| -0.077 | -0.065 | -0.065 | 20 |
| -0.077 | -0.066 | -0.065 | 30 |

## Photoelastic Properties - Piezooptic Coefficients

Blokh and Lutsiv-Shumskii have reported on the dispersion of $\pi_{44}$ in 45° cuts of single crystal ADP at 20°C and 0.4 to 0.7 micron. The value is about $5 \times 10^{-12}$ $m^2/N$ and shows practically no change in this spectral range.

TABLE 7. PIEZOOPTIC COEFFICIENTS OF ADP AT ROOM TEMPERATURE

| $10^{-12}$ $m^2/N$ | | | | | | | | |
|---|---|---|---|---|---|---|---|---|
| $\pi_{11}-\pi_{12}$ | $\pi_{11}$ | $\pi_{33}$ | $\pi_{12}$ | $\pi_{13}$ | $\pi_{31}$ | $\pi_{44}$ | $\pi_{66}^E$ | Reference |
| 0.8 | 7.6 ± 0.5 | -35 ± 2 | 6.8 ± 0.5 | -33 ± 2 | 11.9 ± 0.7 | -6.2 ± 0.1 | -16.5 ± 0.5 | Ziauddin and Narasimhamurty |
| | | -35.7 | | -37.3 | | | | *Achyuthan and Breazeale |
| 0.7 | 8.6 | | 7.9 | | 12.3 | -5.8 | -12.2 | Carpenter |
| 0.78 | | | | | | \|5.78\| | \|19.6\| | Deviot |
| 1.04 | | | | | | | -18.15 | West and Makas |

* Data taken at 5461 Å; wavelength not reported for other data.

TABLE 8. PIEZOOPTIC COEFFICIENT $\pi_{66}$ OF ADP AS A FUNCTION OF TEMPERATURE AT 5460 Å [Landolt-Börnstein, citing Carpenter].

| $\pi_{66}^E$ | $\pi_{66}^D$ | $\pi_{66}^P$ | T(°C) |
|---|---|---|---|
| -14.0 | -7.8 | -7.5 | -120 |
| -14.4 | -8.5 | -8.3 | -110 |
| -14.9 | -9.4 | -9.1 | -100 |
| -15.0 | -9.8 | -9.6 | -90 |
| -14.8 | -10.0 | -9.6 | -80 |
| -14.6 | -10.1 | -9.7 | -70 |
| -14.4 | -10.0 | -9.7 | -60 |
| -14.1 | -10.0 | -9.7 | -50 |
| -13.9 | -10.0 | -9.7 | -40 |
| -13.7 | -9.9 | -9.7 | -30 |
| -13.5 | -9.7 | -9.6 | -20 |
| -13.0 | -9.6 | -9.5 | -10 |
| -12.9 | -9.6 | -9.4 | 0 |
| -12.8 | -9.6 | -9.4 | 10 |
| -12.6 | -9.6 | -9.4 | 20 |
| -12.6 | -9.6 | -9.5 | 30 |

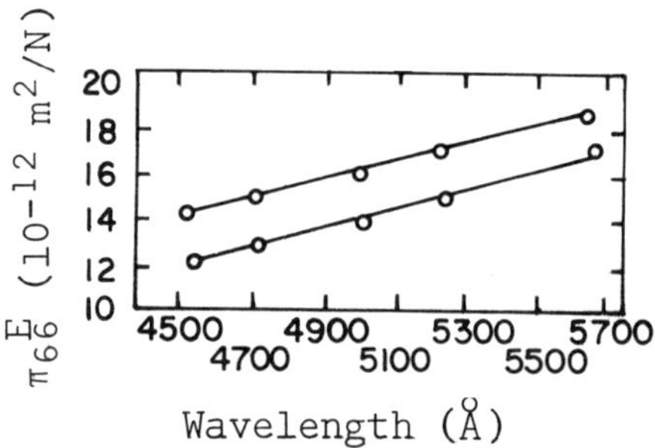


Fig. 11. Piezooptic coefficient $\pi_{66}^E$ of two ADP crystals as a function of wavelength [Popova and Rips].

## Photoelastic Properties - Elastic Constants

TABLE 9. ELASTIC STIFFNESS CONSTANTS OF ADP AT ROOM TEMPERATURE [As cited by Landolt-Börnstein].

| $10^{10}$ N/m$^2$ | | | | | | Method | Reference | |
|---|---|---|---|---|---|---|---|---|
| $c_{11}$ | $c_{33}$ | $c_{12}$ | $c_{13}$ | $c_{44}$ | $c_{66}$ | | | |
| 7.57 | 2.96 | -2.43 | 1.30 | 0.87 | 0.61 | Resonant Frequency | 1946 | Mason |
| 6.17 | 3.28 | 0.72 | 1.94 | 0.85 | 0.592 | Light Diffraction | 1946 | Zwicker |
| 6.89 | 3.35 | 0.40 | 1.89 | 0.856 | 0.595 | Pulse Echo | 1950 | Price and Huntington |
| 6.76 | 3.38 | 0.59 | 2.00 | 0.87 | 0.61 | Resonant Frequency | 1957 | Bechmann and Taylor |
| 6.80 | 3.42 | | | 0.862 | 0.602 | Ultrasonic Propagation | 1962 | Aleksandrov and Ryabinkin |
| 6.877 | 3.402 | 0.406 | 2.038 | 0.862 | 0.601 | | 1964 | Haussühl |

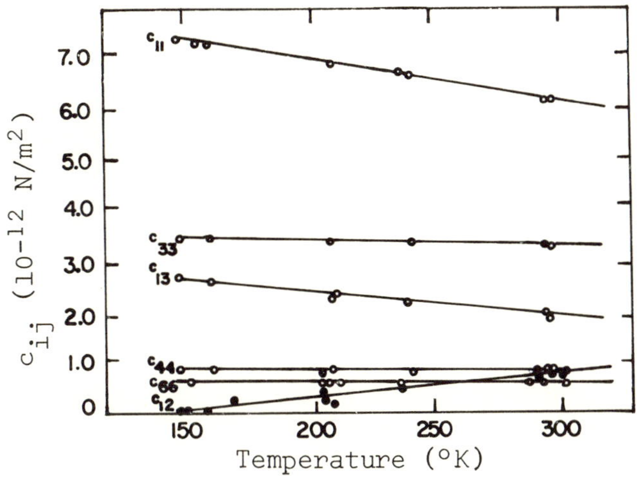


Fig. 12. Elastic stiffness constants of ADP as a function of temperature [Zwicker].

TABLE 10. ELASTIC COMPLIANCE CONSTANTS OF ADP AT ROOM TEMPERATURE.

| $10^{-12}$ $m^2/N$ | | | | | | Method | Reference | |
|---|---|---|---|---|---|---|---|---|
| $s_{11}$ | $s_{33}$ | $s_{12}$ | $s_{13}$ | $s_{44}$ | $s_{66}$ | | | |
| 17.5 | 43.5 | 7.5 | -11.5 | 115 | 164 | Resonant Frequency | 1946 | Mason |
| 20.0 | 45.7 | 1.7 | -12.9 | 117.0 | 185 | Light Diffraction | 1946 | Zwicker |
| 18.1 | 43.5 | 1.9 | -11.8 | 115.3 | 164.6 | Resonant Frequency | 1951 | Bechmann |

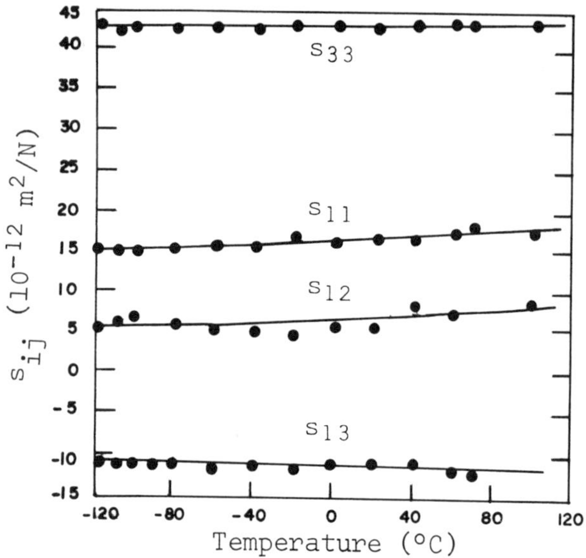


Fig. 13. Elastic compliance constants of ADP as a function of temperature [Mason].

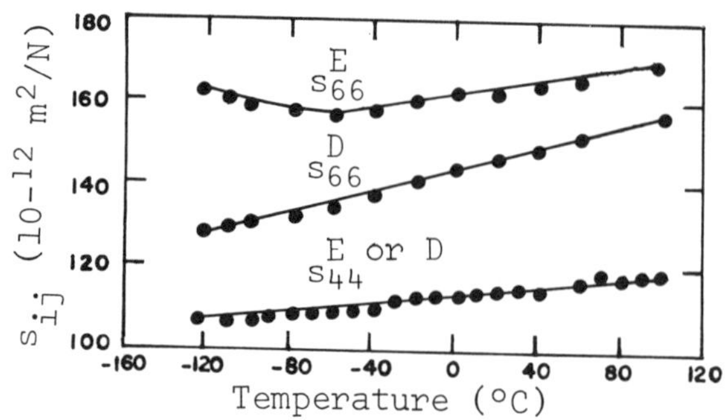


Fig. 14. Elastic compliance constants of ADP as a function of temperature [Mason].

## Piezoelectric Properties

TABLE 11. PIEZOELECTRIC CONSTANTS OF ADP
[As cited by Landolt-Börnstein].

| $10^{-12}$ C/N | | $10^{-1}$ C/m$^2$ | | $10^{-2}$ m$^2$/C | | $10^8$ N/C | | | |
|---|---|---|---|---|---|---|---|---|---|
| $d_{14}$ | $d_{36}$ | $e_{14}$ | $e_{36}$ | $g_{14}$ | $g_{36}$ | $h_{14}$ | $h_{36}$ | T(°C) | Reference |
| | 43.2 | | 2.54 | | 35.1 | | 22.4 | 100 | Mason |
| | 44.2 | | 2.61 | | 34.8 | | 22.6 | 80 | |
| | 45.3 | | 2.71 | | 34.8 | | 23.0 | 60 | |
| | 47.3 | | 2.87 | | 35.25 | | 23.6 | 40 | |
| 1.7 | 49.3 | | 3.01 | | 35.55 | | 24.3 | 20 | |
| | 51.7 | | 3.18 | | 36.0 | | 25.1 | 0 | |
| | 53.7 | | 3.33 | | 35.7 | | 25.3 | -20 | |
| | 56.7 | | 3.55 | | 36.3 | | 26.2 | -40 | |
| | 60.0 | | 3.77 | | 36.6 | | 26.8 | -60 | |
| | 66.0 | | 4.13 | | 37.5 | | 28.0 | -80 | |
| | 69.0 | | 4.30 | | 37.8 | | 28.8 | -100 | |
| | 80.7 | | 4.97 | | 39.0 | | 30.0 | -110 | |
| | 87.0 | | 5.13 | | 39.6 | | 30.6 | -120 | |
| | 90.0 | | 5.20 | | 39.6 | | 30.9 | -122 | |
| 1.76 | -48.3 | 0.153 | -2.94 | | | | | 20 | Bechmann and Taylor |
| | | 0.147 | 3.19 | 0.309 | 37.5 | 0.27 | 26.3 | 0 | Van Dyke and Gordon |
| 1.51 | -48.9 | | | | | | | 20 | Spitzer |
| 1.3 | 48 | | | | | | | RT | Jaffe and Smith |

Sliker has reported the following piezoelectric coupling constants for ADP:

$$k_{14} = 0.006$$

$$k_{36} = 0.33$$

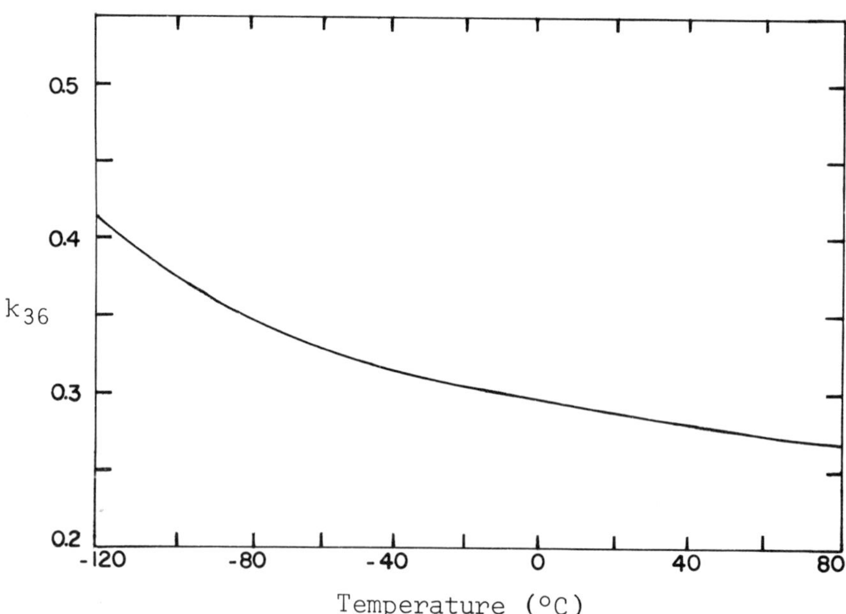


Fig. 15. Piezoelectric coupling constant of ADP as a function of temperature [Mason].

Dielectric Properties - Dielectric Constant

A large number of values for the dielectric constant of ADP, a nonlinear dielectric material, are quoted in the literature; however, in many cases the frequency and/or the temperature of the measurements are not reported. Representative room temperature data, reported by Von Hippel and Belyaev et al., are given in Table 12.

Representative data for the temperature dependence of the dielectric constant are presented in Figure 16. As with other isomorphs of KDP, $\varepsilon_a$ is larger than $\varepsilon_c$ at room temperature but, unlike the others, $\varepsilon_a$ has the more pronounced temperature variation.

TABLE 12. DIELECTRIC CONSTANT OF ADP AT ROOM TEMPERATURE

| Frequency (Hz) | $\varepsilon_c^{(A)}$ | $\varepsilon_c^{(B)}$ | $\varepsilon_a^{(A)}$ | $\varepsilon_a^{(B)}$ |
|---|---|---|---|---|
| $10^2$ | 16.0 ± 0.5 | 16.4 | 55.8 ± 1.5 | 56.4 |
| $10^3$ | 15.9 ± 0.5 | 16.0 | 57.0 ± 1.5 | 56.0 |
| $10^4$ | 15.5 ± 0.5 | 15.4 | 56.0 ± 1.5 | 55.9 |
| $10^5$ | 15.3 ± 0.5 | 14.7 | 55.8 ± 1.5 | 55.9 |
| $10^6$ | - | 14.3 | - | 55.9 |
| $10^7$ | - | 14.3 | - | 55.9 |
| $10^8$ | - | - | - | 55.9 |
| $3 \times 10^8$ | - | 14.3 | - | 55.9 |
| $9.8 \times 10^8$ | 15.0 ± 0.5 | - | 55.5 ± 1.5 | - |
| $9.4 \times 10^9$ | 14.7 ± 0.5 | - | 55.3 ± 1.5 | - |
| $10^{10}$ | - | 13.7 | - | - |
| $3.96 \times 10^{10}$ | 14.0 ± 0.5 | - | 55.0 ± 1.5 | - |

$\varepsilon_c = \varepsilon_{33}/\varepsilon_o$ and $\varepsilon_a = \varepsilon_{11}/\varepsilon_o$, where $\varepsilon_o$ is the permittivity of free space.

(A) Belyaev et al.

(B) Von Hippel

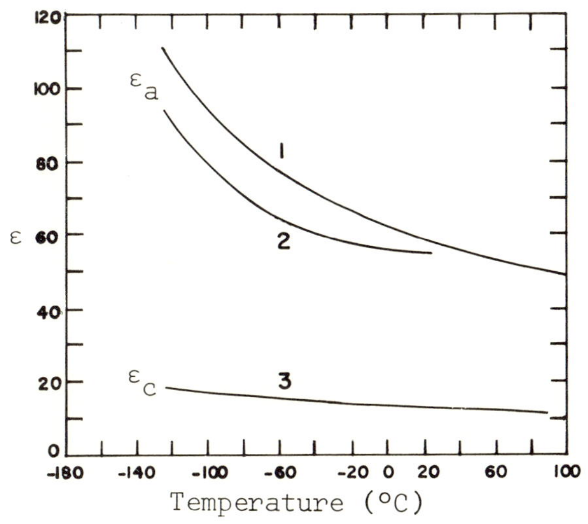


Fig. 16. Dielectric constant of ADP as a function of temperature.

| Curve | f(Hz) | Reference |
|---|---|---|
| 1 | $9.2 \times 10^9$ | Kaminow |
| 2 | $10^4$ | Pepinsky |
| 3 | $9.2 \times 10^9$ | Kaminow |
| | $10^4$ | Pepinsky |
| | $10^3$ | Mason |

TABLE 13. LOSS TANGENT OF ADP AT ROOM TEMPERATURE.

| Frequency (Hz) | $\tan \delta_c^{(A)}$ | $\tan \delta_c^{(B)}$ | $\tan \delta_a^{(B)}$ |
|---|---|---|---|
| $10^2$ | 0.1 | 0.2400 | 0.0400 |
| $10^3$ | 0.065 | 0.0240 | 0.0046 |
| $10^4$ | 0.018 | 0.0070 | 0.00046 |
| $10^5$ | 0.005 | 0.0070 | <0.0005 |
| $10^6$ | - | 0.0060 | <0.0005 |
| $10^7$ | - | 0.0010 | <0.0005 |
| $10^8$ | - | - | <0.0005 |
| $3 \times 10^8$ | - | 0.0005 | <0.0010 |
| $9.8 \times 10^8$ | 0.005 | - | - |
| $9.4 \times 10^9$ | 0.041 | - | - |
| $10^{10}$ | - | 0.0050 | - |
| $3.96 \times 10^{10}$ | 0.08 | - | - |

(A) Belyaev et al.

(B) Von Hippel

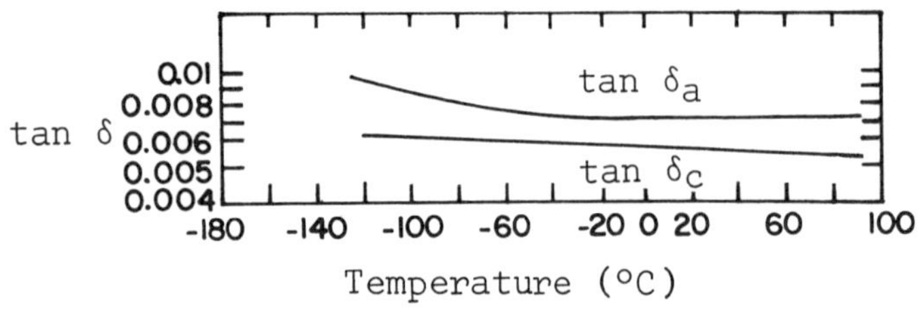


Fig. 17. Loss tangent of ADP as a function of temperature at 9.2 GHz [Kaminow].

Dielectric Properties - Electrical Resistivity

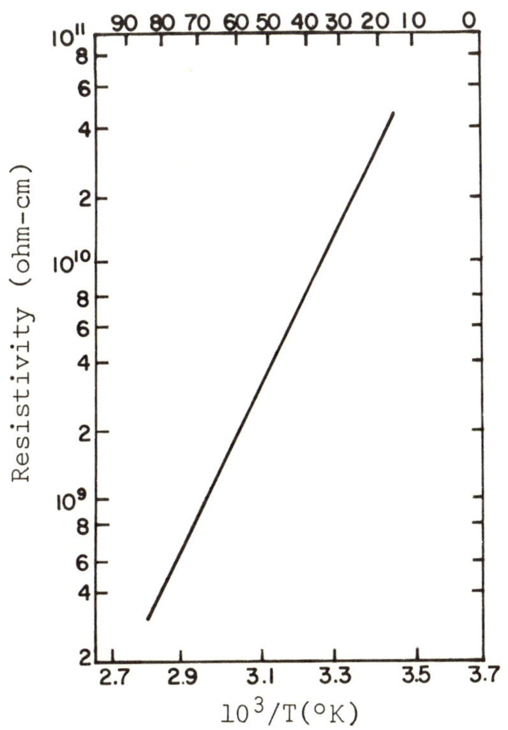


Fig. 18. Electrical resistivity of ADP as a function of temperature [Mason].

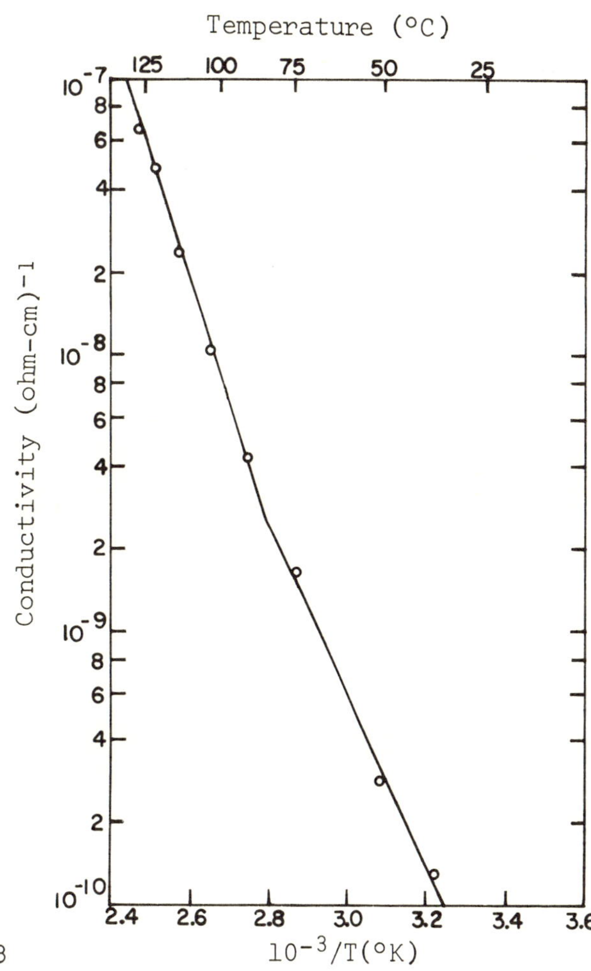


Fig. 19. Electrical conductivity of ADP as a function of reciprocal temperature $\sigma = \sigma_1 \exp(W_1/kT) + \sigma_2 \exp(-W_2/kT)$, where

$T$ = absolute temperature in °K
$k$ = Boltzmann constant
$\sigma_1 = 6.46 \times 10^3$ (ohm-cm)$^{-1}$
$\sigma_2 = 1.54$ (ohm-cm)$^{-1}$
$W_1$ = 20.4 kcal/mole
$W_2$ = 14.6 kcal/mole

[Murphy].

## AMMONIUM DIHYDROGEN PHOSPHATE (ADP)

### Antiferroelectric Properties

Although ammonium dihydrogen phosphate is an isomorph of KDP, at low temperatures it behaves quite differently than KDP with respect to its dielectric properties and its phase transition. Whereas KDP becomes ferroelectric at low temperatures, the $NH_4$ salts undergo a phase transition apparently to an antiferroelectic phase with no net spontaneous polarization in any of the crystal directions. Moreover, the dielectric constants $\varepsilon_a$ and $\varepsilon_c$ rise only slightly as the temperature falls, and drop suddenly on crossing the critical temperature $T_c$. Not much is known regarding the nature of the antipolar phase of ADP, since macroscopic crystals shatter completely during the phase transition. Känzig has reviewed the antiferroelectric nature of ADP.

The antiferroelectric transition temperature is found to be 147.0°K on cooling and 148.9°K on warming, with $T_c$ = 147.9 ± 1.0°K [Stephenson and Zettlemoyer]. Using a DTA method, Loiacono has obtained a Curie temperature value of 148.5°K on heating and 142.1°K on cooling.

The pressure dependence of the Curie temperature has been measured by Skalyo et al. on single crystals. Pressure was increased to 10 kbar and the temperature was changed at a constant rate through the ferroelectric transition. Dependence was linear and the values given are:

| | |
|---|---|
| $T_{c,\ P=0}$ | 151.2 |
| $dT_c/dP$ | -3.39 °K/kbar |

### Thermal Properties

While ADP will heat shock with rapid changes in temperature, it can withstand uniform temperatures up to 90°C [Burnett]; above this temperature, it begins to emit ammonia. Although the maximum safe operating temperature of ADP has been quoted as 80°C [Sliker], Burnett notes that in order to avoid distortion caused by the elasto-optic effect, operating temperature of the crystal, as used in electrooptic devices, should not exceed 40°C.

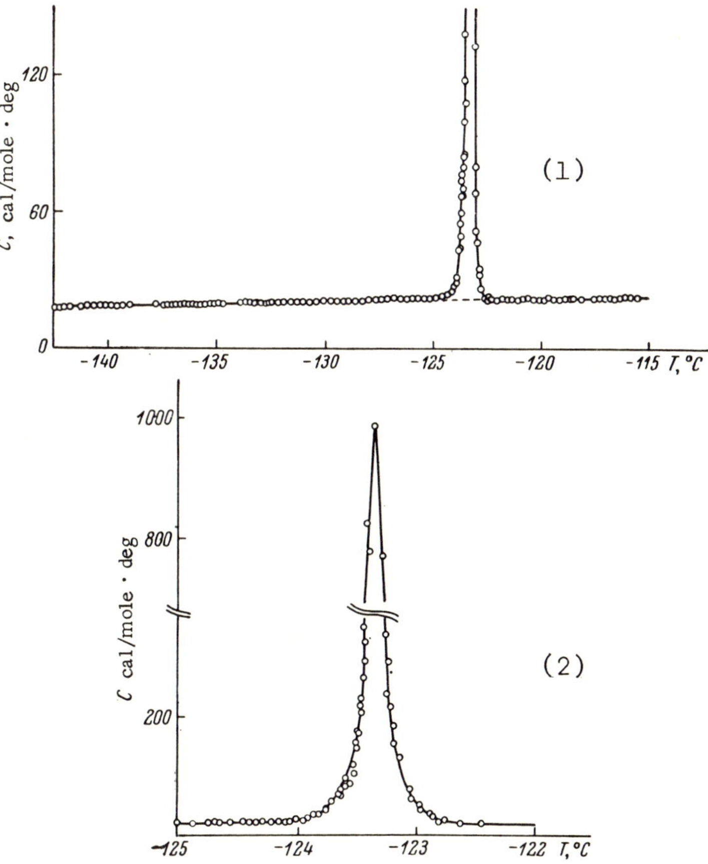


Fig. 19b. Specific heat of ADP as a function of temperature.

Amin and Strukov have reported the specific heat of ADP as a function of temperature over the range, 131 to 158°K. Figure 19b shows their results in the vicinity of the transition temperature.

## Thermal Properties - Thermal Expansion

Cook has surveyed the literature on the thermal coefficient of expansion of ADP, as shown in Table 14. Boiko and Golovnin measured the thermal expansion of single crystals by x-ray diffraction at 110°K to 270°K and show the volume coefficient of thermal expansion.

TABLE 14. THERMAL COEFFICIENT OF EXPANSION OF ADP.

| Temperature Range (°C) | Expansion Coefficient ($10^{-6}$/°C) along a-axis | along c-axis | Reference |
|---|---|---|---|
| -50 to +50 | 32.0 | 4.2 | Cook see Fig. 20 |
| -60 to +80 | 34.0 | 5.3 | Mason see Fig. 21 |
| 24 to 134 | 39.3 | 1.9 | Deshpande and Khan |
| 20 | 37 | 4 | Haussühl |

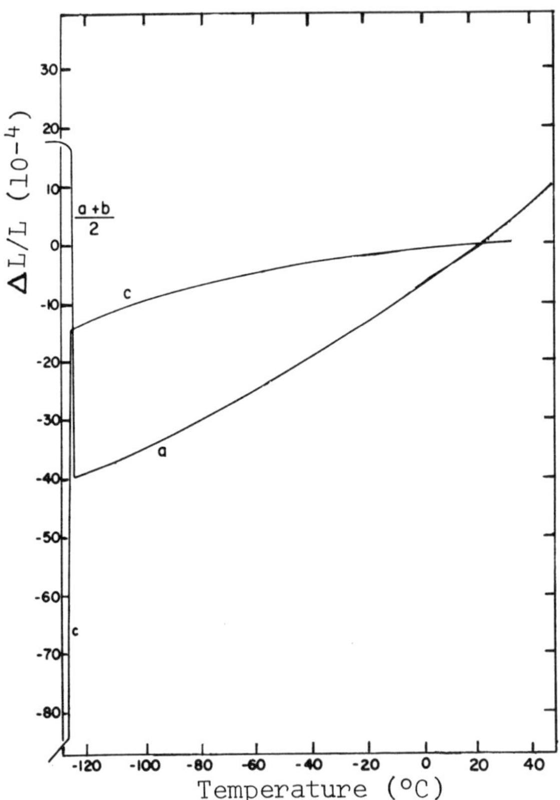


Fig. 20. Thermal expansion of ADP as a function of temperature. The a and c transition temperatures are slightly offset to increase the readability of the graph [Cook].

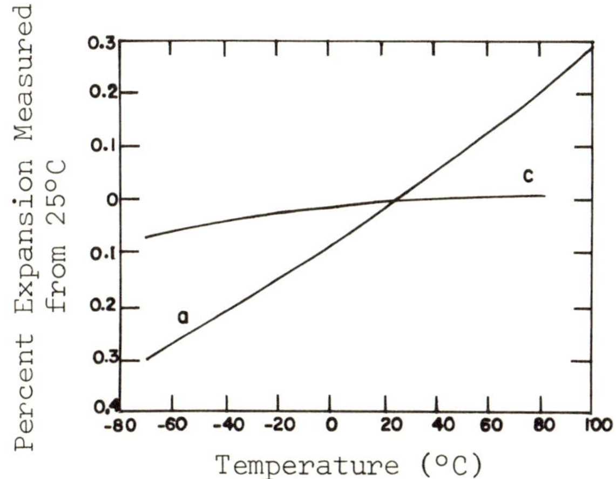


Fig. 21. Thermal expansion of ADP as a function of temperature [Mason].



## Thermal Properties - Thermal Conductivity

TABLE 15. THERMAL CONDUCTIVITY OF ADP [McCarthy and Ballard].

| Orientation | T(°C) | Thermal Conductivity (mW/cm°K) |
|---|---|---|
| a-axis | 40 | 12.5 |
| | 69 | 13.4 |
| c-axis | 42 | 7.1 |
| | 66 | 7.1 |

AMMONIUM DIHYDROGEN PHOSPHATE (ADP)

ACHYUTHAN, K. and M.A. BREAZEALE. Photoelastic Constants of Ammonium Dihydrogen Phosphate (ADP). OPTICAL SOC. OF AMERICA, J., v. 51, 1961. p. 914-915.

ADHAV, R.S. Linear Electro-Optic Effects in Tetragonal Phosphates and Arsenates. OPTICAL SOC. OF AMERICA, J., v. 59, no. 4, Apr. 1969. p. 414-418.

ALEKSANDROV, K.S. and L.N. RYABINKIN. The Elastic Properties of Ammonium Dihydrogen Phosphate and the Laval-Raman Elasticity Theory. SOVIET PHYS.-DOKL., v. 7, no. 2, 1962. p. 99-101.

AMIN, M. and B.A. STRUKOV. Effects of Deuteration on the Specific Heat of Ammonium Dihydrogen Phosphate Crystals. SOVIET PHYS. SOLID STATE, v. 12, no. 7, Jan. 1971. p. 1616-1618.

APKARYANTS, P.A. and A.S. SONIN. Induced Electrooptic Effect in the Antiferroelectric $NH_4H_2PO_4$ (ADP). SOVIET PHYS.-SOLID STATE, v. 11, no. 1, July 1969. p. 148-149.

MICHIGAN UNIV. WILLOW RUB LAB. Optical Materials for Infrared Instrumentation. By: BALLARD, S.S. et al. State-of-the-Art Rept. No. 2389-11-S, Jan. 1959. Contract No. Nonr 1224-12. AD 217 367.

BECHMANN, R. Contour Modes of Square Plates Excited Piezoelectrically and Determination of Elastic and Piezoelectric Coefficients. PHYS. SOC., PROC., B64, 1951. p. 323-337.

BELYAEV, L.M. et al. Dielectric Constant of Crystals Having an Electro-Optical Effect. SOVIET PHYS.-SOLID STATE, v. 6, no. 8, Feb. 1965. p. 2007-2008.

BJORKHOLM, J.E. Relative Measurement of the Optical Nonlinearities of KDP, ADP, $LiNbO_3$, and $\alpha$-$HIO_3$. IEEE J. QUANTUM ELECTRONICS, v. QE-4, Nov. 1968. p. 970-972. Also, Errata v. QE-5, May 1969. p. 260.

BJORKHOLM, J.E. and A.E. SIEGMAN. Accurate cw Measurements of Optical Second-Harmonic Generation in Ammonium Dihydrogen Phosphate and Calcite. PHYS. REV., v. 154, no. 3, Feb. 1967. p. 851-860.

BLOKH, O.G. Dispersion of r-63 for Crystals of ADP and KDP. SOVIET PHYS.-CRYST., v. 7, no. 4, Jan.-Feb. 1963. p. 509-511.

BLOKH, O.G. et al. Temperature and Wavelength Dependence of the r-41 Electro-Optical Coefficient of Ammonium Dihydrogen Phosphate and Potassium Dihydrogen Phosphate Crystals. ACAD. OF SCI., USSR, BULL., PHYS. SER., v. 33, no. 2, 1969.

BLOKH, O.G. and L.F. LUTSIV-SHUMSKII. Photoelastic Effect in 45 Degree X Cuts of Potassium Dihydrogen Phosphate and Ammonium Dihydrogen Phosphate. SOVIET PHYS. SOLID STATE, v. 12, no. 1, July 1970. p. 256-257.

BLOKH, O.G. and L.F. LUTSIV-SHUMSKII. Temperature and Wavelength of the Electrooptical Coefficient $r_{63}$ in ADP Crystals. SOVIET PHYS.-CRYST., v. 12, no. 3, Nov.-Dec. 1967. p. 380-382.

## AMMONIUM DIHYDROGEN PHOSPHATE (ADP)


BLUMENTHAL, R.H. Design of a Microwave-Frequency Light Modulator. IRE PROC., v. 50, 1962. p. 452-456.

BOIKO, A.A. and V.A. GOLOVNIN. Thermal Expansion of ADP and DADP in the Region of Antiferroelectric Phase Transition. SOVIET PHYS. CRYST., v. 15, no. 1, July-Aug. 1970. p. 153-154.

BURNETT, G.D. Light Modulation with Piezoelectric Crystals. ELECTRONIC INDUSTRIES, Nov. 1962. p. 90-95.

CARPENTER, R.O. The Electro-Optic Effect in Uniaxial Crystals of the Dihydrogen Phosphate Type. III. Measurement of Coefficients. OPTICAL SOC. OF AMERICA, J., v. 40, no. 4, Apr. 1950. p. 225-229.

CARPENTER, R.O. Photoelastic Constants of Ammonium Dihydrogen Phosphate and Potassium Dihydrogen Phosphate. OPTICAL SOC. OF AMERICA, J., v. 44, no. 5, 1954. p. 425.

COOK, W.R. Jr. Thermal Expansion of Crystals with $KH_2PO_4$ Structure. J. OF APPLIED PHYS., v. 38, no. 4, 1967. p. 1637-1642.

DESHPANDE, V.T. and A.A. KHAN. X-ray Determination of the Thermal Expansion of Ammonium Dihydrogen Phosphate. ACTA CRYST., v. 16, 1963. p. 936-938.

DEVIOT, B. Elasto-Optical Constants of $PO_4H_2(NH_4)$ and $SO_4Ni$, $6H_2O$ (In Fr.). J. OF PHYS. ET LE RADIUM, v. 16, no. 2, 1955. p. 162.

DIXON, R.W. Photoelastic Properties of Selected Materials and Their Relevance for Applications to Acoustic Light Modulators and Scanners. J. OF APPLIED PHYS., v. 38, no. 13, Dec. 1967. p. 5149-5153.

DOWLEY, M.W. Parametric Fluorescence in Ammonium Dihydrogen Phosphate and Potassium Dihydrogen Phosphate Excited by a 2573 Angstrom CW Pump. OPTO-ELECTRONICS, v. 1, no. 4, Nov. 1969. p. 179-181.

FRANCOIS, G.E. CW Measurements of the Optical Nonlinearity of Ammonium Dihydrogen Phosphate. PHYS. REV., v. 143, no. 2, Mar. 1966. p. 597-600.

HAUSSÜHL, S. Elastic and Thermoelastic Properties of $KH_2PO_4$, $KH_2AsO_4$, $NH_4H_2PO_4$, and $RbH_2PO_4$. Z. FUER KRYSTALLOGRAPHIE, v. 120, 1964. p. 401-414.

JERPHAGNON, J. and S.K. KURTZ. Optical Nonlinear Susceptibilities, Accurate Relative Values for Quartz, Ammonium Dihydrogen Phosphate, and Potassium Dihydrogen Phosphate. PHYS. REV. B, Ser. 3, v. 1, no. 4, Feb. 15, 1970. p. 1739-1742.

KAMINOW, I.P. Microwave Dielectric Properties of $NH_4H_2AsO_4$, and Partially Deuterated $KH_2PO_4$. PHYS. REV., v. 138, no. 5A, May 1965. p. A1539-A1543.

KÄNZIG, W. Ferroelectrics and Antiferroelectrics. SOLID STATE PHYS., v. 4, 1957. N.Y., Academic Press. p. 1-197.

KOETSER, H. Measurement of $r_{63}$ for ADP up to Electric Breakdown. ELECTRONICS LETTERS, v. 3, no. 2, Feb. 1967. p. 54-55.

LANDOLT-BÖRNSTEIN. NUMERICAL DATA AND FUNCTIONAL RELATIONSHIPS IN SCIENCE AND TECHNOLOGY, New Series, v. 1, Elastic, Piezoelectric, Piezooptic and Electro-optic Constants of Crystals. Group III: Crystal and Solid State Physics. Ed. by: Hellwege, K.-H. and A.M. Hellwege. Berlin, Ger.: Springer-Verlag, 1966.

LEY, J.M. Low-Voltage Light-Amplitude Modulation. ELECTRONICS LETTERS, v. 2, no. 1, Jan. 1966. p. 12-13.

LEY, J.M. Variation of $r_{41}$ with Wavelength for ADP. ELECTRONICS LETTERS, v. 3, no. 4, Apr. 1967. p. 145.

LOIACONO, G.M. A DTA Study of the Ferroelectric Transition in Potassium Dihydrogen Phosphate Type Crystals. MAT. RES. BULL., v. 5, no. 9, Sept. 1970. p. 775-782.

MASON, W.P. The Elastic, Piezoelectric, and Dielectric Constants of KDP and ADP. PHYS. REV., v. 69, no. 5-6, Mar. 1946. p. 173-194.

MASON, W.P. Elasto-Electric Constants of $KH_2PO_4$ Type Crystal. Table III. PHYSICAL ACOUSTICS, v. 1, Pt. A. Academic Press, N.Y., 1964. p. 181.

McCARTHY, K.A. and S.S BALLARD. New Data on the Thermal Conductivity of Optical Crystals. OPTICAL SOC. OF AMERICA, J., v. 41, 1951. p. 1062-1063.

McMAHON, D.H. and A.R. FRANKLIN. Laser Focusing Effects on Second Harmonic Generation in ADP. APPLIED PHYS. LETTERS, v. 6, no. 1, Jan. 1, 1965. p. 14-16.

MILLER, R.C. et al. Quantitative Studies of Optical Harmonic Generation in CdS, $BaTiO_3$, and $KH_2PO_4$ Type Crystals. PHYS. REV. LETTERS, v. 11, no. 4, Aug. 15, 1963. p. 146-149.

MORLOT, G. et al. Absorption Spectra of Ammonium Dihydrogen Phosphate in the Infrared Range (In Fr.). PHYS. STATUS SOLIDI, B, v. 49, no. 1, 1972. p. K47-K52.

MURPHY, E.J. Conduction in Single Crystals of Ammonium Dihydrogen Phosphate. J. OF APPLIED PHYS., v. 35, Sept. 1964. p. 2609-2614.

NAMBA, S. High Voltage Measurement by ADP Crystal Plate. REV. OF SCI. INSTRUMENTS, v. 32, 1961. p. 595-597.

OTT, J.H. and T.R. SLIKER. Linear Electro-Optic Effects in $KH_2PO_4$ and Its Isomorphs. OPTICAL SOC. OF AMERICA, J., v. 54, no. 12, Dec. 1964. p. 1442-1444.

PENN STATE COLLEGE. Ferroelectric and High Dielectric Crystals: Dielectric Measurements of Crystals. By: PEPINSKY, R. Tech. Rept. No. 3, Feb. 1952.

PERFILOVA, V.E. The Quadratic Electrooptic Effect in Single Crystals of Ammonium Dihydrogen Phosphate. SOVIET PHYS.-CRYST., v. 12, no. 4, 1968. p. 621-622.

PERFILOVA, V.E. and A.D. SONIN. The Quadratic Electro-Optical Effect in KDP Group Crystals. ACAD. OF SCI. USSR, BULL. PHYS. SER., v. 31, no. 7, July 1967. p. 1154-1157.

PHILLIPS, R.A. Temperature Variation of the Index of Refraction of ADP, KDP and Deuterated KDP. OPTICAL SOC. OF AMERICA, J., v. 56, May 1966. p. 629-632.

POPOVA, A.S. and L.B. RIPS. The True Electro-Optic Effect in Crystals of $NH_4H_2PO_4$. SOVIET PHYS.-CRYST., v. 10, no. 3, Nov.-Dec. 1965. p. 347-348.

PRICE, W.J. and H.B. HUNTINGTON. Acoustical Properties of Anisotropic Materials. ACOUSTICAL SOC. OF AMERICA, J., v. 22, 1950. p. 32-37.

SILVERSTEIN, L. and M. SUCHER. Determination of the Pockels Electro-Optic Coefficient in ADP at 5.5 GigaHertz. ELECTRONICS LETTERS, v. 2, no. 12, Dec. 1966. p. 437-438.

SKALYO, J., Jr. et al. The Pressure Dependence of the Transition Temperature in KDP and ADP. J. OF PHYS. AND CHEM. OF SOLIDS, v. 30, no. 8, Aug. 1969. p. 2045-2051.

CLEVITE CORP. Reference Data on Linear Electro-Optic Effects. Engineering Memorandum 64-10. By: SLIKER, T.R. May 15, 1964. 9pp.

STEPHENSON, C.C. and A.C. ZETTLEMOYER. The Heat Capacity of Ammonium Dihydrogen Phosphate from 15 to 300°K. The Anomaly at the Curie Temperature. AMERICAN CHEM. SOC., J., v. 66, 1944. p. 1405-1408.

TREVELYAN, B. The Practical Design of a Laser Modulator Using 45 Degree Cut ADP Crystals. J. OF SCI. INSTRUMENTS, (J. OF PHYS. E), Ser. 2, v. 2, 1969. p. 425-428.

VAN DER ZIEL and N. BLOEMBERGEN. Temperature Dependence of Optical Harmonic Generation in $KH_2PO_4$ Ferroelectrics. PHYS. REV., v. 135, 1964. p. 1662-1669.

VASILEVSKAYA, A.S. The Electro-Optical Properties of Crystals of KDP Type. SOVIET PHYS.-CRYST., v. 11, no. 5, Mar.-Apr. 1967. p. 644-647.

VASILEVSKAYA, A.S. and A.S. SONIN. Electro-Optical and Elasto-Optical Properties of Deuterated Ammonium Dihydrogen Phosphate Crystals. SOVIET PHYS. SOLID STATE, v. 8, no. 11, May 1967. p. 2756-2757.

VASILEVSKAYA, A.S. and A.S. SONIN. The Relation of Structure to Electrooptic and Elastooptic Properties in Crystals of the KDP Group. SOVIET PHYS. CRYST., v. 14, no. 4, Jan.-Feb. 1970. p. 611-613.

## AMMONIUM DIHYDROGEN PHOSPHATE (ADP)

VISHNEVSKII, V.N. et al. On the Temperature Dependence of Double-Refraction Dispersion in Ammonium Dihydrogen Phosphate Crystals. OPT. AND SPECTRO., v. 18, no. 5, May 1965. p. 468-469.

VISHNEVSKII, V.N. and I.V. STEFANSKII. Temperature Dependence of the Dispersion of the Refractivity of ADP and KDP Single Crystals. OPT. I SPECTRO., v. 20, no. 2, Feb. 1966. p. 195-196.

DIELECTRIC MATERIALS AND APPLICATIONS. Chap. V: Tables of Dielectric Materials. Ed. by: VON HIPPEL, A.R. Wiley, N.Y., 1954. p. 291-425.

WARD, J.F. Absolute Measurement of an Optical-Rectification Coefficient in Ammonium Dihydrogen Phosphate. PHYS. REV., v. 143, Mar. 1966. p. 569-574.

WARD, J.F. and G.H.C. NEW. Optical Rectification in Ammonium Dihydrogen Phosphate and Quartz. ROYAL SOC. OF LONDON, PROC., A, v. 299A, no. 1457, June 1967. p. 238-263.

WEST, C.D. and A.S. MAKAS. Some Photoelastic Constants of Crystals Ammonium Dihydrogen Phosphate and Potassium Dihydrogen Phosphate. AMERICAN MINERALOGIST, v. 35, 1950. p. 130.

WIENER, E. et al. Antiferroelectric Transitions in Ammonium Dihydrogen Phosphate and Ammonium Dihydrogen Arsenate Studied by Infrared Absorption. J. OF CHEM. PHYS., v. 52, no. 6, Mar. 15, 1970. p. 2891-2900.

WIENER, E. et al. Proton Dynamics in Potassium Dihydrogen Phosphate Type Ferroelectrics Studied by Infrared Absorption. J. OF CHEM. PHYS., v. 52, no. 6, Mar. 15, 1970.

YAMAZAKI, M. and T. OGAWA. Temperature Dependence of the Refractive Indices of $NH_4H_2PO_4$, $KH_2PO_4$ and Partially Deuterated $KH_2PO_4$. OPTICAL SOC. OF AMERICA, J., v. 56, no. 10, Oct. 1966. p. 1407-1408.

ZERNIKE. F., Jr. Refractive Indices of Ammonium Dihydrogen Phosphate and Potassium Dihydrogen Phosphate Between 2000 Angstroms and 1.5 Microns. OPTICAL SOC. OF AMERICA, J., v. 54, Oct. 1964. p. 1215-1220.

ZIAUDDIN, M. and T.S. NARASHIMHAMURTY. Photoelastic Constants of Ammonium Dihydrogen Phosphate. OPTICAL SOC. OF AMERICA, J., v. 57, Nov. 1967. p. 1392-1393.

ZWICKER, B. Elastic Studies in $NH_4H_2PO_4$ and $KH_2PO_4$ (In Ger.). HELV. PHYS. ACTA, v. 19, 1946. p. 523-549.

## BARIUM SODIUM NIOBATE

### Introduction

Barium sodium niobate ($Ba_2NaNb_5O_{15}$) has been shown to be an outstanding material for electrooptic and nonlinear optic applications, and is particularly useful for the second harmonic generation of 0.53 radiation from 1.06 microns and for the parametric conversion of 0.53 micron radiation to longer wavelengths. The material was discovered in 1967 by researchers at the Bell Telephone Laboratories and was found to be stable to intense visible radiation (laser beams) and to have a number of useful nonlinear, elastooptic and piezoelectric properties. The discovery of this nonlinear crystal made possible the high-power generation of continuous, coherent green light through the conversion of infrared laser radiation. This, in turn, made a practical solid-state source of green laser light possible. When operated as a harmonic generator with a neodymium-doped yttrium-aluminum garnet laser cavity, the new crystal generates about 210 mW at 0.532-micron radiation. It is hoped that work with this crystal eventually will lead to a continuously operating tunable parametric oscillator. Barium sodium niobate is phase matchable without double refraction; its nonlinear coefficients are approximately twice those observed in lithium niobate, and it is stable under intense laser radiation. A most important practical feature is the fact that the phase-match temperatures are quite reproducible from crystal to crystal, the variations being less than 10°C.

### Chemical and Physical Properties

The composition $Ba_2NaNb_5O_{15}$ (also written as: $Ba_{0.8}Na_{0.4}Nb_2O_6$) is reported to be a stoichiometric compound with all sites filled in the crystal lattice structure.

| Density | 5.39 g/cm$^3$ | Van Uitert et al. |
|---|---|---|

### Crystallography

Van Uitert and his coworkers at Bell Telephone Laboratories have conducted extensive crystal growth studies (using Czochralski techniques) on barium sodium niobate and other mixed and related niobates. Melt conditions, pulling rates and furnace conditions are detailed in their paper. A wide range of thermal gradient conditions, as well as chemical compositions, were investigated in order to obtain good boules and reduce growth striae. These authors also report that the grown barium sodium niobate crystals require a certain amount of processing before they are ready for device use. To obtain crystals having good optical quality: (1) moderate thermal gradient conditions should be employed, (2) the melt should be centered in a uniformly heated, high thermal mass furnace that provides radial and vertical gradients that just suffice to permit controlled crystal growth, (3) close temperature control and low crystal rotation rates during growth are desirable, (4) platinum crucibles in air or in atmospheres that are predominantly argon or nitrogen can be utilized, (5) annealing in oxygen or air in the 800 to 900°C range may be required subsequent to pulling when the latter atmospheres are employed.

Barium sodium niobate tends to grow as an ordered crystal from the stoichiometric melt; there are unique sites for both Ba(A1) and Na(A2) in the proper number and ratio (2:1) to meet the electrostatic requirement for a filled structure. Linares and Sigsway have also conducted Czochralski crystal pulling studies on this material; they varied the partial pressures, pulling rates, rotation rates, oxygen and nitrogen atmospheres and used platinum and iridium crucibles. Two types of defect phenomenon were present: whisker-like inclusions and growth rings. The growth and preparation of large single crystals of barium sodium niobate, also involving the Czochralski method, have been detailed by Rice et al. at Union Carbide Corp.

Preliminary x-ray and optical crystal analyses [Geusic et al.] indicate that at room temperature the crystal is orthorhombic and belongs to the point group mm2. At approximately 260°C, the crystal undergoes a structural transformation. Above 260°C, it is tetragonal. This structural transformation manifests itself in producing a high degree of microtwinning in the as-grown crystal. The twinning is such that all crystallites have their c axes in the same direction but the a and b axes are interchanged. The microtwinning is easily removed by heating the crystal above 300°C with a compressive stress applied along one of the tetragonal a axes. On cooling below 300°C, the axis along which the compressive stress is applied becomes the intermediate or b axis of the orthorhombic phase. Once detwinned in this manner, the barium sodium niobate is a uniform single crystal and can be handled, cut and polished normally without undue concern of the crystal reverting to the twinned state of the as-grown crystal. Table 1 lists the various phases known to exist for this crystal.

TABLE 1. CRYSTAL STRUCTURE OF BARIUM SODIUM NIOBATE [Van Uitert et al.].

| Temperature Range | Phase | Symmetry | Space Group |
|---|---|---|---|
| RT to 260°C | Ferroelectric | Orthorhombic | Cmm2 ($C_{2v}^{11}$) |
| 260° to 560°C | Ferroelectric | Tetragonal | P4bm ($C_{4v}^{2}$) |
| > 560°C ($T_c$) | Paraelectric | (centrosymmetric) | |

The lattice constants of barium sodium niobate have been reported as follows:

| Formula | $a_o$ | $b_o$ | $c_o$ | | Symmetry | Reference |
|---|---|---|---|---|---|---|
| $NaBa_2Nb_5O_{15}$ | 17.59 | 17.61 | 7.982 | Å | orthorhombic | Giess et al. |
| $Na_{1.74}Ba_{4.13}Nb_{10}O_{30}$ | | | | | 2 formula weight/cell G=5.413 g/cm$^3$ | Jamieson et al. |
| | 17.59182 | 17.6256 | 3.9949 | | orthorhombic | |
| | 17.59182 | 17.6256 | 7.0909 | | orthorhombic G=5.4076 g/cm$^3$ | Barns |
| | 17.626 | 17.592 | 3.995 | | | Van Uitert et al. |

These authors also have presented a representative picture of the tungsten bronze structure (tetragonal unit cell consisting of 10 Nb octahedrons) of barium sodium niobate (Figure 1). They note that this arrangement provides space for up to four cations in 9-coordinated tricapped prismatic A1 sites, two cations in somewhat 12-coordinated cubo-octahedral A2 sites, and four cations in relatively small 3-coordinated planar trigonal C sites.

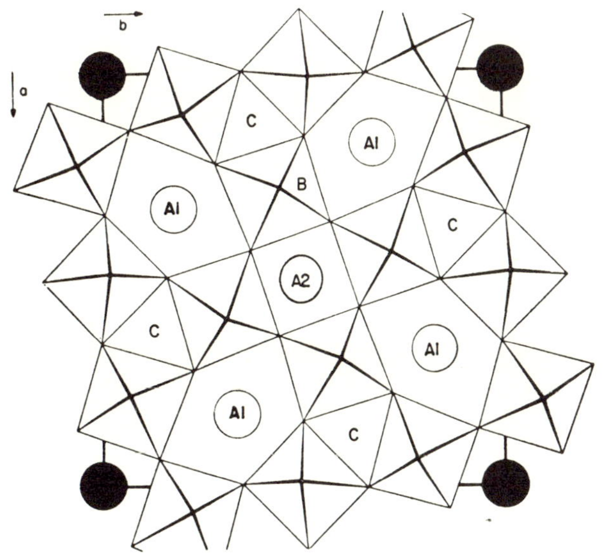


Fig. 1. A representation of the structure of $Ba_2NaNb_5O_{15}$. Ba occurs in A1 sites, Na in A2 sites and Nb in B sites. The C sites are empty. [Van Uitert et al.]

| Site Designation | A1 | A2 | C | B |
|---|---|---|---|---|
| Sites Available | 4 | 2 | 4 | 10 |

## Optical Properties - Refractive Index

Barium sodium niobate is transparent in the range 0.4 to 5 microns [Geusic et al. Refractive index data at 30°C are presented in Table 2. In the temperature range 260 to 560°C, the crystal is uniaxial. The birefringence is reported to be negligible above $T_C$ (560°C).

TABLE 2. REFRACTIVE INDEX OF BARIUM SODIUM NIOBATE AT 30°C [Singh et al.].

| λ (nm) | $n_x$ | $n_y$ | $n_z$ |
|---|---|---|---|
| 457.9 | 2.4284 | 2.4266 | 2.2931 |
| 476.5 | 2.4094 | 2.4076 | 2.2799 |
| 488.0 | 2.3991 | 2.3974 | 2.2727 |
| 496.5 | 2.3920 | 2.3903 | 2.2678 |
| 501.7 | 2.3879 | 2.3862 | 2.2649 |
| 514.5 | 2.3786 | 2.3767 | 2.2583 |
| 532.1 | 2.3672 | 2.3655 | 2.2502 |
| 632.8 | 2.3222 | 2.3205 | 2.2177 |
| 1064.2 | 2.2580 | 2.2567 | 2.1700 |

$dn_{x,y}/dT = -2.5 \times 10^{-5}/°C$ $dn_z/dT = 8 \times 10^{-5}/°C$ at 1.064μ

There are four types of optical aberration in barium sodium niobate which tend to restrict the usefulness of crystals for electrooptic or nonlinear applications. The most persistent defects are growth striations which appear to some extent in all crystals. Rice et al. reported that the optical behavior of a-axis crystals is superior to c-axis crystals since the light is propagated normal to the striations and thus suffers less beam divergence. A second more subtle optical defect is a refractive index variation parallel to the growth direction which is observable as bands parallel to the growth axis. A third type of defect is a slow variation of refractive index which produces interference fringes when a crystal with flat and parallel faces is used as a Fabry-Perot etalon in a laser beam. A fourth type of defect, seen only occasionally, is a gross localized refractive index anomaly which may appear quite different when viewed under different polarizations.

## Optical Properties - Nonlinear Optical Behavior

Geusic et al. report that the nonlinear optical coefficients of barium sodium niobate are approximately three times those of $LiNbO_3$. They note that the material at room temperature does not have the serious problem of optically induced refractive index inhomogeneities which has been observed in several nonlinear materials. In addition, barium sodium niobate has unique and reproducible phase-match temperatures (80-90°C). These authors report the following nonlinear coefficients relative to $d_{36}^{2\omega}$ of KDP:

$$d_{33}^{2\omega} = 34.4 \pm 2$$
$$d_{31}^{2\omega} = 29 \pm 6 \qquad \text{at 1.06 microns}$$
$$d_{32}^{2\omega} = 33.9 \pm 3$$

The results of Byer et al. as shown in Figure 2, indicate that the nonlinear coefficient $d_{31}^{2\omega}$ is constant from room temperature to about 300°C, and then breaks sharply and decreases to zero at the Curie temperature of approximately 560°C. The crystals were striated and remained so over the full temperature range examined; however, above 300°C a significant reduction of crystal strain was observed.

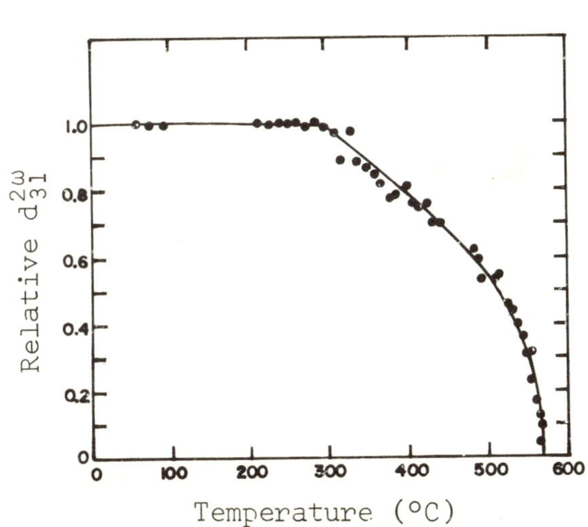


Fig. 2. Relative nonlinear coefficient $d_{31}^{2\omega}$ as a function of temperature at 1.06 micron [Byer et al.].

## Electrooptic Properties

Barium sodium niobate has five independent nonzero electrooptic coefficients: $r_{13}$, $r_{23}$, $r_{33}$, $r_{42}$, and $r_{51}$. Singh et al. using 6330 Å light and a low (60 Hz) frequency obtained the following values:

| $r_{13}$ | $r_{23}$ | $r_{33}$ | $r_{42}$ | $r_{51}$ | |
|---|---|---|---|---|---|
| 15 | 13 | 48 | 92 | 90 | $10^{-12}$ m/V |

Nash et al. used 90 MHz frequency and 6328, Å light and obtained:

$$r_{13} = r_{23} = 6.2 \times 10^{-12} \text{ m/V}, \; r_{33} = 24.4 \times 10^{-12} \text{ m/V}$$

| Light Direction | Half-wave Voltage | Electrooptic Coefficient |
|---|---|---|
| y-axis | 1.72 kV | $\|n_z^3 r_{33} - n_x^2 r_{13}\| = 3.7 \times 10^{-10}$ m/V |
| x-axis | 1.57 kV | $\|n_z^3 r_{33} - n_y^3 r_{13}\| = 4.0 \times 10^{-10}$ m/V |

The above data are reported by Geusic et al. for an electric field applied along the z-axis and the light polarized at 45° to the z-axis; at 6328Å. Additionally they report the following:

$$|n_z^3 r_{33}| = 6.2 \pm 0.4 \times 10^{-10} \text{ m/V}$$
$$|n_x^3 r_{13}| = 2.3 \pm 0.4 \times 10^{-10} \text{ m/V}$$
$$|n_y^3 r_{23}| = 1.7 \pm 0.4 \times 10^{-10} \text{ m/V}$$

Indications are that $r_{13}$, $r_{23}$ and $r_{33}$ have the same relative sign. The above results are consistent with the later work of Rice et al.

The half-wave voltage versus temperature, with an electric field along the c axis and light propagating along the a axis, is shown in Figure 3.

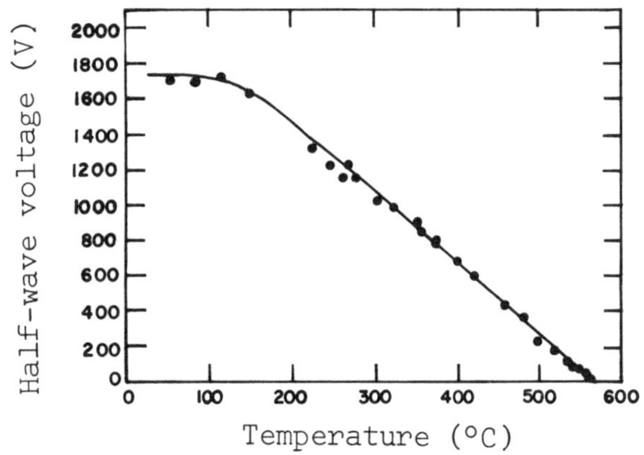


Fig. 3. Half-wave voltage as a function of temperature at 6328 Å [Byer et al.].

## Photoelastic Properties - Elastooptic Coefficients

There are twelve possible nonzero elastooptic coefficients for barium sodium niobate. Spencer and Van Uitert have investigated the elastooptic behavior of $Ba_2NaNb_5O_{15}$ by Bragg diffraction techniques at 100 MHz. The coefficients $p_{23}$, $p_{32}$ and $p_{33}$ were small but readily observable; the largest value obtained was for $p_{31} = 0.14$ to 0.2.

## Photoelastic Properties - Elastic Constants

TABLE 3. ELASTIC CONSTANTS OF $Ba_2NaNb_5O_{15}$

| $c_{ij}^E$ | $10^{11}$ N/m$^2$ | $s_{ij}^E$ | $10^{-12}$ m$^2$/N |
|---|---|---|---|
| $c_{11}$ | 2.39 | $s_{11}$ | 5.30 |
| $c_{12}$ | 1.04 | $s_{12}$ | -1.98 |
| $c_{13}$ | 0.50 | $s_{13}$ | -1.20 |
| $c_{22}$ | 2.47 | $s_{22}$ | 5.14 |
| $c_{23}$ | 0.52 | $s_{23}$ | -1.25 |
| $c_{33}$ | 1.35 | $s_{33}$ | 8.33 |
| $c_{44}$ | 0.65 | $s_{44}$ | 15.4 |
| $c_{55}$ | 0.66 | $s_{55}$ | 15.2 |
| $c_{66}$ | 0.76 | $s_{66}$ | 13.2 |

## Piezoelectric Properties

Spencer and Van Uitert have shown barium sodium niobate to be strongly piezoelectric with low acoustic loss at high frequency (acoustic Q $\sim 10^5$ at 500 MHz). The piezoelectric properties of $Ba_2NaNb_5O_{15}$ have been studied further by Warner and coworkers. Taking measurements at 23°C and following the IRE convention to designate the lattice directions ( $c < a < b$), these authors report the following piezoelectric constants and coupling factors:

| | | | |
|---|---|---|---|
| $d_{15}$ | $4.2 \times 10^{-11}$ C/N | $e_{15}$ | 2.8 C/m$^2$ |
| $d_{24}$ | 5.2 | $e_{24}$ | 3.4 |
| $d_{31}$ | -0.7 | $e_{31}$ | -0.4 |
| $d_{32}$ | -0.6 | $e_{32}$ | -0.3 |
| $d_{33}$ | 3.7 | $e_{33}$ | 4.3 |

$k_{15} = 0.21 \pm 0.05$

$k_{24} = 0.25 \pm 0.05$

$k_t = 0.57 \pm 0.01$

$k_{31} = 0.14$

$k_{32}$ 0.13

## BARIUM SODIUM NIOBATE

The high coupling factor $k_t = 0.57$ is effectively independent of temperature from 0° to 100°C; it is the highest value yet reported for a single-crystal ferroelectric material usable at and above room temperature. Since barium sodium niobate has good mechanical handling characteristics, a relatively low dielectric constant ($\varepsilon_{33}/\varepsilon_o$), and a pure mode of vibration, plates of this material are attractive as high-frequency longitudinal mode transducers.

### Dielectric Properties

Dielectric constant data are presented in Table 4. Details of the dielectric constant versus temperature near the Curie point as shown in Figure 4; no dielectric anomaly was observed at the orthorhombic-tetragonal near 300°C.

TABLE 4. DIELECTRIC CONSTANTS OF BARIUM SODIUM NIOBATE

| Constant Stress (T)<br>Constant Strain (S) | $\varepsilon_{33}/\varepsilon_o$ | $\varepsilon_{11}/\varepsilon_o$ | $\varepsilon_{22}/\varepsilon_o$ | T(°C) | Reference |
|---|---|---|---|---|---|
| S | 32 | 222 | 227 | 23 | Warner et al., 15 |
| T | 51 | 235 | 247 | | |
| | 51 | 246 | 242 | 30 | Geusic et al., 3 |
| | 48,000 | 191 | 191 | 560 | |

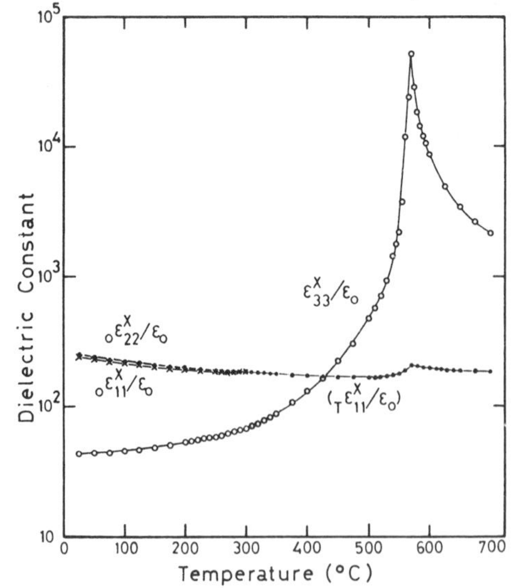


Fig. 3a Temperature characteristics of dielectric constants of $Ba_2NaNb_5O_{15}$ [Yamada et al.].

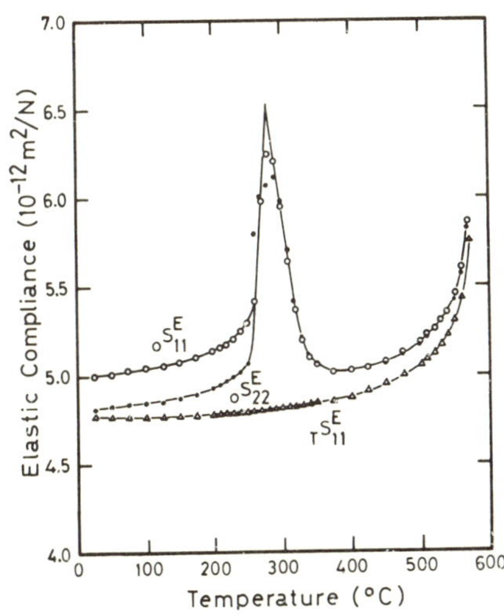


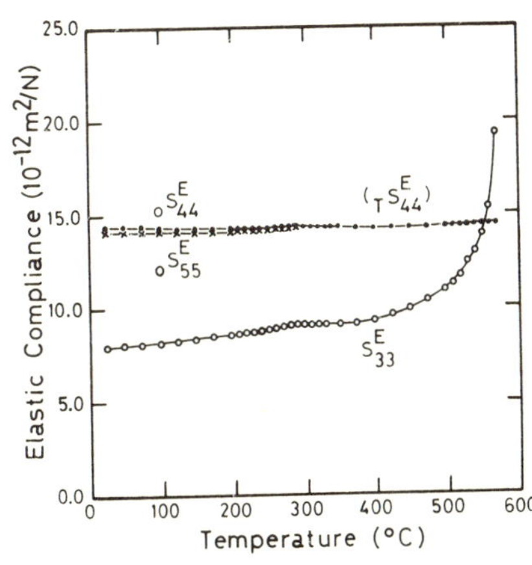


Fig. 3b Elastic compliance coefficients of $Ba_2NaNb_5O_{15}$ as a function of temperature [Yamada et al.].

ELASTIC AND PIEZOELECTRIC VALUES
FOR $Ba_2NaNb_5O_{15}$ AT 20°C.

| | | |
|---|---|---|
| $_os_{11}{}^E = 5.00 \times 10^{-12} m^2/N$ | | $_od_{15} = 3.2 \times 10^{-11} C/N$ |
| $_os_{22}{}^E = 4.81$ | | $_od_{24} = 4.5$ |
| $_os_{33}{}^E = 7.85$ | and | $_od_{31} = -0.68$ |
| $_os_{44}{}^E = 14.3$ | | $_od_{32} = -0.69$ |
| $_os_{55}{}^E = 14.0$ | | $_od_{33} = 3.4.$ |

[Yamada et al.]

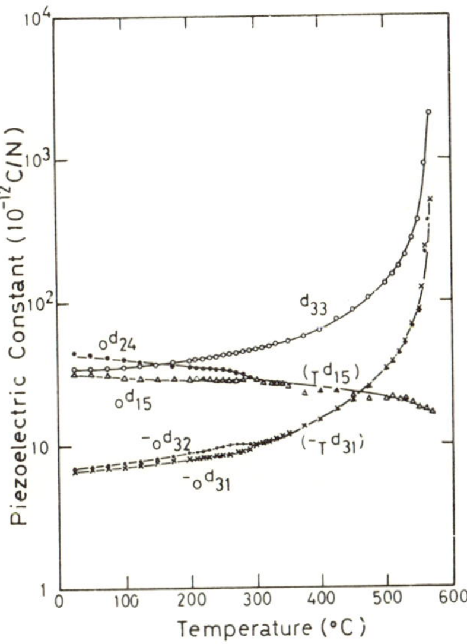


Fig. 3c Piezoelectric constants of $Ba_2NaNb_5O_{15}$ as a function of temperature [Yamada et al.].

Nash et al. report two values for the dielectric constant $\varepsilon_{33}$: at 1 kHz $\varepsilon_{33}^{T}$ = 41.8, at 20-100 MHz $\varepsilon_{33}^{S}$ = 29.5.

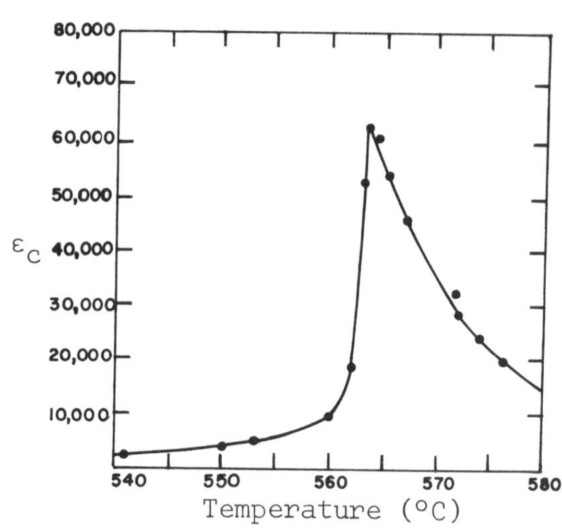


Fig. 4. Dielectric constant along the c axis as a function of temperature at 10 kHz [Rice et al.].

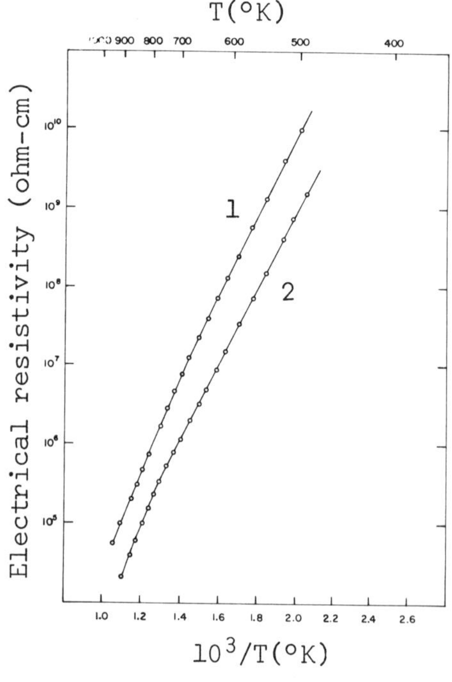


Fig. 5. Electrical resistivity of barium sodium niobate as a function of reciprocal temperature.

1. crystal grown using an Ir crucible under Ar cover.
2. crystal grown using a Pt crucible in air.

The activation energy for conduction (Q) was calculated using the relation $p \sim \exp(Q/kT)$ for crystal 1:

| | |
|---|---|
| Q = 1.00 eV | 510-700°K |
| Q = 1.33 eV | 700-900°K |

[Van Uitert et al.].

## Ferroelectric Properties

The Curie temperature of $Ba_2NaNb_5O_{15}$ has been determined as $T_c$ = 585°C by Singh et al. and found to be sensitive to chemical composition by Rice et al.

Pulled boules of barium sodium niobate consist of many ferroelectric domains. The domain structure in both a- and c-axis crystals consists of stacked concave disks in which the polarization is oppositely directed (Figure 6). The domains tend to be normal to the growth direction of the pulled crystal, about two to ten microns apart.

To be useful as an active optical component a crystal must first be converted to a single ferroelectric domain by an appropriate poling procedure. Both a- and c-axis crystals can be completely poled by the same procedure which consists of applying a field in excess of 100 V/cm along the c direction near the Curie temperature. The field can be applied a few degrees below the Curie point for a sufficient period or the crystal can be cooled through the Curie point under field [Rice et al.]. Rice and Fay have studied the characteristics of partially poled $Ba_2NaNb_5O_{15}$.

The room-temperature spontaneous polarization in $Ba_2NaNb_5O_{15}$ has been measured directly by Camlibel using a pulsed field method and an aqueous solution of LiCl as the electrode material. A value of $P_s$ = 0.40 ± 0.01 C/m$^2$ is reported.

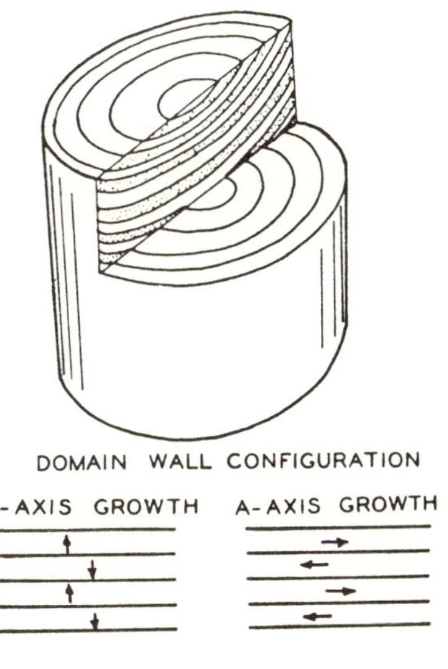


Fig. 6. A representation of the domain structure of $Ba_2NaNb_5O_{15}$ in a pulled boule [Van Uitert et al.].

## Thermal Properties

The compound melts congruently at 1437°C according to Giess et al.

COEFFICIENT OF THERMAL EXPANSION [Singh et al.].

| $\Delta T$ (°C) | $\alpha_a$ ($10^{-6}$ deg$^{-1}$) | $\alpha_c$ ($10^{-6}$ deg$^{-1}$) |
|---|---|---|
| 50–200 | 10.4 | 11.4 |
| 200–250 | 14 | 16 |
| 250–300 | 16.6 | 10.4 |
| 300–350 | 8.3 | 0.298 |
| 350–400 | 8.3 | −0.955 |
| 400–450 | 8.3 | −17 |
| 450–500 | 8.3 | −31 |
| 500–550 | 8.3 | −49 |
| 550–600 | 60 | −7.15 |

## Vendor Sources

High optical quality single crystals of barium sodium niobate are now commercially available from the Isomet Corporation, 433 Commercial Avenue, Palisades Park, N.J. 07650. Various sizes are available and blanks are fabricated from boules which have been annealed, poled, detwinned, and x-ray oriented. Entrance and exit faces are polished flat to λ/5 and parallel to 10 seconds.

## Applications

Warner et al. have noted that barium sodium niobate is attractive for applications as high-frequency longitudinal mode transducers. Other potential applications include optical harmonic generators and optical parametric oscillator devices as well as optical switching.

Geusic et al., at Bell Telephone Laboratories, reported on a preliminary experiment with a 2.8 mm-thick crystal inside the cavity of a 1.064-micron Nd:YAG laser which resulted in the generation of 210 mW of 0.532-micron radiation. Non-optimal coupling prevented them from obtaining higher power. These workers in a later paper [Geusic et al.] reported on the development of a continuous 0.532-micron solid state source using this new optical nonlinear material within the cavity of a 1.064-micron YAG:Nd laser. A power of 1.1W in the $TEM_{oo}$ mode at 0.532 microns was produced by SHG which equalled the available 1.064-micron $TEM_{oo}$ output of the basic YAG:Nd laser employed. This performance, according to these investigators, is nearly three orders of magnitude greater than what had previously been reported using $LiNbO_3$. Additionally, barium sodium niobate crystals have been reported to suffer no optical damage even under intense laser radiation at 0.53 microns.

## BARIUM SODIUM NIOBATE

Smith et al. have used this crystal to attain low threshold cw optical parametric oscillation. The threshold was measured to be 45 mW of multimode power at 0.532 microns. The efficiency was found to be 1% with 300 mW of pump power. In another paper [Smith et al.] these workers reported on their observation of optical parametric oscillation in barium sodium niobate with a measured pulsed threshold power of 13W for a pump wavelength of 0.532 microns. The overall efficiency in their experiments was approximately 0.1%. By varying the temperature of the crystal from 88.4°C to 96°C, the tunability of the signal was achieved from 1.056 microns to 0.948 microns, corresponding to idler wavelengths from 1.072 to 1.214 microns.

Byer and coworkers have presented potential tuning curves for an argon-pumped barium sodium niobate parametric oscillator. The tuning curves are smooth and continuous through the phase transition at 300°C. They report that it should be possible to electrooptically tune an optical parametric oscillator at about 1000 $cm^{-1}$ by using this crystal at a temperature of 550°C, where the half-wave voltage is about 100V and the optical nonlinearity is still reasonably large.

The conversion efficiency for second harmonic generation has been found to approach 100% under optimum conditions and continuous parametric oscillation. The half-wave retardation voltage for optical switching was found to be conveniently low, the field distance product being about 1500V in the optimum direction.

## BARIUM SODIUM NIOBATE


ABELL, J.S. et al. A Dilatometric Study of the Orthorhombic-Tetragonal Phase Transition in Barium Sodium Niobate. J. OF MATERIALS SCIENCE, v. 6, 1971. p. 1084-1092.

BARNS, R.L. Barium Niobate ($Ba_{4+x}Na_{2-2x}Nb_{10}O_{30}$); Crystallographic Data and Thermal Expansion Coefficient. J. APPL. CRYST., v. 1, 1968. p. 290-292.

BONNER, W.A. et al. Effects of Changes in Melt Composition on Crystal Growth of Barium Sodium Niobate. MAT. RES. BULL., v. 5, 1970, P. 243-252.

BYER, R.L. et al. Nonlinear Optical Properties of $Ba_2NaNb_5O_{15}$ in the Tetragonal Phase. J. OF APPLIED PHYS., v. 40, no. 1, Jan. 1969. p. 444-445.

CAMLIBEL, I. Spontaneous Polarization Measurements in Several Ferroelectric Oxides Using a Pulsed-Field Method. J. OF APPLIED PHYS., v. 40, no. 4, Mar. 1969. p. 1690-1693.

GEUSIC, J.E. et al. The Nonlinear Optical Properties of $Ba_2NaNb_5O_{15}$. APPLIED PHYS. LETTERS, v. 11, no. 9, Nov. 1967. p. 269-271. Also Errate, APPLIED PHYS. LETTERS, v. 12, no. 6, Mar. 1968. p. 224.

GEUSIC, J.E. et al. Continuous 0.532 micron Solid-State Source Using $Ba_2NaNbO_{15}$. APPLIED PHYS. LETTERS, v. 12, no. 9, May 1968. p. 306-308.

GIESS, E.A. et al. Alkali Strontium-Barium-Lead Niobate Systems with a Tungsten Bronze Structure: Crystallographic Properties and Curie Points. AMERICAN CERAMIC SOC., J., v. 52, no. 5, May 1969. p. 276-281.

JAMIESON, P.B. et al. Ferroelectric Tungsten Bronze-Type Crystal Structures. II. Barium Sodium Niobate $Ba_{(4+x)}Na_{(2-2x)}Nb_{10}O_{30}$. J. OF CHEM. PHYS., v. 50, no. 10. May 1969. p.4352-4362.

LINARES, R.C. and R.L. SIGSWAY. Some Optical Defects in Barium Sodium Niobate. MAT. RES. BULL., v. 3, no. 10, Oct. 1968. p. 825-830.

NASH, F.R. et al. Measurements of Second-Harmonic Generation and the Variations in the Free and Clamped Values of the Dielectric Constants and Electro-Optic Coefficients in Barium Sodium Niobate. J. OF APPLIED PHYS., v. 43, no. 1, Jan. 1972. p. 1-14,

RICE, R.R. and H. Fay. Comparison of the Electro-Optic and Pyroelectric Effects in Partially Poled alpha-axis $Ba_2NaNb_5O_{15}$ Crystals. J. OF APPLIED PHYS., v. 40, no. 2, Feb. 1969. p. 909-910.

RICE, R.R. et al. Characteristics of $Ba_2NaNb_5O_{15}$ for Optical Switching and Harmonic Generation. ELECTROCHEM. SOC., J., v. 116, no. 6, June 1969. p. 839-843.

SINGH, S. et al. Optical and Ferroelectric Properties of Barium Sodium Niobate. PHYS. REV., B, Ser. 3, v. 2, no. 7, Oct. 1970. p. 2709-2724.

SMITH, R.G. et al. Continuous Optical Parametric Oscillation in $Ba_2NaNb_5O_{15}$. APPLIED PHYS. LETTERS, v. 12, no. 9, May 1969. p. 308-310.

SMITH, R.G. et al. Low-Threshold Optical Parametric Oscillator Using $Ba_2NaNb_5O_{15}$. J. OF APPLIED PHYS., v. 39, no. 8, July 1968. p. 4030-4032.

SPENCER, E.G. and L.G. VAN UITERT. Elastic Properties of $Ba_2NaNb_5O_{15}$. PHYS. LETTERS, v. 27A, no. 9, Sept. 1968. p. 626-627.

VAN UITERT, L.G. et al. Some Characteristics of Niobates Having Filled Tetragonal Tungsten Bronze-Like Structures. MAT. RES. BULL., v. 3, no. 1, Jan. 1968. p. 47-58.

VAN UITERT, L.G. et al. Growth of $Ba_2NaNb_5O_{15}$ Single Crystals for Optical Applications. IEEE J. OF QUANTUM ELECTRONICS, v. QE-4, no. 10, Oct. 1968. p. 622-627.

VAN UITERT, L.G. et al. Some Characteristics of Barium, Strontium, Sodium Niobates. MAT. RES. BULL., v. 4, no. 1, Jan. 1969. p. 63-74.

WARNER, A.W. et al. Piezoelectric Properties of $Ba_2NaNb_5O_{15}$. APPLIED PHYS. LETTERS, v. 14, no. 1, Jan. 1969. p. 34-35.

WARNER, A.W. et al. Elastic and Piezoelectric Constants of $Ba_2NaNb_5O_{15}$. J. OF APPLIED PHYSICS, v. 40, no. 11, Oct. 1969. p. 4353-4356.

YAMADA, T. et al. Elastic Anomaly of $Ba_2NaNb_5O_{15}$. J. OF APPLIED PHYS., v. 41, no. 10, Sept. 1970. p. 4141-4147.

# BISMUTH GERMANIUM OXIDE

## Introduction

Two stoichiometric compositions are known for this compound; $Bi_{12}GeO_{20}$ with a molar ratio of 6 $Bi_2O_3$:1 $GeO_2$ and $Bi_4Ge_3O_{12}$ with 2 $Bi_2O_3$:3 $GeO_2$.

Bismuth germanate, $Bi_{12}GeO_{20}$, has been grown as large single crystals, using the Czochralski technique, by several investigators, including Ballman and Lauer, Aldrich et al., Dickinson et al. It is strongly piezoelectric, optically active, photoconductive and shows a small linear electrooptic effect. A practical advantage of this material is its low melting point, (918°C), which allows the preparation of mechanically sound crystals more easily than other electro-optic crystals with higher melting points. The phase diagram has been published by Speranskaya and Arshakuni.

The other modification has been grown and reported by Nitsche and is isomorphous with Eulytite, ($Bi_4Si_3O_{12}$). The effective electro-optic coefficient of this bismuth germanate, expressed as $n_o^3 r_{41}$, is one fifth that of the higher bismuth compound.

## Chemical and Physical Properties

| | | |
|---|---|---|
| Formula | $Bi_{12}GeO_{20}$ | $Bi_4Ge_3O_{12}$ |
| Molecular Weight | 2900.35 | 1245.8 |
| Density | 9.232 g/cm$^3$ (Bernstein) | 7.120 g/cm$^3$ (Kuzminov et al.) |
| Hardness | 4.5 Mohs (Safonov et al.) | |
| Microhardness | 244 kg/mm$^2$ | 315 kg/mm$^2$ (Kuzminov et al.) |
| Colour | yellow, transparent (Ballman) | yellow tinge, transparent (Dickinson) |
| Formation Temperature | 822°C | 790°C (Kuzminov et al.) |
| Melting Point | 918°C | 1040°C (Kuzminov et al.) |
| Crystal Symmetry | cubic (Abrahams et al.) | cubic (Nitsche) |
| Space Group | I23 Z2 (Abrahams et al.) | I43d Z4 (Nitsche) |
| Lattice Parameters $a_o$ | 10.1455 Å (Abrahams et al.) | 10.527 Å (Nitsche) |

## Optical Properties

| | | |
|---|---|---|
| Transmission | 90% at 0.45-7.5μ (Safonov et al.) | 90% at 0.4-2μ<br>60% at 2-6μ (Nitsche) |
| Refractive Index | 2.55 at 0.51 (Venturini et al.) | 2.07 (Nitsche) |
| Photoconductivity Maximum | 0.5μ (Douglas and Zitter) | |

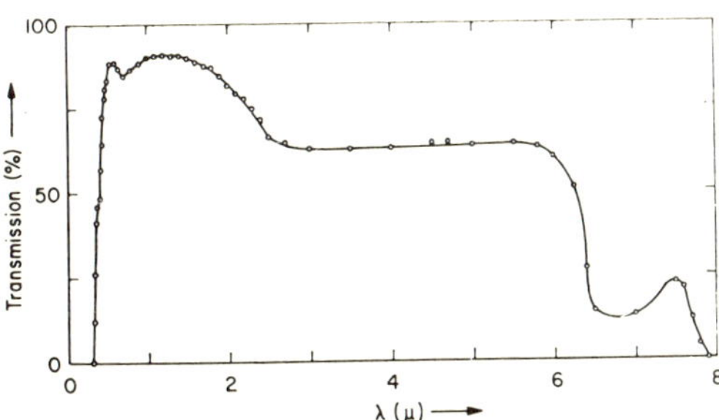


Fig. 1. Optical Transmission of single crystals of $Bi_4Ge_3O_{12}$, 0.5 mm thick.

[Nitsche]

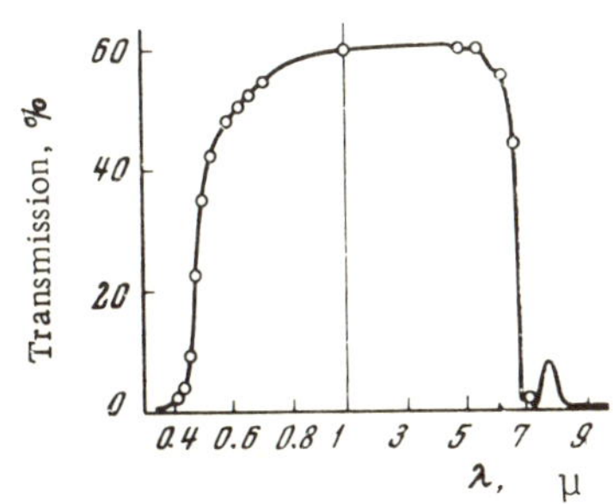


Fig. 2. Optical Transmission of single crystals of $Bi_{12}GeO_{20}$

[Safonov et al.]

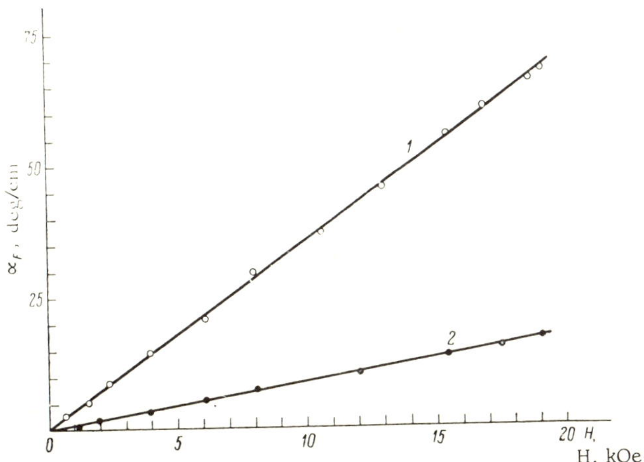


Fig. 3. Specific Rotation of the polarization plane on the external magnetic field in $Bi_{12}GeO_{20}$ at 20°C.

[Pisarev et al.]

## BISMUTH GERMANIUM OXIDE

### Electrooptic Properties

Lenzo and coworkers have shown that the application of an electric field induces linear birefringence (the usual electrooptic effect) as well as a change in the optical activity (circular birefringence). Modulation has been observed at frequencies up to 500 MHz. Neither effect can be measured directly; separation of the two phenomena show that the dc linear electrooptic coefficient for the two bismuth germanides is:

| | $Bi_{12}GeO_{20}$ | | $Bi_4Ge_3O_{12}$ |
|---|---|---|---|
| $n_o^3\ r_{41}$ | = 53.3 x $10^{-12}$ m/V at 6660 Å | (Lenzo et al.) | 9.14 x $10^{-12}$ m/V (Nitsche) |
| $r_{41}$ | = 3.4 x $10^{-12}$ m/V | (Lenzo et al.) | 1.03 x $10^{-12}$ m/V (Nitsche) |

When, in addition to the biasing electric field, a side light is applied (photoexcitation), the polarization state is changed again and can be analyzed to separate the photoinduced linear birefringence from the photoinduced change in optical activity [Lenzo et al.]. In an electric-field-biased crystal 1.5 mm thick the plane of polarization can be rotated by 90 degrees using only a weak monochromatic lamp [Spencer et al.].

### Elastooptic Coefficients

The expected elastooptic coefficients for a crystal in the cubic point group 23 are $p_{11}$, $p_{12}$, $p_{13}$ and $p_{44}$. Since bismuth germanium oxide is strongly optically active, the individual elastooptic coefficients cannot be measured using the normal ultrasonic procedures; in practice, an incident linearly polarized optical beam rotates as it passes the elastic-wave region and no single elastooptic coefficient is involved in the diffraction. Venturini et al. have reported the following effective elastooptic coefficients for $Bi_{12}GeO_{20}$:

| Acoustic wave Properties | Effective p |
|---|---|
| Longitudinal wave velocity in the [110] crystallographic direction = 3.42 x $10^5$ cm/sec | 0.115 |
| Transverse wave velocity in the [110] direction with motion in the [100] direction = 1.77 x $10^5$ cm/sec | 0.031 |

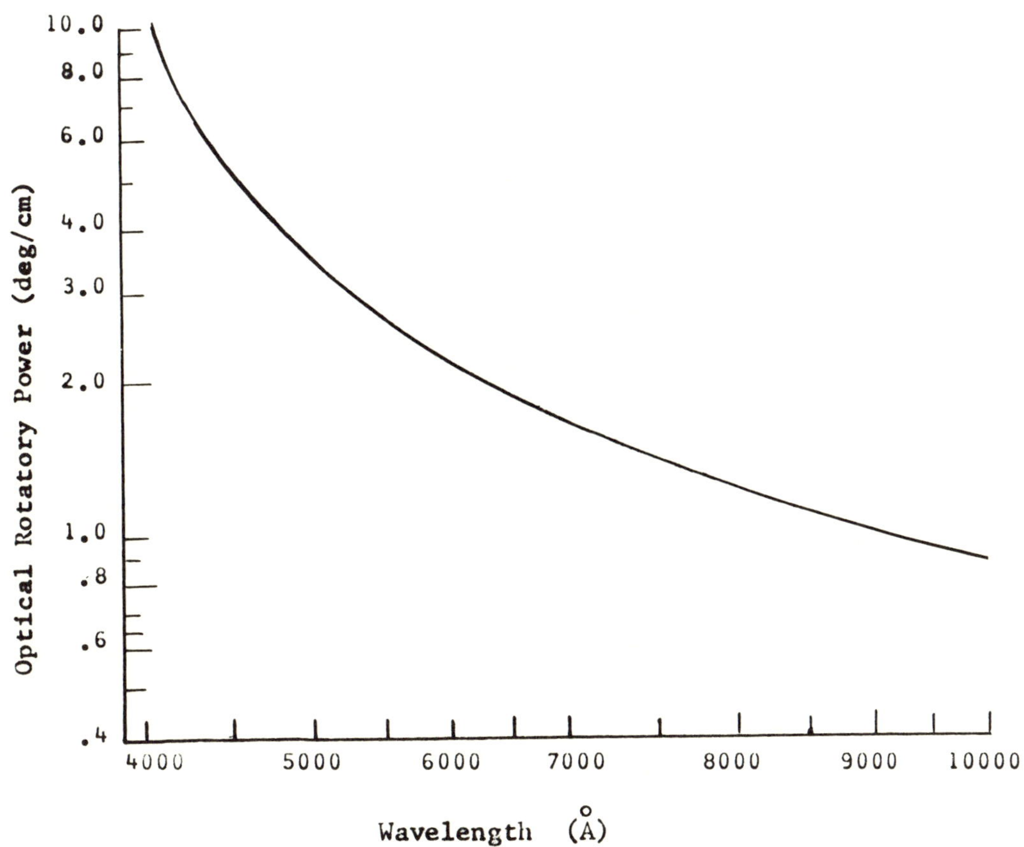


Fig. 4. Specific rotation of the polarization plane as a function of wavelength in $Bi_{12}GeO_{20}$ at 20°C [Feldman et al., Batog et al.].

## BISMUTH GERMANIUM OXIDE

The ultrasonic attenuation characteristics of bismuth germanium oxide have been reported by Spencer et al., further work on the acoustic surface wave properties of $Bi_{12}GeO_{20}$ has been done by Kraut et al. and Pratt et al. This compound also shows low-frequency oscillations which are light and electric-field dependent (Lenzo). Very recently Dickinson et al. and Lauer have reported photoluminescence spectra for this material at 6358 Å and 9537 Å. Dickinson, by doping the compound with certain rare earths, has been able to shift the luminescence peaks, further into the infrared which gives a possibility of a 1.54 micron crystal laser.

The elastic, piezoelectric and dielectric constants of $Bi_{12}GeO_{20}$ have been reported by Onoe et al., Bairamov et al., Kraut et al. and Slobodnik and Ṣethares. These last named tested over 45 samples employing frequency measurements of the acoustic resonances of thin plates. They also measured the piezoelectric constant by three different methods. Their values are given below; on single crystals at 300°K with Kraut's value for comparison. Kraut et al. also found that the temperature coefficients of the elastic constants are negative.

| Elastic Constants | | Value: Slobodnik & Sethares | Value: Kraut et al. | Units |
|---|---|---|---|---|
| $c_{11}$ | | 1.28 | 1.2848 | $10^{11}$ N/m$^2$ |
| $c_{12}$ | | 0.305 | 0.2942 | |
| $c_{44}$ | | 0.255 | 0.2552 | |
| **Piezoelectric Properties** | | | | |
| Electromechanical Coupling Constant | k | 0.22 | | |
| Piezoelectric Constant | $e_{14}$ | 0.99 | 0.983 | C/m$^2$ |
| **Dielectric Properties** | | | | |
| Relative Dielectric Constant | $\varepsilon_{11}^S/\varepsilon_o$ | 38.7 | 38 | |
| Dielectric Coeff. | $\varepsilon_{11}^S$ | $34.2 \times 10^{-11}$ F/m | | |
| Dissipation Factor | | $10^{-4}$ | | |
| Resistivity | | $10^{10}$-$10^{11}$ ohm-cm (Aldrich et al.), Kuzminov et al.) | | |
| Magnetic Susceptibility | $\chi_g$ | $-1.8 \times 10^{-7}$ cgs (Pisarev et al.) at 300°K | | |

ABRAHAMS, S.C. et al. Crystal Structure of Piezoelectric Bismuth Germanium Oxide $Bi_{12}GeO_{20}$. J. OF CHEM. PHYS., v. 47, no. 10, Nov. 15, 1967. p. 4034-4041.

ALDRICH, R.E. et al. Electrical and Optical Properties of $Bi_{12}SiO_{20}$. J. OF APPL. PHYS., v. 42, no. 1, Jan. 1971. p. 493-494.

BAIRAMOV, B. Kh. et al. Scattering of Light by Phonons in $Bi_{12}GeO_{20}$. SOVIET PHYS.-SOLID STATE, v. 13, no. 11, May 1972. p. 2827-2831.

BALLMAN, A.A. Growth and Properties of Piezoelectric Bismuth Germanium Oxide $Bi_{12}GeO_{20}$. J. OF CRYSTAL GROWTH, v. 1, no. 1, 1967. p. 37-40.

BATOG, V.N. et al. Nonlinear Optical Properties of Single Crystals of the Sillenite Type. SOVIET PHYS.-CRYST., v. 16, no. 5, Mar.-Apr. 1972. p. 914-915.

BATOG, V.N. et al. The Optical Activity of Bismuth Compounds. SOVIET PHYS.-CRYST., v. 14, no. 5, Mar.-Apr., 1970. p. 803-804.

BERNSTEIN, J.L. The Unit Cell and Space Group of Piezoelectric Bismuth Germanium Oxide ($Bi_{12}GeO_{20}$). J. OF CRYSTAL GROWTH, v. 1, 1967. p. 45-46.

DICKINSON, S.K. et al. Czochralski Synthesis and Properties of Rare-Earth-Doped Bismuth Germanate. MAT. RES. BULL., v. 7, 1972. p. 181-192.

DOUGLAS, G.G. and R.N. ZITTER. Transport Processes of Photoinduced Carriers in Bismuth Germanium Oxide ($Bi_{12}GeO_{20}$). J. OF APPLIED PHYS., v. 39, no. 4, Mar. 1968. p. 2133-2135.

FELDMAN, A. et al. Optical Activity and Faraday Rotation in Bismuth Oxide Compounds. APPLIED PHYS. LETTERS, v. 16, no. 5, Mar. 1, 1970. p. 201-202.

KRAUT, E.A. et al. Acoustic Surface Waves on Metallized and Unmetallized $Bi_{12}GeO_{20}$. Acoustic Surface Waves on Metallized and Unmetallized $Bi_{12}GeO_{20}$. APPLIED PHYS. LETTERS, v. 17, no. 7, Oct. 1, 1970. p. 271-272.

KUZMINOV, Yu.S. et al. Preparation and Physicochemical Properties of Two Bismuth Germanates. SOVIET PHYS.-CRYSTALLOGRAPHY, v. 14, no. 2, Sept.-Oct., 1969. p. 297-299.

LENZO, P.V. et al. Optical Activity and Electrooptic Effect in Bismuth Germanium Oxide ($Bi_{12}GeO_{20}$). APPLIED OPTICS, v. 5, Oct. 1966. p. 1688-1689.

LENZO, P.V. et al. Photoactivity in Bismuth Germanium Oxide. PHYS. REV. LETTERS, v. 19, no. 11, Sept. 11, 1967. p. 641-644.

NITSCHE, R. Crystal Growth and Electro-Optic Effect of Bismuth Germanate $Bi_4(GeO_4)_3$. J. OF APPLIED PHYS., v. 36, Aug. 1965. p. 2358-2360.

ONOE, M. et al. Elastic and Piezoelectric Characteristics of Bismuth Germanium Oxide $Bi_{12}GeO_{20}$. IEEE TRANS. ON SONICS AND ULTRASONICS, v. SU-14, no. 4, Oct. 1967. p. 165-167.

## BISMUTH GERMANIUM OXIDE

PISAREV, R.V. et al. Faraday Effect in Bismuth Germanate and Silicate. SOVIET PHYS.-SOLID STATE, v. 12, no. 5, Nov. 1970. p. 1241-1242.

SAFONOV, A.I. et al. Production and Optical Properties of $Bi_{12}GeO_{20}$ Single Crystals. SOVIET PHYS.-CRYST. v. 14, no. 1, July-Aug., 1969. p. 131.

AIR FORCE CAMBRIDGE RES. LAB. L.G. HANSCOM FIELD, BEDFORD, MASS. Measurement of the Elastic, Piezoelectric, and Dielectric Constants of $Bi_{12}GeO_{20}$. By: A.J. SLOBODNIK, Jr. and J.C. SETHARES. AFCRL-71-0570, Nov. 10, 1971, Phys. Sci. Res. Papers, no. 467. 32p.

SPENCER, E.G. et al. Ultrasonic Properties of Bismuth Germanium Oxide. APPLIED PHYS. LETTERS, v. 9, no. 8, Oct. 1966. p. 290-291.

SPENCER, E.G. et al. Dielectric Materials for Electrooptic, Elastooptic, and Ultrasonic Device Application. IEEE PROC., v. 55, no. 12, Dec. 1967. p. 2074-2108.

SPERANSKAYA, E.I. and A.A. ARSHAKUNI. The Bismuth Oxide-Germanium Dioxide System. RUSSIAN J. OF INORGANIC CHEM., v. 9, no. 2, Feb. 1964. p. 226-229.

VENTURINI, E.L. et al. Elasto-Optic Properties of $Bi_{12}GeO_{20}$, and $Sr_xBa_{1-x}Nb_2O_6$. J. OF APPLIED PHYS., v. 40, no. 4, Mar. 1969. p. 1622-1624.

# CALCIUM PYRONIOBATE

## Introduction

Calcium pyroniobate, $Ca_2Nb_2O_7$, is closely related chemically and structurally to $LiNbO_3$ and $LiTaO_3$. A linear electrooptic effect has been observed in this material which is sufficiently strong to be of device interest. A relatively small amount of data is available on the crystal.

## Chemical and Physical Properties [Rowland et al.]

| | | |
|---|---|---|
| Chemical Formula | $Ca_2Nb_2O_7$ | ($2CaO \cdot Nb_2O_5$) |
| Molecular Weight | 1511.9 | |
| Density | 4.39 g/cm$^3$ | |

## Crystallography

| | |
|---|---|
| Crystal Symmetry | Monoclinic |
| Space Group | $P2_1$ |
| Point Group | 2 or $C_2$ |
| Lattice Constants [Rowland et al.] | a = 13.36 Å<br>b = 5.50 Å<br>c = 7.70 Å<br>$\beta$ = 98°25' |

A crystal of class $C_2$ has a single twofold axis of symmetry which may be taken as the crystallographic b axis. The crystallographic a and c axes lie in the plane normal to the b axis. The angle $\beta$ between the a and c axes is, in general, not 90°. The (100) plane in calcium pyroniobate is a natural cleavage plane. The direction normal to the (100) plane is the acute bisector of the optic axes, which lie in the plane of the a and c axes [Holmes et al.].

It is reported that single crystals of calcium pyroniobate have been grown at Standard Telephone Laboratories, Harlow, England, using the Czochralski technique. The melt was contained in an iridium crucible, under an atmosphere of argon and oxygen, and growth rates of between 3 and 12 mm/hour were used while rotating the seed at about 25 revs/min. Boules with well developed (100) mirror facets have been grown and, with careful control, ribbon crystals typically 10 cm long x 1 cm wide x 0.1 cm thick have been obtained. The ribbon faces are (100) planes along which occasional growth steps about 1 micron high occur. Growth in the (100) direction is difficult, as would be expected, but by growing at less than 4mm per hour, boules of good quality have been grown, exhibiting (001) and (011) facets and thus a roughly hexagonal cross-section. All the boules which have been grown are polysynthetically twinned parallel to the (100) plane and exhibit psuedo-orthorhombic symmetry.

## Optical Properties [Rowland et al.]

| | |
|---|---|
| Color | Pale Yellow |
| Optical Character | Biaxial negative |
| Refractive Index | $n_x = 1.97$ |
| | $n_y = 2.16$ |
| | $n_z = 2.17$ |

## Electrooptic Properties

There are eight allowed linear electrooptic coefficients for $Ca_2Nb_2O_7$, including three principal axis coefficients ($r_{12}$, $r_{22}$, $r_{32}$) and five skew coefficients ($r_{52}$, $r_{41}$, $r_{61}$, $r_{43}$, $r_{63}$). Observed values of these coefficients are given in Table 1. Holmes et al. repeated their measurements at 77°K and at room temperature and 5890 Å; no essential variations from the values given in Table 1 were observed.

TABLE 1. ELECTROOPTIC COEFFICIENTS FOR CALCIUM PYRONIOBATE AT 6328 Å AND ROOM TEMPERATURE.

| Coefficient | Value ($10^{-12}$ m/V) | Half-wave Voltage (V) | Reference |
|---|---|---|---|
| $r_{12}^S$ | 6.7 ± 0.4 | | Rosner & Turner f = 60 MHz |
| $r_{22}^S$ | 25.5 ± 1.5 | | |
| $r_{32}^S$ | 6.4 ± 0.4 | | |
| $r_{41}^S$ | 2.7 ± 1 | | |
| $r_{52}^S$ | <0.6 | | |
| $r_{63}^S$ | 0.9 | | |
| $r_{c1}^S$ | 20.4 ± 1.8 | | |
| $r_{c1}^T$ | 12.3 | 5080 | Holmes et al. f = 3 GHz, dc |
| $r_{c3}^T$ | 13.7 | 4550 | |
| $r_{c3}^S$ | 13.0 | 4700 | |
| $r_{c3}^S$ | 19.0 ± 1.9 | | Rosner & Turner |

$r_{c1} = r_{22} - (n_x/n_y)^3\, r_{12}$, $\quad (n_x/n_y)^3 = 1.01$ Spencer et al.

$r_{c3} = r_{22} - (n_z/n_y)^3\, r_{32}$, $\quad (n_z/n_y)^3 = 0.76$

Spencer et al. state that electro-optic values are independent of frequency up to 3 GHz.

$r_{c3} = r_{22} - 1.01\, r_{32} = 4.1 \times 10^{-7}$ cm/stat V $= 13.7 \times 10^{-12}$ m/V

$r_{c1} = r_{22} - 0.76\, r_{12} = 3.7 \times 10^{-7}$ cm/stat V $= 12.3 \times 10^{-12}$ m/V

Although the size of the electrooptic coefficients ($r_{22}$ in particular) would make $Ca_2Nb_2O_7$ a competitor of such materials as $LiNbO_3$ for electrooptic modulators, Rosner and Turner point out that the crystal has the following disadvantages:

1. The crystal is difficult to grow.
2. The optical quality is not good; scattering of light, plus absorption, caused a loss of about 10 dB in 3.5 mm.
3. The crystal has micaceous cleavage which makes it difficult to polish well.
4. Optically induced index inhomogeneities are generated even by low power density laser beams at 0.63 microns.

## Piezoelectric Properties

Calcium pyroniobate is piezoelectric with $k_{33}$ = 0.30 [Ballman].

## Dielectric Properties

Rosner and Turner report that the relative dielectric constants $\varepsilon_2$ and $\varepsilon_3$ are both approximately 45 at 50 MHz.

## Ferroelectric Properties

Calcium pyroniobate does not exhibit ferroelectric characteristics [Ballman].

## Thermal Properties

Rowland et al. report that $Ca_2Nb_2O_7$ melts congruently at 1575°C. Later work by Jongejan, indicates a melting point of 1571°C. This investigation proposes a phase diagram for the calcium oxide-niobium oxide system.

## CALCIUM PYRONIOBATE

BALLMAN, A.A. Growth of Piezoelectric and Ferroelectric Materials by the Czochralski Technique. AMERICAN CERAM. SOC., J., v. 48, no. 2, Feb. 1965. p. 112-113.

EMMENEGGER, F.P. and H. ROETSCHI. Dielectric Properties of Some Niobates and Tungstates. J. OF PHYS. AND CHEM. OF SOLIDS, v. 32, no. 4, Apr. 1971. p. 787-790.

HOLMES, C.H. et al. The Electrooptic Effect in Calcium Pyroniobate. APPLIED OPTICS, v. 4, May 1965. p. 551-553.

JONGEHAN, A. Phase Relationships in the High-Lime Part of the System Calcium Oxide-Niobium Oxide. J. OF LESS COMMON METALS, v. 19, no. 3, Nov. 1969. p. 193-202.

ROSNER, R.D. and E.H. TURNER. Electrooptic Coefficients in Calcium Pyroniobate. APPLIED OPTICS, v. 7, no. 1, Jan. 1968. p. 171-174.

ROWLAND, J.F. et al. The Crystallography of Compounds in the Calcium Oxide-Niobium Pentoxide System. ADVANCES IN X-RAY ANALYSIS, Plenum Press Inc., N.Y., 1960. p. 97-106.

SPENCER, E.G. et al. Dielectric Materials for Electrooptic, Elastooptic, and Ultrasonic Device Applications. IEEE PROC., v. 55, no. 12, Dec. 1967. p. 2074-2108.

# CUPROUS CHLORIDE

## Introduction

About 1963, cuprous chloride (CuCl) was found to have a number of advantageous electrooptic characteristics as compared to the KDP family of electrooptic materials: it has a low optical loss over a wide spectral range from the visible through the infrared, a low refractive index and a low dielectric constant as well as a low loss tangent at microwave frequencies. This means that a wider range of infrared frequencies may be modulated (CuCl transmits to 20 microns), greater angular aperture, lower modulating power, and a greater ease of use in both cavity-type and travelling-wave type modulators. Cuprous chloride modulators, requiring only modest input power, have been demonstrated experimentally and operated continuously with bandwidths of tens of MHz at microwave frequencies. The major problem hindering its full development and use has been the (difficulty in producing) good strain-free crystals, a problem arising from the phase transformation.

## Chemical and Physical Properties

| | | |
|---|---|---|
| Chemical Formula | CuCl | |
| Molecular Weight | 98.99 | |
| Density (25°C) | 4.137 g/cm$^3$ | [Smakula] |
| Microhardness | 11 kg/mm$^2$ | |

Cuprous chloride belongs to the I-VII class of compounds and has been reported to be hygroscopic. The crystal is relatively insoluble in water, dissolving only to the extent of 1.52 parts per 100 parts of water. CuCl is unstable in air, forming, with time, a green cupric chloride, $CuCl_2$, cuprous hydroxide or a complex oxychloride.

## Crystallography

| | | |
|---|---|---|
| Crystal Symmetry | cubic | |
| Point Group | $\bar{4}3m$ or $T_d$ | |
| Lattice Constant | 5.418 ± 0.002 Å | [Donnay] |

Because of the promise that CuCl shows as a light modulator, numerous efforts have been made by various investigators [e.g., Blattner et al. at RCA] to grow crystals showing a high degree of optical perfection. All large crystals of this material grown in the past have been highly strained because the material goes through a phase change from hexagonal to cubic. Cuprous chloride crystallizes in the zincblende lattice (two face-centered cubic lattices displaced from each other by one-quarter of a body diagonal) up to a temperature of 407°C, and in a wurtzite (hexagonal) lattice from 407°C to its melting point at 422°C. Crystals grown at temperatures below 407°C crystallize directly in the zincblende phase, and show no strains. Techniques for growing crystals below 407°C include deposition from a vapor phase onto a substrate and growing from solution; however, crystals grown by these techniques are presently too small for useful application. Crystals with a volume of a few cubic centimeters have been grown directly from the melt. However, these crystals form first in the wurtzite phase, and convert to the zincblende phase when cooled below 407°C. Because the material is a solid at 407°C, the phase transition is in general not complete and, as a result, large crystals grown from the melt are generally badly strained. Some of the strain can be relieved by annealing at temperatures below 407°C.

To overcome the wurtzite-to-zincblende phase inversion at 407°C, some investigators [e.g., Soga et al.] have grown crystals from a flux melt in chich the melting temperature is depressed. Fluxes such as KCl, $SrCl_2$, $PbCl_2$, and $BaCl_2$ have been explored. The single crystals grown by the flux method have a comparatively small optical strain as shown in Table 1. The authors report that they grew a strain-free single crystal cube of 5 mm on each edge by this flux method in a Bridgman apparatus.

TABLE 1. CHARACTERISTICS OF CuCl SINGLE CRYSTALS GROWN BY THE FLUX METHOD [Soga et al.].

| Flux, adding 2 mole % to CuCl | Optical strain, per mm | Optical density | Specific resistance (ohm-cm) | Breakdown field (kV/mm) |
|---|---|---|---|---|
| $SrCl_2$ | 0.005 | 0.20 | $5 \times 10^8$ | 1.45 |
| $BaCl_2$ | 0.007 | 0.19 | $6 \times 10^8$ | 0.83 |
| $PbCl_2$ | - | 0.58 | $1.5 \times 10^8$ | - |
| KCl | - | 1.1 | $1 \times 10^6$ | - |
| RbCl | - | 0.47 | $9 \times 10^6$ | - |
| Growing without flux | 0.02 | 0.12 | $3.5 \times 10^8$ | 0.58 |

## CUPROUS CHLORIDE

A study of the application of the Czochralski technique to the growth of CuCl crystals from a flux has been described by Wilcox and Corley. Purification methods (e.g. wet chemical and zone-refining), crystal imperfections and mechanical properties are discussed by these authors. Purification of the starting materials is necessary for the growth of transparent cuprous chloride crystals. Chemical methods and zone-refining were both successfully employed for purification, but problems arose with tube breakage during zone melting. Ingots were easily pulled from the pure melt but were invariably highly twinned and polycrystalline because of passage through the wurtzite-to-zincblende phase transition.

Eden reported on the CuCl crystal-preparation program at Texas Instruments. The cupric ion (a common impurity in cuprous halides) absorbs light of wavelength around 0.9 micron. This ion is responsible for the blue coloring of most CuCl crystals, and must be eliminated by zone refining processes. Eden was unsuccessful in obtaining good crystals because of the bad strains found in the samples.

Armington and coworkers, at the Air Force Cambridge Research Laboratories, made a study of some factors which influence the growth of cuprous chloride in silica gel. They found that the yield of clear crystals produced by this method is small (between 5 and 10 percent) in relation to the total number of crystals formed in the gel. Most crystals were attacked in air, particularly if they were moist. Some crystals, however, remained stable in the air indefinitely. Where large crystals are desired, there are certain difficulties that must be overcome in their present system. A second and more complicated problem related to the growth of large crystals is the limitation of nucleation sites. Crystals of 6 to 7 mm were easily obtained by the silica gel method.

CuCl crystals have also been grown from aqueous HCl solutions and by sublimation [Kaifu et al.].

CUPROUS CHLORIDE

Optical Properties - Refractive Index

TABLE 2. REFRACTIVE INDEX OF CUPROUS CHLORIDE

| Wavelength (μ) | n Chemla et al. | n Smakula | n Alonas et al. | n Kaifu & Komatsu | Temp.(°K) 300 |
|---|---|---|---|---|---|
| | single cry. | single cry. grown from melt | single cry., high quality surfaces | transparent, polycrystal-line cubic films | |
| 0.4047 | 2.1535 | | | | |
| 0.4078 | 2.1410 | | | | |
| 0.4358 | 2.0720 | | | | |
| 0.4678 | 2.0336 | | | | |
| 0.4800 | 2.0234 | | | | |
| 0.5086 | 2.0042 | | | | |
| 0.5461 | 1.987 | | | | |
| 0.5791 | 1.9760 | | | | |
| 0.5893 | | | | 1.973 | |
| 0.5896 | 1.9726 | | | | |
| 0.6438 | 1.9584 | | | | |
| 0.6563 | | | | 1.955 | |
| 0.7699 | 1.9411 | | | | |
| 2.5 | | 1.90 | | | |
| 3.0 | | | 1.92 | | |
| 3.3 | | 1.90 | | | |
| 5.0 | | 1.90 | | | |
| 6.7 | | 1.89 | | | |
| 10.0 | | 1.88 | | | |
| 14.3 | | 1.86 | | | |
| 16.7 | | 1.78 | | | |
| 18.2 | | 1.73 | | | |
| 20.0 | | 1.72 | | | |
| 20.8 | | 1.72 | | | |
| 21.7 | | 1.68 | | | |
| 22.7 | | 1.64 | | | |
| 23.8 | | 1.61 | | | |
| 25.0 | | 1.57 | | | |
| 26.3 | | 1.49 | | | |
| 27.8 | | 1.46 | | | |
| 29.4 | | 1.43 | | | |
| 30.0 | | | 1.65 | | |
| 31.3 | | 1.39 | | | |

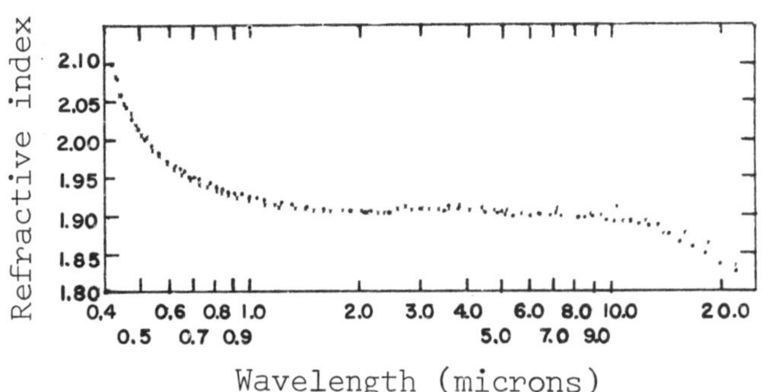


Fig. 1. Refractive index of single crystal CuCl as a function of wavelength

• 54.13 microns thick

× 14.99 microns thick

[Feldman and Horowitz].

Dispersion ($\lambda$, $\mu$)

$$n = 1.9 + \frac{2.45 \times 10^{-2}}{\lambda^2}$$ Jerphagnon et al.

$$n^2 = 3.580 + \frac{[3.16 \times 10^{-2}]\lambda^2}{\lambda^2 - 0.16} + \frac{9.3 \times 10^{-2}}{\lambda^2}$$ Feldman and Horowitz

## Optical Properties - Transmission

Sterzer et al., Alonas et al., and Chemla et al. report that cuprous chloride is transparent [80% transmission (Wittke et al., McCarthy)] in the range from 0.4 to 20.5 microns. In this range the transmission is constant to within a few percent. Absorption data are presented in Table 3.

TABLE 3. ABSORPTION COEFFICIENT OF CUPROUS CHLORIDE.

| Wavelength (μ) | ABSORPTION COEFFICIENT ($cm^{-1}$) | | | Temp.(°K) |
|---|---|---|---|---|
| | Chemla et al. | Smakula | Sueta et al. | 300 |
| 0.5 | 0.6 | | | |
| 1.0 | 0.5 | | | |
| 5.0 | 0.5 | | | |
| 10.0 | 0.4 | 0.01 | | |
| 10.6 | | | 0.02 | |
| 14.3 | | 0.11 | | |
| 16.0 | 0.4 | | | |
| 16.7 | | 0.29 | | |
| 18.2 | | 1.15 | | |
| 19.0 | 0.7 | | | |
| 20.0 | 1.0 | 4.39 | | |
| 20.8 | | 6.64 | | |
| 21.0 | 4.4 | | | |
| 21.7 | | 10.5 | | |
| 22.7 | | 15.9 | | |
| 23.8 | | 22.4 | | |
| 25.0 | | 29.9 | | |
| 26.3 | | 40.1 | | |
| 27.8 | | 53.6 | | |
| 29.4 | | 80.4 | | |
| 31.3 | | 218.0 | | |

## Electrooptic Properties

The electrooptic properties of natural copper chloride crystals were first investigated by West and later used to modulate light by Sterzer et al. Because of its cubic crystal structure ($\bar{4}3m$), CuCl has three identical non-zero electrooptic coefficients: $r_{41} = r_{52} = r_{63}$.

The properties of the cubic crystal class $\bar{4}3m$ are of particular importance to electrooptic applications:

1. They respond to fields transverse to the light path. This mode of operation not only permits reduction in the voltage by the use of a light path large compared with the distance over which the field is impressed, but also avoids the necessity of using either transparent electrodes (which are not operable at high frequencies) or electrodes with holes (which operate only with fringing fields).

2. They possess not natural birefringence and, therefore, the problems of alignment and beam parallelism are minimized as compared with problems found in the use of naturally birefringent crystals such as KDP.

TABLE 4. STATIC, UNCLAMPED ELECTROOPTIC PROPERTIES OF CuCl.

| Wavelength (Å) | $n_o^3 r_{41}^T$ ($10^{-12}$ m/V) | Half-wave Voltage (kV) | Reference |
|---|---|---|---|
| 5461 | 44 | Longitudinal 6.2<br>Transverse (minimum)<br>E ⊥ (111) plane 7.2<br>E ⊥ (110) plane 6.2 | Sterzer et al. |
| 5250<br>6750 | 29<br>34 | Melt grown and annealed (low values due to crystal stresses and disorientation) | Belyaev et al. |
| 10600 | 23 | Single crystals grown by crystallographic transition control<br>$\rho = 10^{-8}$ ohm-cm<br>$\alpha = 0.02\ cm^{-1}$ at 10.6μ | Sueta et al. |

Sterzer et al. point out that, since the difference between the clamped and unclamped values of the half-wave voltage $V_{1/2}$ for zincblende is negligible, and since CuCl crystallizes in the zincblende lattice, it seems likely that this difference for CuCl is also small.

Piezo-optical Coefficients

| Value ($10^{-5}$/bar) | $\lambda = 0.5893\mu$ | $\lambda = 0.6328\mu$ | |
|---|---|---|---|
| $\pi_{11} - \pi_{12}$ | -0.125 | -0.122 | 300°K |
| $\pi_{44}$ | -0.0064 | -0.0059 | |

Dielectric Properties

The optical dielectric constant, $\varepsilon_\infty = 3.70$. This value was obtained by Alonas et al. from optical measurements at 3 to 30μ on single crystals with carefully prepared surfaces. The static dielectric constant, $\varepsilon_o = 7.5$, the same value obtained by Sterzer et al. from measurements at the C-band.

The variation of the dielectric constant of CuCl is shown in Table 5 and Figure 3. Although these data indicate that ε at 25°C decreases with increase in frequency and approaches a constant value of 8.3 above $10^{10}$ Hz, lower values of 7.7

and 7.5 at $10^9$-$10^{10}$ Hz have been reported, respectively, by Kaminow and Turner and Sterzer et al. At -190°C, ε= 7.4 and is frequency-independent. The temperature coefficient of the dielectric constant at 25°C and $10^7$ Hz is: $(1/\varepsilon)(d\varepsilon/dT) = 52 \times 10^{-5}/°C$ [Smakula]. The influence of moisture on the dielectric constant is shown in Figure 4.

TABLE 5. DIELECTRIC CONSTANT OF CUPROUS CHLORIDE AT ROOM TEMPERATURE [Belyaev et al.]

| Frequency (Hz) | ε |
|---|---|
| $10^2$ | 10.0 ± 0.5 |
| $10^3$ | 9.8 |
| $10^4$ | 9.2 |
| $10^5$ | 8.8 |
| $9.8 \times 10^8$ | 8.6 |
| $9.4 \times 10^9$ | 8.4 |
| $3.96 \times 10^{10}$ | 8.3 |

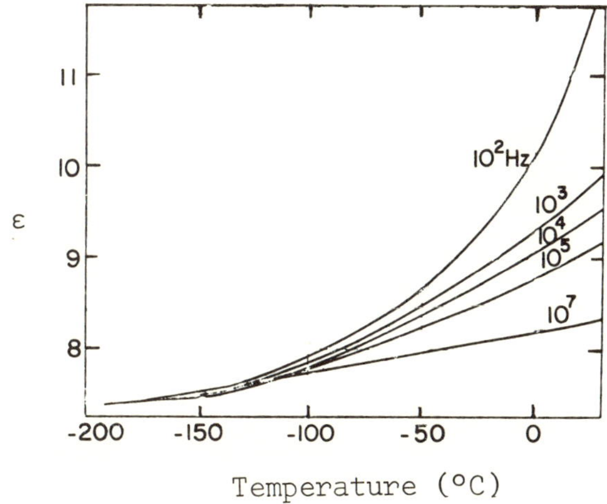


Fig. 3. Dielectric constant of CuCl as a function of temperature measured at several frequencies [Smakula].

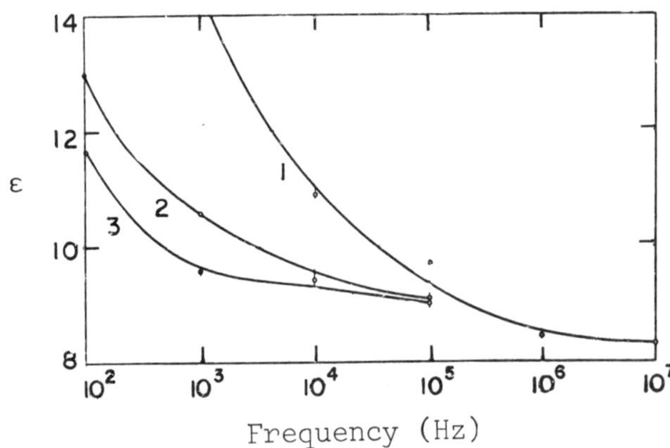


Fig. 4. Dielectric constant of CuCl samples with different moisture contents as a function of frequency at 25°C.

1. Before drying.
2. After drying for half an hour at 100°C.
3. After drying for one hour at 100°C

[Smakula].

The following loss tangent values have been reported for CuCl:

| | |
|---|---|
| $\tan \delta = 0.001$ at 6 GHz | [Sterzer et al.] |
| $\tan \delta = 0.002$ at 9.3 GHz | [Kaminow and Turner] |

The electrical resistivity and breakdown of CuCl have been reported by Soga and coworkers and are detailed in Table 1. Gentile evaluated single-crystal cuprous chloride crystals for light modulator applications and observed electrical breakdown to occur during room temperature operation at relatively low dc fields (3.4 kV/cm). The cause was ascribed to copper precipitated as either filaments with dendritic branches or as networks.

## CUPROUS CHLORIDE

### Thermal Properties

| Property | T(°C) | Value | Reference |
|---|---|---|---|
| Phase transformation temperature (cubic to hexagonal) | | 407°C | [Sterzer et al.] |
| Melting Point | | 422°C | |

| Linear Thermal Expansion Coefficient | T(°C) | $(1/\ell)(d\ell/dT)$ $10^{-6}/°C$ | Reference |
|---|---|---|---|
| Single crystal | 25 | 13.6 | [Smakula] |
| Pressed pellets | 30-75 | 7.27 | |
| Pressed pellets | 40-140 | 10 | |
| Powder (X-ray technique) | 20-120 | 17 | [Lawn] |

| Property | Value | Reference |
|---|---|---|
| Maximum Safe Operating Temperature | > 200°C | [Sliker] |
| Thermal EMF | 0.9 mV/°C at 20°C | [Mogilevskii & Usmanov, Hsueh & Christy] |

## CUPROUS CHLORIDE


ALONAS, P. et al. Dielectric Properties of CuCl at 300°K in the 3-30 micron Region. APPLIED OPTICS, v. 8, no. 12, Dec. 1969. p. 2557-2559.

AIR FORCE CAMBRIDGE RES. LABS., BEDFORD, MASS. SOLID STATE SCI. LAB. A Study of Some Factors which Influence the Growth of Cuprous Chloride in Silica Gel, by ARMINGTON, A.F. et al. July 1967. 17 p. AD 659 135.

BELYAEV, L.M. et al. Dielectric Constant of Crystals Having an Electro-Optical Effect. SOVIET PHYS.-SOLID STATE, v. 6, no. 8, Feb. 1965. p. 2007-2008.

BELYAEV, L.M. et al. Electro-Optical Properties of Copper Chloride and Bromide Crystals. SOVIET PHYS.-SOLID STATE, v. 6, no. 12, June 1965. p. 2988.

BLATTNER, D.J. et al. A Research Program on the Utilization of Coherent Light. RCA Interim Rept., no. 6, Jan. 20, 1963. 28 p. AD 296 145.

CHEMLA, D. et al. Nonlinear Properties of Cuprous Halides. IEEE J. QUANTUM ELECTRONICS, v. QE-7, no. 3, Mar. 1971. p. 126-132.

DONNAY, J.D.H. Ed. CRYSTAL DATA, DETERMINATIVE TABLES. 2nd Ed., American Crystallographic Assn., ACA Monograph No. 5, 1963.

TEXAS INSTRUMENTS, INC. Solid State Techniques for Modulation and Demodulation of Optical Waves, by: EDEN, D.D. Final Tech. Rept. ECOM-03250-F. Sept. 1966. 283 p. AD 489 390.

FELDMAN, A. and D. HOROWITZ. Refractive Index of Cuprous Chloride. OPTICAL SOC. OF AMERICA, J., v. 59, no. 11, Nov. 1969. p. 1406-1408.

GENTILE, A.L. Electric Breakdown Mechanism in Cuprous Chloride Single Crystals. APPLIED PHYS. LETTERS, v. 9, Sept. 1966. p. 237-239.

GOTO, T. and M. UETA. Single Crystals of Cuprous Halides and Their Exciton Emissions. PHYS. SOC. OF JAPAN, J., v. 22, no. 4, Apr. 1967. p. 1123-1124.

GOTO, T. et al. Exciton Luminescence of CuCl, CuBr and CuI Single Crystals. PHYS. SOC. OF JAPAN, J., v. 24, no. 2, Feb. 1968. p. 314-327.

HADNI, A. et al. Far-Infrared-Active Phonon Processes in CuCl. J. CHEM. PHYS., v. 49, no. 1, July 1968. p. 471-473.

HSUEH, Y.W. and R.W. CHRISTY. Thermoelectric Power of CuCl Containing $CdCl_2$. Dartmouth College, Dept. of Physics, Tech. Report No. 4, July 1963.

JERPHAGNON, J. et al. Second Harmonic Generation in Cuprous Chloride. ACAD. DES SCIENCES, COMPTES RENDUS, B. v. 265, no. 19, Nov. 1967. p. 1032-1033. (In Fr.)

KAIFU, Y. et al. Some Optical Properties of CuCl Single Crystals. PHYS. SOC. OF JAPAN, J., v. 22, no. 2, Feb. 1967. p. 517-524.

KAIFU, Y. and T. KOMATSU. Refractive Index of CuCl. PHYS. SOC. OF JAPAN, J., v. 25, no. 2, Aug. 1968. p. 644.

KAMINOW, I.P. and E.H. TURNER. Electrooptic Light Modulators. IEEE PROC., v. 54, Oct. 1966. p. 1374-1390.

LAWN, B.R. The Thermal Expansion of Silver Iodide and the Cuprous Halides. ACTA CRYST., v. 17, no.11, Nov. 1964, p. 1341-1347.

McCARTHY, D.E. The Reflection and Transmission of Infrared Materials: Pt. III. Spectra from 2 to 50 microns. APPLIED OPTICS, v. 4, no. 3, Mar. 1965. p. 317-320.

MOGILEVSKII, B.M. and O.U. USMANOV. Thermoelectric Properties of Silver and Copper Halides in Solid and Liquid Phases. SOVIET ELECTROCHEM., v. 3, no. 9, Sept. 1967, p. 1002-1004.

SCHWAB, C. and P. ROBINC. Photoelastic Properties of Cuprous Halides. OPTICS COMMUNICATIONS, v. 4, no. 4, Dec. 1971. p. 304-306.

CLEVITE CORP. Reference Data on Linear Electro-Optic Effects. Engineering Memorandum 64-10, by: SLIKER, T.R. May 15, 1964. 9 pp.

M.I.T. CRYSTAL PHYSICS LAB. A Study of the Physical Properties of High-Temperature Single Crystals, by: SMAKULA, A. Rept. No. AFCRL-67-0645. Sept. 1967. 124 pp. AD 663 734.

SOGA, M. et al. A Method of Growing CuCl Single Crystals with Flux. ELECTROCHEM. SOC., J., v. 114, no. 4, Apr. 1967. p. 388-390.

STERZER, F. et al. Cuprous Chloride Light Modulators. OPTICAL SOC. OF AMERICA, J., v. 54, no. 1, Jan. 1964. p. 62-68.

SUETA, T. et al. Modulation of 10.6 micron Laser Radiation by CuCl. IEEE, PROC., v. 58, no. 9, Sept. 1970. p. 1378-1379.

WEST, C.D. Electrooptic and Related Properties of Crystals with the Zinc Blende Structure. OPTICAL SOC. OF AMERICA, J., v. 43, no. 1. 1953. p. 335.

AEROSPACE CORP. EL SEGUNDO, CALIF. LABS. DIV. Czochralski Growth of CuCl, by: WILCOX, W.R. and R.A. CORLEY. Rept. No. Tr-1001 (9320-13)-3, Apr. 1967. 15 p. AD 813 027.

RCA. Solid State Laser Explorations, by: WITTKE, J.P. et al. Tech. Rept. AFAL-TR-64-334. Jan. 1965.

# LITHIUM NIOBATE

Lithium niobate possesses a very unique combination of properties and characteristics: 1) ferroelectric with a high Curie point, 2) large nonlinear optical coefficient, 3) large birefringence, 4) strong piezoelectric effect, 5) excellent acoustic properties, and 6) large electrooptic effect. This unusual dielectric material has been extensively investigated only quite recently.

Midwinter made an assessment of lithium niobate in 1967 for nonlinear optics. It has proved to be an efficient nonlinear optical material of prime importance in the visible and near-infrared regions of the spectrum for phase-matched second harmonic generation and difference frequency generation. Recently, Spencer et al. and Wemple and DiDomenico, at Bell Telephone Laboratories, have reviewed this crystalline material and contrasted it with other dielectric and ferroelectric materials for electrooptic, elastooptic, ultrasonic, and nonlinear optical device applications. Abrahams, Nassau and coworkers, also of Bell Telephone Laboratories, have reviewed the growth, structure and properties of lithium niobate. Carruthers et al. and Byer et al. report the growth of high quality crystals and phase equilibrium.

More recently, Bergman et al. and Fay et al. have shown that the Curie temperature, birefringence, and phase-matching temperature of lithium niobate are strong functions of melt stoichiometry. It seems probable that other electrooptic, acoustic, and piezoelectric properties of this material are also functions of stoichiometry.

## Chemical and Physical Properties

| | | |
|---|---|---|
| Chemical Formula | $LiNbO_3$ | |
| Molecular Weight | 147.85 | |
| Density (23°C) | 4.628 g/cm$^3$ | [Abrahams et al.] |
| (25°C) | 4.64 | [Nassau et al.] |
| Solubility | Insoluble in water and dilute acids | |

## Crystallography

Single crystal lithium niobate was first prepared using the Czochralski technique by Ballman and later by Nassau and co-workers at Bell Telephone Laboratories. Parfitt and Robertson also used this technique and varied the growth conditions to obtain various domain structures. $LiNbO_3$ has also been grown by the flux method using lithium carbonate and single-crystal lithium chloride in a platinum crucible in the temperature range 1250° to 1300°C [Smolenskii et al.].

An observed characteristic of $LiNbO_3$ is its susceptibility to the introduction of color centers. If $LiNbO_3$ is grown or annealed in an $O_2$-deficient atmosphere, it becomes colored anywhere from light brown to black, depending on the degree of $O_2$-deficiency. The color can be removed by annealing in $O_2$ at elevated temperatures (1100°C). Levinstein et al. have found that the color can be removed by field annealing at temperatures as low as 400°C with a field of 1000 V/cm. The color-free region forms near the negative electrode and progresses at constant temperature and field at a rate such that the distance of the boundary between clear and colored material from the negative electrode varies as the square of the time. The effect of annealing on the transmittance of lithium niobate crystals is shown in Figure 2.

Nassau and co-workers have conducted a considerable amount of work on the preparation of single domain $LiNbO_3$ crystals and the study of their dislocations and etching behavior. They cite the following particular set of conditions which appears satisfactory, but not always sufficient, for the growth of single-domain $LiNbO_3$:

(1) 0.5 at % $MoO_3$ added to the melt,
(2) growth direction 20-40° from the c-axis, e.g. perpendicular to (11·12),
(3) negative dipole end of crystal facing the melt,
(4) flat melt-crystal interface,
(5) seed free from twin boundaries, and very stable temperature so that twins will not form,
(6) steady, slow pulling rate, e.g. 1/2-3/4 in./hr.

These authors reported that single domain crystals can be prepared by two additional methods: 1) growth in an electric field, and 2) poling at elevated temperatures. In their crystal growing evaluation, these authors noted that the following defects can be present in pulled $LiNbO_3$: solid phase inclusions, gaseous inclusions, prominent low angle grain boundaries, twin planes, and dislocations; all of these, except the last, can be avoided by careful growth. Niizeki et al. also have studied the growth features (ridges, etched hillocks) and crystal structure of $LiNbO_3$ grown from the melt.

At the time the ferroelectric properties of $LiNbO_3$ were first noted by Matthias and Remeika in 1949, this crystal was thought to have the ilmenite ($FeTiO_3$) crystal structure (space group $R\bar{3}$). The detailed structure of lithuum niobate at room temperature has been unamibguously established now by X-ray and neutron diffraction studies of single domain crystals [Abrahams et al.]. Lithium niobate crystallizes into the noncentro-symmetric space group R3c and has six formula units per hexagonal and two per equivalent rhombohedral unit cell. It also has been well established now that $LiNbO_3$ is a ferroelectric crystal with a Curie temperature of about 1210°C, close to its melting point of about 1260°C. The transformation of the ferroelectric phase into the paraelectric phase is depicted in Figure 1. The following summarizes the various crystallographic parameters of lithium niobate:

Space Group

| | | | |
|---|---|---|---|
| Below $T_c$ | Rhombohedral | R3c or $C_{3v}^6$ | Abrahams et al. |
| Above $T_c$ | Rhombohedral | $R\bar{3}c$ or $D_{3v}^6$ | Ismailzade et al. |

Point Group

| | | |
|---|---|---|
| Below $T_c$ | 3m or $C_{3v}$ | Niizeki et al. |
| Above $T_c$ | $\bar{3}2/m$ or $D_{3d}$ | |

Lattice Constants at room temperature

| Hexagonal | | Rhombohedral | | Reference |
|---|---|---|---|---|
| $a_H$ (Å) | $c_H$ (Å) | $a_{Rh}$ (Å) | $\alpha_{Rh}$ | |
| 5.147 | 13.857 | 5.492 | 55° 53' | Shapiro et al. |
| 5.154 | 13.865 | 5.497 | 55° 55' | Ismailzade |
| 5.14829 | 13.8631 | 5.4944 | 55° 52' | Abrahams et al. |

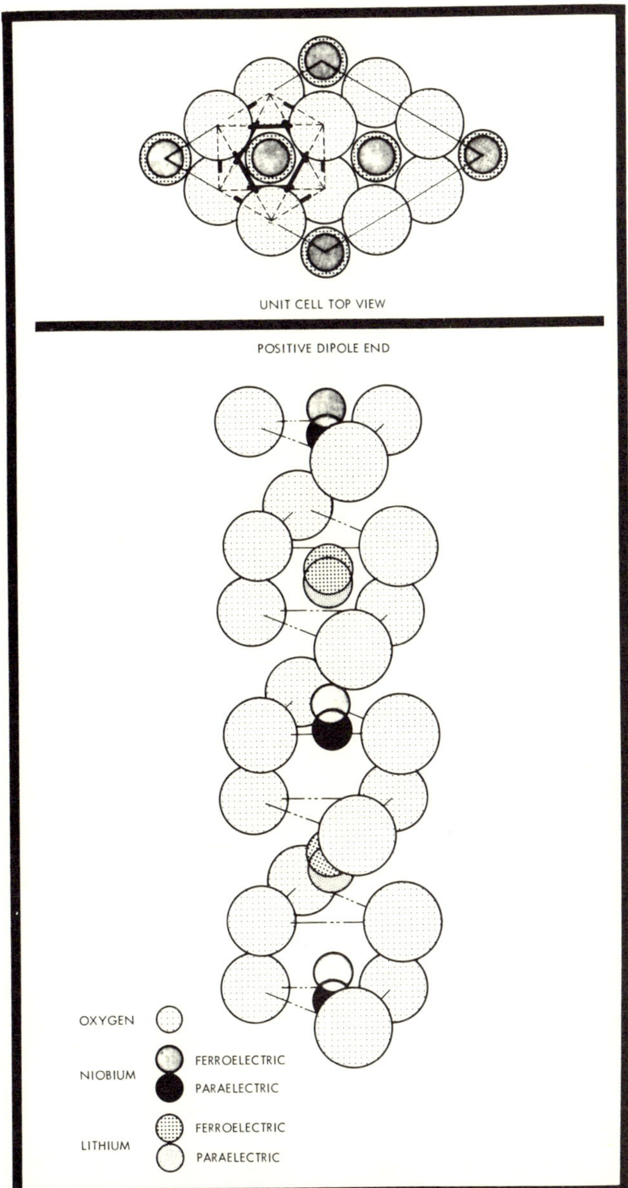




Fig. 1. A three-dimensional view of a lithium niobate crystal. The top view of a unit cell illustrates the ions seen along the direction of the polar axis. Negatively-charged oxygen ions form the corners of an octahedron. The smaller, positively-charged lithium and niobium ions occupy the space within the octahedrons in an ordered fashion (bottom view). When the temperature is above the Curie point, the lithium ions lie within the plane of the oxygen ions; the niobium ions are midway between planes. Hence, the crystal has no charge (paraelectric). With the temperature below the Curie point, both the lithium and niobium ions move in the same direction, resulting in a positive dipole at one end (ferroelectric). When a crystal is initially grown, however, the positive dipole ends in half of the domains point in one direction; dipoles in the remaining domains are oriented in the opposite direction. To switch all of the dipoles into the same direction, the crystal is placed in a dc field at temperatures approaching the Curie point. [Laudise].

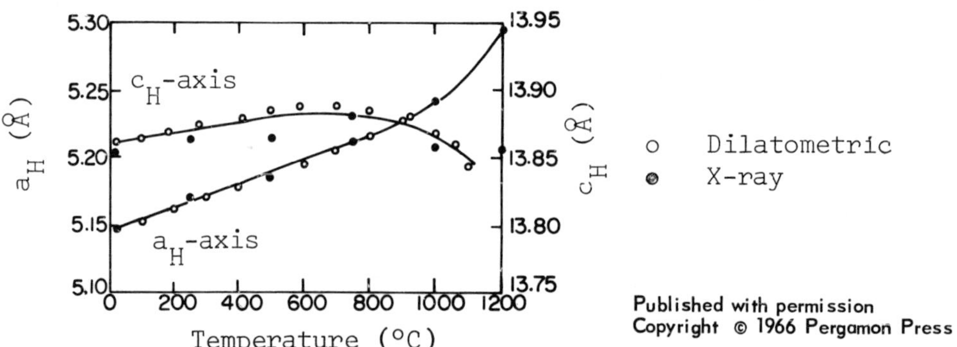


Fig. 2. Lattice constants of $LiNbO_3$ as a function of temperature [Abrahams et al.].

## Optical Properties - Refractive Index

The refractive indices of lithium niobate are of interest for the design of optical parametric amplifiers and other nonlinear devices where the achievement of phase-matching of the light waves is important. Midwinter has shown that the refractive indices of $LiNbO_3$ are sensitively dependent on the crystal chemical composition, with departures from stoichiometry or impurity content producing large changes in the phase-matching temperatures for a given nonlinear process.

The most quoted refractive index data for lithium niobate are those given by Boyd et al. for multidomain material, with $n_o = 2.2967$ and $n_e = 2.2082$ at 6000 Å The results of more recent refractive index measurements by Boyd et al. on $LiNbO_3$ typical of recent high-purity crystals to which 0.5% by weight MgO has been added to prevent cracking, are given in Table 1. These values are good to 2 parts in $10^4$ on a relative basis, though probably only to twice that amount on an absolute basis. Barker and Loudon have shown that the refractive index in the transparent region of lithium niobate can be explained in terms of a single uv oscillation term and several infrared terms (Figure 3).

Vinogradov et al. have measured the ordinary refractive index, $n_o$ at 120 to 150 GHz and report a value of 7.2; Irisova and Kozlov report birefringence of 1.57.

TABLE 1. REFRACTIVE INDICES OF $LiNbO_3$ [Boyd et al.]

| Wavelength | T = 25°C | | T = 80°C | |
|---|---|---|---|---|
| (microns) | $n_e$ | $n_o$ | $n_e$ | $n_o$ |
| 0.42 | 2.3038 | 2.4144 | 2.3090 | 2.4170 |
| 0.45 | 2.2765 | 2.3814 | 2.2814 | 2.3836 |
| 0.50 | 2.2446 | 2.3444 | 2.2498 | 2.3462 |
| 0.55 | 2.2241 | 2.3188 | 2.2276 | 2.3199 |
| 0.60 | 2.2083 | 2.3002 | 2.2118 | 2.3013 |
| 0.65 | 2.1964 | 2.2862 | 2.1993 | 2.2865 |
| 0.70 | 2.1874 | 2.2756 | 2.1900 | 2.2758 |
| 0.80 | 2.1741 | 2.2598 | 2.1766 | 2.2600 |
| 0.90 | 2.1647 | 2.2487 | 2.1671 | 2.2490 |
| 1.00 | 2.1580 | 2.2407 | 2.1601 | 2.2407 |
| 1.20 | 2.1481 | 2.2291 | 2.1503 | 2.2293 |
| 1.40 | 2.1410 | 2.2208 | 2.1426 | 2.2208 |
| 1.60 | 2.1351 | 2.2139 | 2.1372 | 2.2138 |
| 1.80 | 2.1297 | 2.2074 | 2.1318 | 2.2074 |
| 2.00 | 2.1244 | 2.2015 | 2.1265 | 2.2011 |
| 2.20 | 2.1187 | 2.1948 | 2.1211 | 2.1947 |
| 2.40 | 2.1138 | 2.1882 | 2.1156 | 2.1881 |
| 2.60 | 2.1080 | 2.1814 | 2.1099 | 2.1812 |
| 2.80 | 2.1020 | 2.1741 | 2.1037 | 2.1738 |
| 3.00 | 2.0955 | 2.1663 | 2.0972 | 2.1660 |
| 3.20 | 2.0886 | 2.1580 | 2.0903 | 2.1577 |
| 3.40 | 2.0814 | 2.1493 | 2.0830 | 2.1490 |
| 3.60 | 2.0735 | 2.1398 | 2.0746 | 2.1396 |
| 3.80 | 2.0652 | 2.1299 | 2.0669 | 2.1298 |
| 4.00 | 2.0564 | 2.1193 | 2.0582 | 2.1193 |

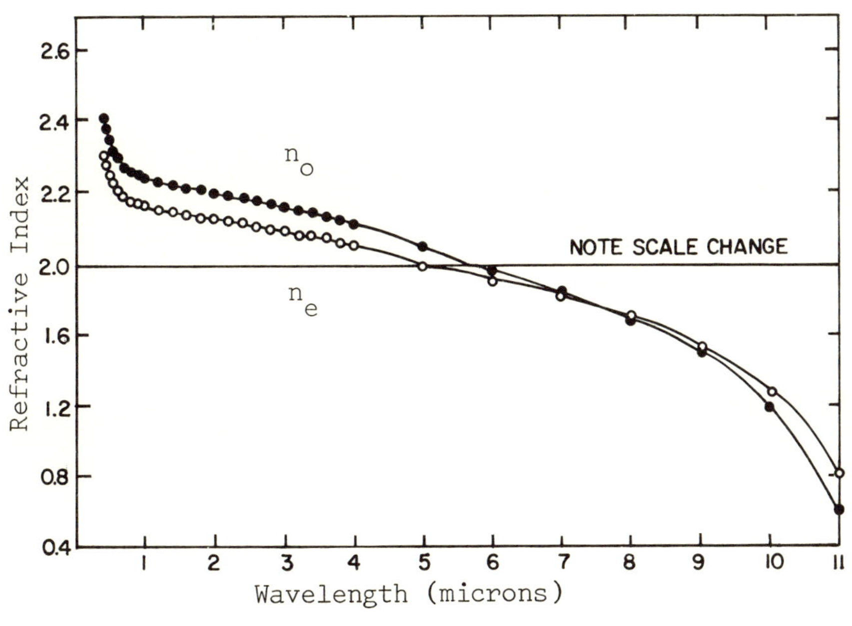


Fig. 3. Refractive indices of $LiNbO_3$ as a function of wavelength at 300°K. The data points for $\lambda<5\mu$ were taken from Boyd et al. The solid curves are least-square fits using the IR phonon modes and one adjustable oscillator in the UV [Barker and Loudon].

● $n_o(E\perp c)$

○ $n_e(E||c)$

— oscillator fit

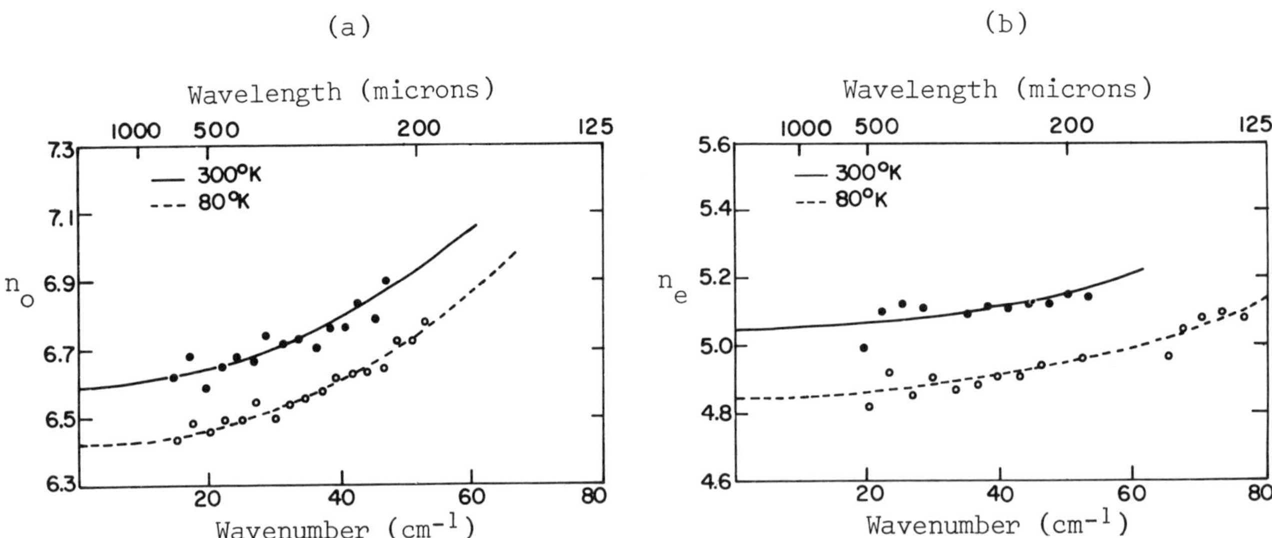


Fig. 4. Far-infrared refractive indices of $LiNbO_3$ as a function of wavenumber at two temperatures.

(a) Ordinary refractive index $n_o$

(b) Extraordinary refractive index $n_e$

[Bosomworth ]



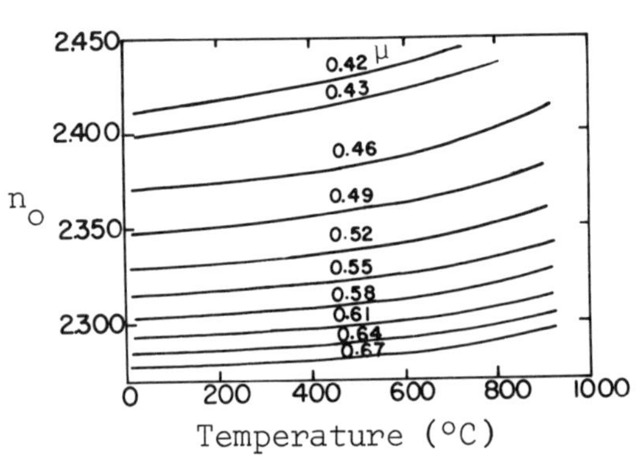


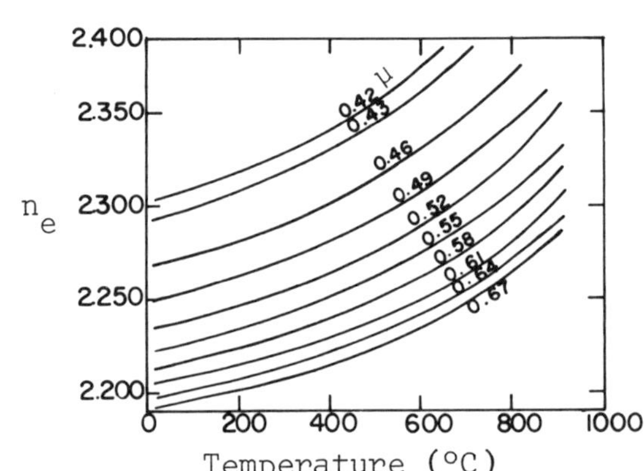


Fig. 5. Refractive indices of $LiNbO_3$ as a function of temperature at several wavelengths. [Iwasaki et al.]

Vedam and Davis have measured the ordinary and extraordinary indices up to 7 kbars at 5893 Å; $dn_o/dP = 0.32 \times 10^{-3}$/kbar and $dn_e/dP = 0.69 \times 10^{-3}$/kbar. The respective temperature coefficients are $1.4 \times 10^{-5}/°K$ and $6.0 \times 10^{-5}/°K$

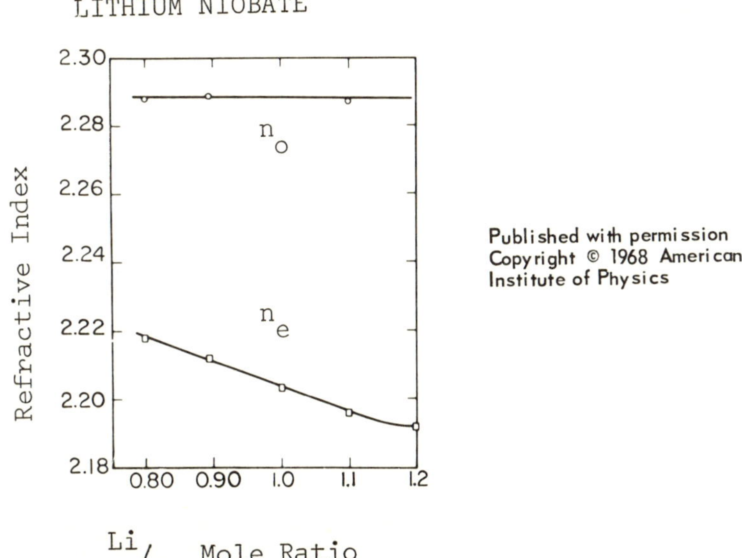




Fig. 6. Refractive indices of $LiNbO_3$ at 6328 Å as a function of melt stoichiometry [Bergman et al.].

Optical Properties - Birefringence

From the data presented by Boyd et al. in Table 1, lithium niobate is optically uniaxial negative, with $n_o$ = 2.3002 and $n_e$ = 2.2083 at 6000 Å and 25°C, giving a birefringence of $n_e - n_o$ = -0.0919.

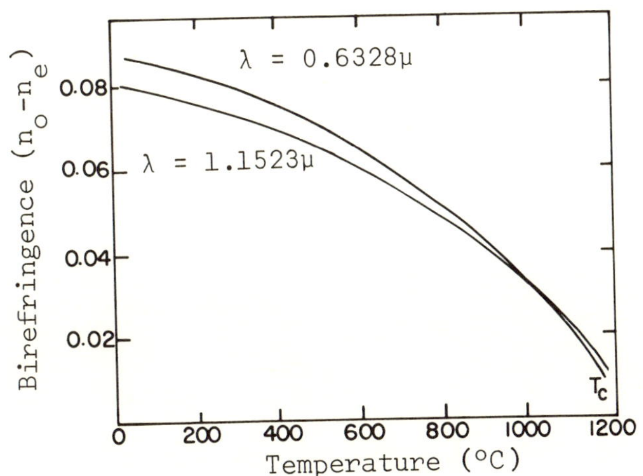




Fig. 7. Birefringence of $LiNbO_3$ as a function of temperature at two wavelengths. At 0.6328μ,

$$d(n_o-n_e)/dT = -2.7 \times 10^{-5}/°C \text{ at } 100°C$$
$$= -9.5 \times 10^{-5}/°C \text{ at } 1000°C$$

[Warner et al.]

Irisova and Kozlov report birefringence, Δn at 300°K and 100 to 150 GHz as 1.57.

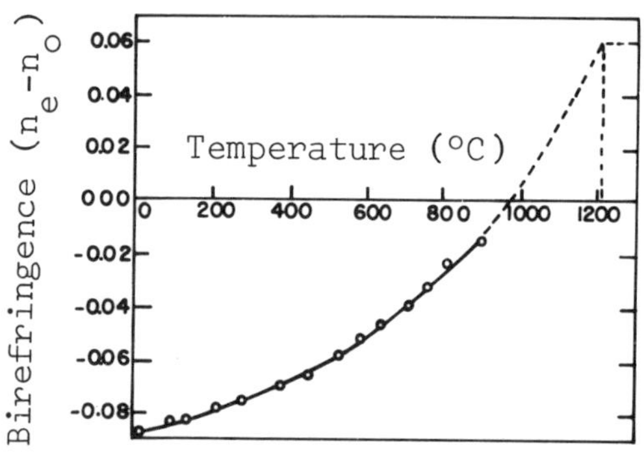


Fig. 8. Birefringence of $LiNbO_3$ as a function of temperature at 0.67 μ, from the data of Figure 5 [Iwasaki et al.].

## Optical Properties - Transmission

The optical transmission of single-domain lithium niobate is shown in Figure 9. Since lithium niobate has large refractive index values, a considerable amount of reflection from the surfaces of the specimen is observed. When allowance is made for reflection losses, the transmission becomes 100% in the central region of Figure 9. Detailed evaluation of the reflectivity by Guseva et al. shows that in a 1 cm thick specimen at λ = 1.06 μ, 5.6% of the radiation is lost by absorption and 14.6% by reflection from one surface, so that the total loss is about 31%. At λ = 0.58 μ, the respective losses are 6.8, 16, and 33%.

To reduce the losses due to reflection, a monolayer coating of quartz (n = 1.46) or magnesium fluoride (n = 1.38) may be applied to the $LiNbO_3$ crystal surface. The effect of a $MgF_2$ coating on the transmission of $LiNbO_3$ is shown in Figure 10. The effect of specimen thickness on the infrared transmission and the effect of the intensity of coloring of the crystal on the visible transmission is shown in Figure 11 and 12, respectively.

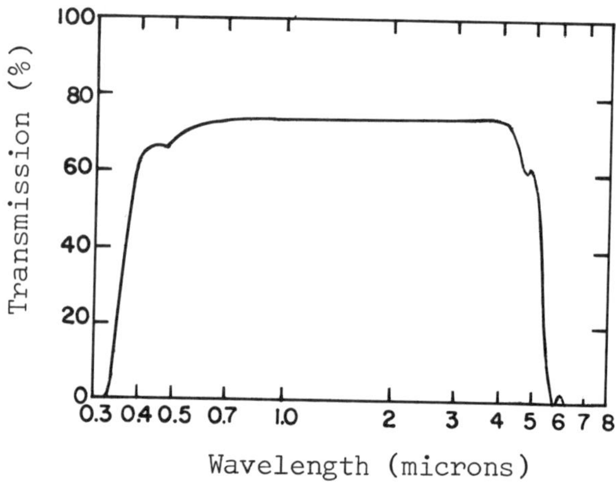


Fig. 9. Transmission of single-domain lithium niobate 0.600 cm thick as a function of wavelength. Transmission is uncorrected for reflection losses [Nassau et al.].

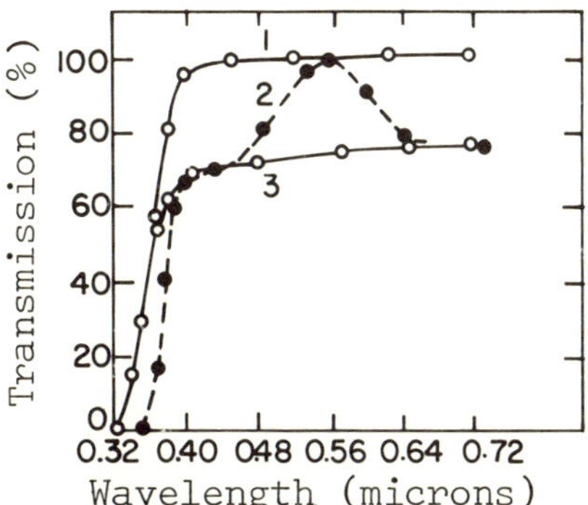


Fig. 10. Transmission of coated and uncoated polydomain lithium niobate as a function of wavelength.

1) Reflection losses taken into account by obtaining data for two specimens of different thicknesses (d ≃ 4mm).
2) Specimen coated with $MgF_2$ to a thickness to insure minimum reflection at $\lambda = 0.56\mu$. (d = 4mm).
3) Reflection not taken into account; uncoated specimen (d = 4.8mm).

[Guseva et al.]

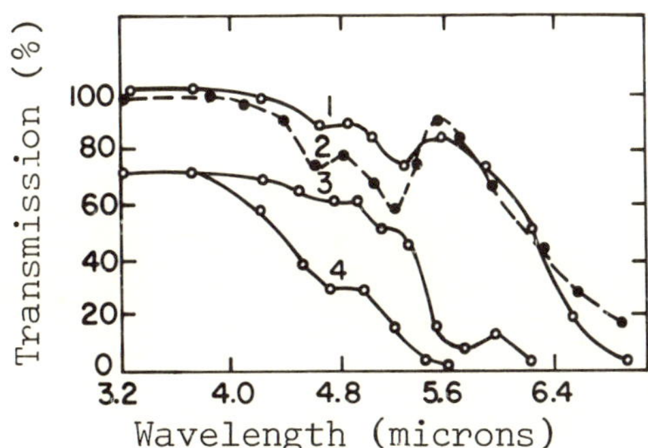




Fig. 11. Transmission of lithium niobate as a function of wavelength for various specimen thicknesses.

1) Corrected for reflection; d = 1.18mm.
2) Corrected for reflection; d = 2.43mm.
3) Uncorrected for reflection; d = 1.95mm.
4) Uncorrected for reflection; d = 6.78mm.

[Guseva et al.]

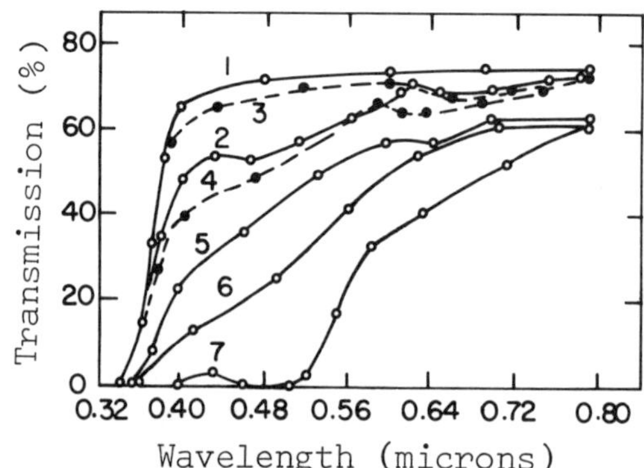


Fig. 12. Transmission of variously colored lithium niobate crystals as a function of wavelength

1) Clear crystal; d = 4.8 mm
2) Yellowish; d = 5.6 mm
3) Crystal 2 after annealing in oxygen
4) Crystal 2 after annealing for 5 min in hydrogen at 500°C
5) Yellowish; d = 2.7 mm
6) Orange; d = 7.27 mm
7) Dark brown; d = 3.1 mm

[Guseva et al.]

<u>Optical Properties - Nonlinear Optical Behavior</u>

Using a neodymium laser (1.06μ) to observe second-harmonic generation in lithium niobate, Boyd et al. and Miller and Savage have measured the following optical nonlinear coefficients relative to $d_{36}^{2\omega}$ for potassium dihydrogen phosphate (KDP); Miller et al. report values for several Li/Nb stoichiometric ratios:

| | | 1.083/4.7 | 0.946/5.6 | 0.852/5.2 |
|---|---|---|---|---|
| $\lvert d_{22}^{2\omega}\rvert$ | = 6.3 ± 0.6 | | | |
| $\lvert d_{31}^{2\omega}\rvert$ | = 11.9 ± 1.7 | -14.5 | -11.6 | -8.4 |
| $d_{33}^{2\omega}$ | = 83 ± 21 | -72.7 | -72.4 | -67.1 |

Absolute values of $d_{22}^{2\omega}$ and $d_{31}^{2\omega}$ are indicated above, since Bjorkholm has found that these coefficients are of opposite sign in $LiNbO_3$. The temperature dependence of $d_{33}^{2\omega}$ is indicated in Figure 13. Bjorkholm also reports $d_{31}^{2\omega}/d_{36}^{2\omega}$ (KDP) = 10.9 ± 1.7 at 1.15 microns.

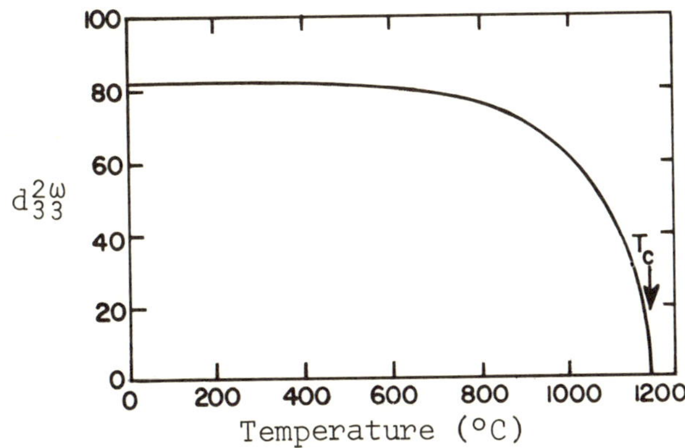


Fig. 13. Nonlinear coefficient $d_{33}^{2\omega}$ for $LiNbO_3$ as a function of temperature [Miller and Savage].



The development at Bell Telephone Laboratories of lithium niobate as a nonlinear optical material has been reviewed by Laudise. Lithium niobate was first found by Boyd et al. to have a high efficiency as a harmonic generator. Furthermore, the birefringence of this material was found to exceed the dispersion in the 1.06-to-0.53-micron range, making it especially useful as a harmonic generator for the 1.06-micron light emitted by the Nd-doped YAG laser.

It was found [Ashkin et al.], however, that the refractive index of lithium niobate changed when the crystal was subjected to laser light at power densities as low as 2 millivolts in a 4-mil-diameter beam. Chen has observed local index changes in poled single crystal $LiNbO_3$; $n_e$ was observed to decrease as much as $10^{-3}$ with a focused Ar laser at 20 mW intensity, while $n_o$ was much smaller. Above approximately 170°C the index inhomogeneity (optical damage) relaxes faster than it is generated. Levinstein et al. have found that this inhomogeneity can be reduced through the use of gold electrodes and an applied field. When lithium niobate is field annealed at 250 V/cm at 700°C using gold or platinum paste electrodes, Au/Pt is observed to diffuse into the sample from the positive electrode along dislocation lines decorating the dislocations; the susceptibility to laser-induced index change is found to be less in those regions in which the dislocations are decorated than in the rest of the crystal. Nevertheless, lithium niobate became a promising nonlinear optical crystal only after it was discovered that the birefringence and, consequently, the phase-matching temperature for second-harmonic generation are a function of melt stoichiometry (see Figure 6). Thus, the phase-matching temperature could be adjusted to occur above 170°C where the crystals could be used without optical damage [Fay et al., Bergman et al.].

## Electrooptic Properties

Peterson et al., first investigated (in 1964) the electrooptic behavior of lithium niobate. In 1966, Lenzo et al. first presented quantitative measurements of the dc electrooptic coefficients of single-domain $LiNbO_3$.

Kaminow and Johnston have given a new insight into the electrooptic behavior of lithium niobate by deriving a simple relationship between an electrooptic coefficient measured at radio-frequencies and the corresponding Raman-scattering efficiencies. These authors have found that the dominant contribution to the electrooptic coefficients $r_{33}$ and $r_{13}$ comes from the lowest-frequency $A_1$ optic mode and to $r_{42}$ and $r_{22}$ from the next lowest E mode, with only a small pure electronic contribution.

Zook et al. and Wemple and co-workers have discussed the relationship between the linear and quadratic electrooptic effects in lithium niobate. Using a directly measured value of the spontaneous polarization ($P_s$ = 0.71 C/m$^2$), the latter authors show that the linear effect is related fundamentally to a biased quadratic effect associated with each $BO_6$ octahedron in $LiNbO_3$.

Lithium niobate belongs to the point group 3m($C_{3v}$). Because of symmetry requirements, only eight of the possible 18 different elements of the electrooptic tensor are non-zero, with only four coefficients required to describe the electro-optic behavior of the crystal;

$$r_{13} = r_{23}$$

$$r_{22} = -r_{12} = -r_{61}$$

$$r_{42} = r_{51}$$

$$r_{33}$$

TABLE 2. ELECTROOPTIC COEFFICIENT OF LITHIUM NIOBATE (in $10^{-12}$ m/V)

| Constant Stress (T) Constant Strain (S) | $r_{33}$ | $r_{13}$ | $r_{22}$ | $r_{42}$ | $n_e^3 r_{33} - n_o^3 r_{13}$ | λ(μ) | T(°C) | Reference |
|---|---|---|---|---|---|---|---|---|
| T | | | 6.7 | | 190 | 0.6328 | R.T. | Lenzo et al. |
| S | 30.8 | 8.6 | 3.4 | 28 | 224 | 0.6328 | R.T. | Turner |
| T | +32.2 | +10.0 | 6.81 | | | 0.6328 | 100 | Zook et al. |
| T | | | 6.7<br>5.4<br>3.1 | | 190<br>180<br>169 | 0.6328<br>1.15<br>3.39 | R.T. | Smakula and Claspy |
| T | | | -6.4 | | | 0.6328 | R.T. | Iwasaki et al. |
| T | | | | 32±2 | | 0.6328 | R.T. | Bernal et al. |
| T | +30.8 | +9.6 | +6.6 | +32.8 | | | | Hulme et al. |

The dc electrooptic coefficients $r_{ij}^T$, corresponding to a free crystal, are related to the electrooptic coefficients $r_{ij}^S$, measured at high modulating frequencies, corresponding to a "clamped" crystal, by

$$r_{ij}^T = r_{ij}^S + p_{ik} d_{jk} \quad , \quad (k = 1 \text{ to } 6)$$

where $p_{ik} d_{jk}$ is the strain contribution to $r_{ij}^T$. Here, $p_{ik}$ are the elastooptic coefficients and $d_{jk}$ the piezoelectric constants. For the lithium niobate point group 3m($C_{3v}$),

$$r_{22}^T = r_{22}^S + (p_{11} - p_{12}) d_{22} - p_{14} d_{15} \quad .$$

Kludzin has recently made a complete set of measurements of the photoelastic constants of single crystals of lithium niobate, using 30 MHz and a helium-neon gas laser at 6328 Å. Measurements are shown in Table 4.

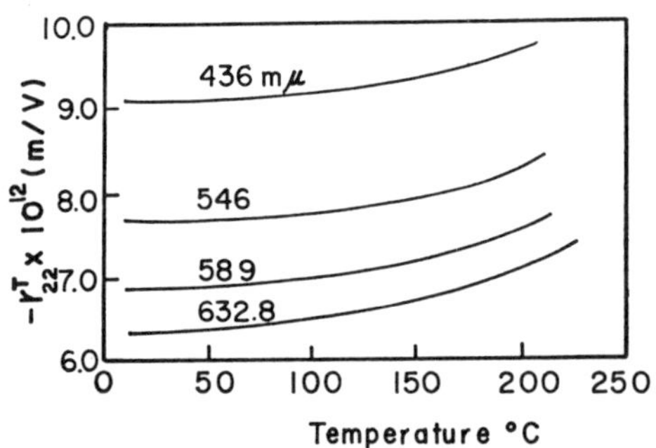


Fig. 14. The electrooptic coefficient of lithium niobate, $r_{22}^T$ as a function of temperature at several wavelengths [Iwasaki et al.].

Turner et al. report measurements at low and high frequencies for several Li/Nb ratios and find no variation with melt composition. Data are given at 6300 Å for the stoichiometric compound.

| 76 MHz | | | Low Frequency |
|---|---|---|---|
| $n_e^3 r_{33}$ | $n_o^3 r_{33}$ | $V_{0.5}$ | $V_{0.5}$ |
| $10^{-10}$ m/V | | kV | kV |
| 3.28 | 1.03 | 2.81 | 3.0 |

TABLE 3. SPATIAL DIFFERENTIATION IN ELECTROOPTIC PROPERTIES OF LITHIUM NIOBATE [Lenzo et al.].

| Light Direction | Field Direction | | | | | |
|---|---|---|---|---|---|---|
| | [100] | | [010] | | [001] | |
| | $r_{22}$ ($10^{-12}$m/V) | (kV) $V_{0.5}$ | $r_{22}$ ($10^{-12}$m/V) | (kV) $V_{0.5}$ | $0.9r_{33}-r_{13}$ ($10^{-12}$m/V) | (kV) $V_{0.5}$ |
| [100] | | | 7.3 | 7.23 | 18 | 2.94 |
| [010] | | | 6.0 | 9.00 | 17 | 3.16 |
| [001] | 6.6 | 4.00 | 6.3 | 4.25 | | |

$V_{0.5}$ = Direct current half-wave voltage at 20°C and 6328 Å

V = Applied Voltage, 1-3 kV

$n_o$ = Ordinary refractive index = 2.286

$n_e$ = Extraordinary refractive index = 2.200

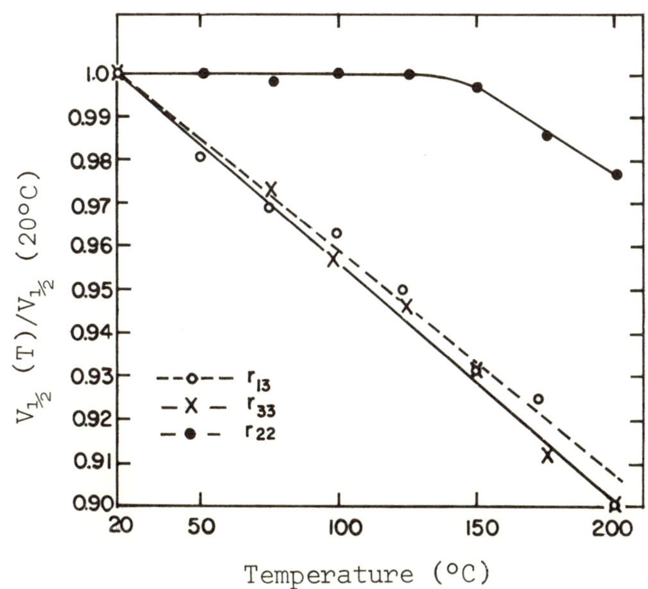


Fig. 15. Half-wave voltage $V_{0.5}$ as a function of temperature normalized with respect to its room temperature value for measuring $r_{13}$, $r_{22}$ and $r_{33}$ in $LiNbO_3$. The unusual temperature dependence of $r_{22}$ above 150°C was checked by repeated measurements on two different crystals[Zook et al.].



Vasilevskaya et al. have also observed an anomaly in $r_{22}$ from 150 to 170°C The following temperature dependences at 100°C may be determined from Figure 15:

$$(dr_{33}^T/dT)/r_{33}^T = 4.9 \times 10^{-4}/°C$$

$$(dr_{13}^T/dT)/r_{13}^T = 4.6$$

$$(dr_{22}^T/dT)/r_{22}^T \sim 0$$

## Photoelastic Properties - Elastooptic Coefficients

TABLE 4. ELASTOOPTIC COEFFICIENTS OF LITHIUM NIOBATE AT 6328 Å

| $p_{11}$ | $p_{12}$ | $p_{13}$ | $p_{14}$ | $p_{31}$ | $p_{33}$ | $p_{41}$ | $p_{44}$ | $p_{66}$ | Reference |
|---|---|---|---|---|---|---|---|---|---|
| 0.036 | 0.072 | 0.092 | | 0.178 | 0.088 | 0.155 | | | Dixon and Cohen |
| | | 0.096 | 0.070 | | 0.093 | | | | Reintjes and Schulz |
| | | | 0.056 | | | 0.050 | 0.062 | 0.048 | Maloney et al. |
| 0.045 | 0.096 | 0.106 | 0.055 | 0.138 | 0.076 | 0.12 | 0.019 | | Kludzin |

## Photoelastic Properties - Piezooptic Coefficients

TABLE 5. DIFFERENTIAL PIEZOOPTIC COEFFICIENTS OF LITHIUM NIOBATE [Spencer et al.]

| Coefficient | Value ($10^{-12}$ $m^2/N$) | Light Direction | Stress Direction |
|---|---|---|---|
| $\pi_{12} - \pi_{11}$ | 1.02 | [001] | [100] |
| $\pi_{33} - 1.12\ \pi_{13}$ | 0.808 | [010] | [001] |
| $\pi_{33} - 1.12\ \pi_{13}$ | 0.484 | [100] | [001] |
| $\pi_{31} - 1.12\ \pi_{11}$ | 0.687 | [100] | [010] |
| $\pi_{11} - \pi_{12}$ | 0.692 | [001] | [010] |

## Photoelastic Properties - Elastic Constants

TABLE 6. ELASTIC STIFFNESS CONSTANTS OF LITHIUM NIOBATE AT 20°C [Warner et al.]

| | $10^{10}$ $N/m^2$ | | | | | | |
|---|---|---|---|---|---|---|---|
| | $c_{11}$ | $c_{12}$ | $c_{13}$ | $c_{14}$ | $c_{33}$ | $c_{44}$ | $c_{66}$ |
| Constant Field (E) | 20.3 | 5.3 | 7.5 | 0.9 | 24.5 | 6.0 | 7.5 |
| Constant Displacement | 21.9 | 3.7 | 7.6 | -1.5 | 25.2 | 9.5 | 9.1 |

TABLE 7. ELASTIC COMPLIANCE CONSTANTS OF LITHIUM NIOBATE AT 20°C

| | $10^{-12}$ $m^2/N$ | | | | | | | | |
|---|---|---|---|---|---|---|---|---|---|
| | $s_{11}$ | $s_{12}$ | $s_{13}$ | $s_{14}$ | $s_{33}$ | $s_{44}$ | $s_{66}$ | $2s_{13}+s_{44}$ | Reference |
| Constant Displacement (D) | 5.20 | -0.44 | -1.45 | 0.87 | 4.89 | 10.8 | 11.3 | 7.9 | Warner et al. |
| Constant Field (E) | 5.78 | -1.01 | -1.47 | -1.02 | 5.02 | 17.0 | 13.6 | 14.1 | |
| | 5.64 | | | -0.84 | 4.94 | | | 13.9 | Yamada et al. |

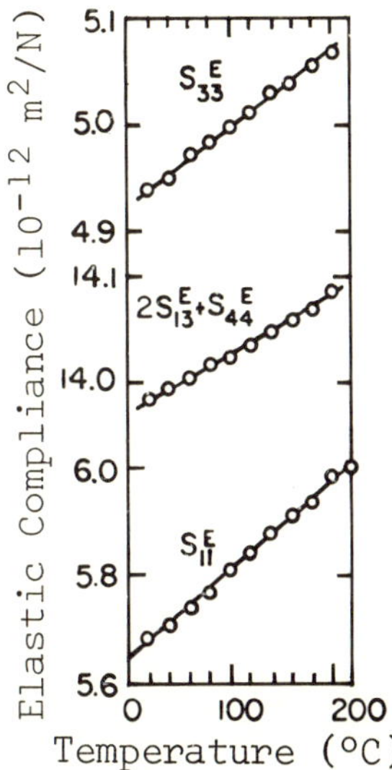


Fig. 16. Elastic compliance constants of lithium niobate as a function of temperature

$$s_{33}^E = 4.94[1 + (T-20)1.5 \times 10^{-4}] \times 10^{-12} \text{ m}^2/\text{N}$$
$$2s_{13}^E + s_{44}^E = 13.9[1 + (T-20)2.0 \times 10^{-4}]$$
$$s_{11}^E = 5.64[1 + (T-20)1.5 \times 10^{-4}]$$

[Yamada et al.]

## Piezoelectric Properties

Four coefficients are required to describe the piezoelectric properties of lithium niobate; the non-zero coefficients of the piezoelectric strain tensor are as follows:

$$d_{15} = d_{24}$$
$$d_{22} = -d_{16} = -d_{21}$$
$$d_{31} = d_{32}$$
$$d_{33}$$

Spencer and co-workers , in reviewing the literature on lithium niobate, found it to have unusual capabilities as an active element of an elastooptic device. They confirmed it to have large piezoelectric coupling coefficients, low microwave-elastic propagation losses, and a large elastooptic coupling coefficient.

Smith and Welsh have recently reported new determinations of the elastic, piezoelectric and dielectric constants of lithium niobate and their temperature dependence in the 0° to 110°C range. They used a combination of ultrasonic phase-velocity, resonance frequency and capacitance measurements. The elastic constant temperature derivatives are all about $-2\text{x}10^{-4}$/°C, the piezoelectric values are all positive in the range 0.8 to $8.9\text{x}10^{-4}$/°C. The dielectric coefficients are $3\text{-}6\text{x}10^{-4}$/°C

TABLE 8. PIEZOELECTRIC CONSTANTS OF LITHIUM NIOBATE AT ROOM TEMPERATURE [Warner et al.]

| | | |
|---|---|---|
| $d_{15}$ = 6.8 x $10^{-11}$ C/N | | $e_{15}$ = 3.7 $C/m^2$ |
| = (7.4)* | | |
| $d_{22}$ = 2.1 | (2.1) | $e_{22}$ = 2.5 |
| $d_{31}$ = -0.1 | (-0.086) | $e_{31}$ = 0.2 |
| $d_{33}$ = 0.6 | (1.62) | $e_{33}$ = 1.3 |
| $h_{15}$ = 9.5 x $10^9$ N/C | | $g_{15}$ = 9.1 x $10^{-2}$ $m^2/C$ |
| $h_{22}$ = 6.4 | | $g_{22}$ = 2.8 |
| $h_{31}$ = 0.8 | | $g_{31}$ = -0.4 |
| $h_{33}$ = 5.1 | | $g_{33}$ = 2.3 |

* ( ) Data of Yamada et al.

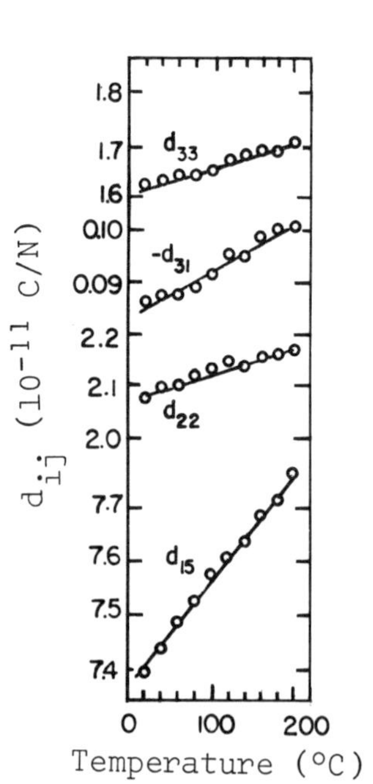


Fig. 17. Piezoelectric constants of lithium niobate as a function of temperature.

$d_{33}$ = 1.62[1 + (T-20)2.9 x $10^{-4}$] x $10^{-11}$ C/N

$d_{31}$ = -0.086[1 + (T-20)11 x $10^{-4}$]

$d_{22}$ = 2.1[1 + (T-20)2.4 x $10^{-4}$]

$d_{15}$ = 7.4[1 + (T-20)2.8 x $10^{-4}$]

[Yamada et al.]

Electromechanical Coupling Factors:

| [Yamada et al.] | [Chkalova et al.] |
|---|---|
| $k_{31} = 2.3\%$ | $k_{31} = 8.7\%$ |
| $k_{22} = 32\%$ | $k_{22} = 24.6\%$ |
| $k_{33} = 47\%$ | $k_{33} = 32.7\%$ |
| | $k_{15} = 44.6\%$ |

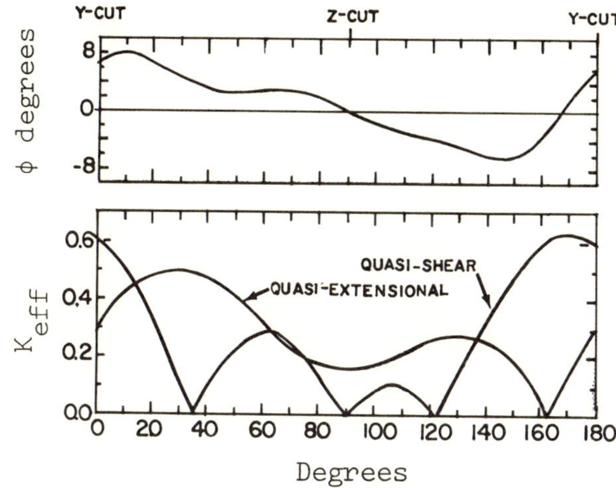


Fig. 18. Effective coupling factors and angle $\phi$ between quasi-extensional wave displacement and plate normal for rotated y-cuts of $LiNbO_3$ [Warner et al.]



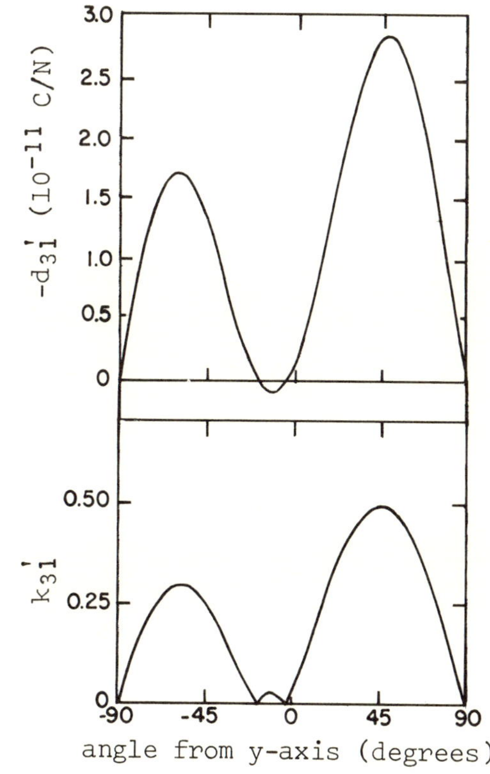


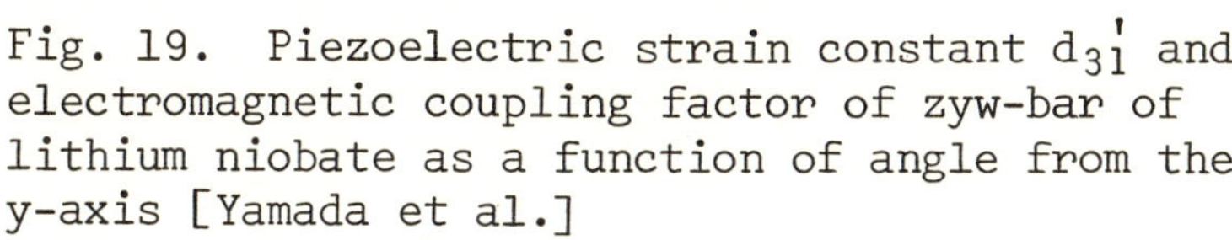
Fig. 19. Piezoelectric strain constant $d_{31}'$ and electromagnetic coupling factor of zyw-bar of lithium niobate as a function of angle from the y-axis [Yamada et al.]

## Dielectric Properties - Dielectric Constant

An overview of the dispersion of the dielectric constant of lithium niobate is shown schematically in Figure 20. At low frequencies the dielectric constant goes through one or more piezoelectric resonances and decreases from the dc (free crystal value $\varepsilon^T$ to the clamped value $\varepsilon^S$ [Ohmachi et al.]. It has been shown by Barker and Loudon, and Axe and O'Kane, that values of $\varepsilon^S$ are in reasonable agreement with values of the "static" dielectric constant $\varepsilon_o$, obtained from infrared reflectivity measurements. This agreement is apparent from the data presented in Table 9.

The low-frequency dispersion is very strong for E⊥c ($\varepsilon_a$), as indicated by the results of Nassau et al. shown in Figure 21. Although these results do not exhibit the strong piezoelectric resonances near 1MHz that are observed by Ohmachi et al. , the observed differences in $\varepsilon^T$ and $\varepsilon^S$ agree with the following room temperature values observed by the latter authors:

$$\varepsilon_a^T - \varepsilon_a^S = 38.5$$
$$\varepsilon_c^T - \varepsilon_c^S = 2.5.$$

The temperature dependence of the dielectric constants $\varepsilon^T$ and $\varepsilon^S$ is illustrated in Figures 22 and 23, respectively.

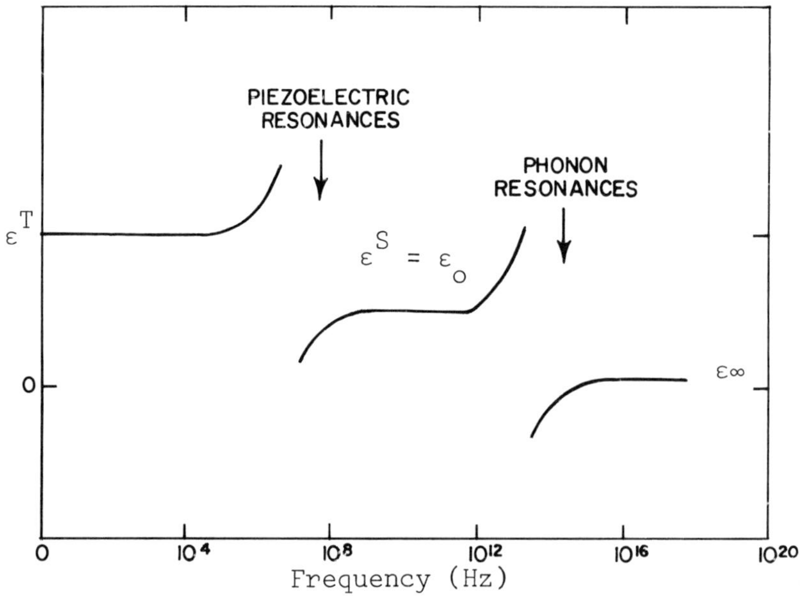


Fig. 20. Schematic of the dielectric constant of lithium niobate as a function of frequency.

TABLE 9. DIELECTRIC CONSTANTS OF LITHIUM NIOBATE AT ROOM TEMPERATURE

| E⊥c | | | | E‖c | | | | Reference |
|---|---|---|---|---|---|---|---|---|
| $\varepsilon^T_a$ | $\varepsilon^S_a$ | $\varepsilon_o(n_o)$ | $\varepsilon_\infty(n_o)$ | $\varepsilon^T_c$ | $\varepsilon^S_c$ | $\varepsilon_o(n_e)$ | $\varepsilon_\infty(n_e)$ | |
| 84 | 44 | | | 30 | 29 | | | Warner et al., |
| 84.6 | | | | 28.6 | | | | Yamada et al., |
| | | 43.5 | 4.98 | | | 25.5 | 4.64 | Bosomworth |
| | | 41.5 | 5.0 | | | 26.0 | 4.6 | Barker and Loudon |

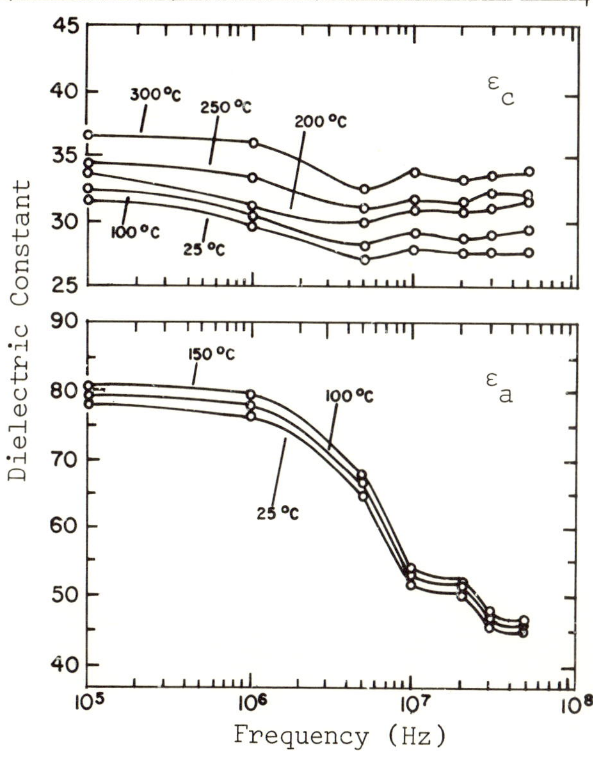


Fig. 21. Dielectric constants of lithium niobate as a function of frequence at several temperatures [Nassau et al.].

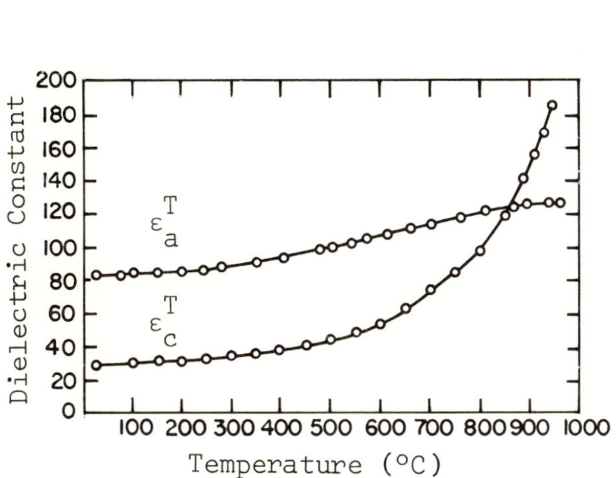


Fig. 22. Dielectric constants of lithium niobate as a function of temperature at $10^5$ Hz. The $\varepsilon^T_c$ curve shows the expected high temperature increase as the Curie temperature is approached [Nassau et al.].

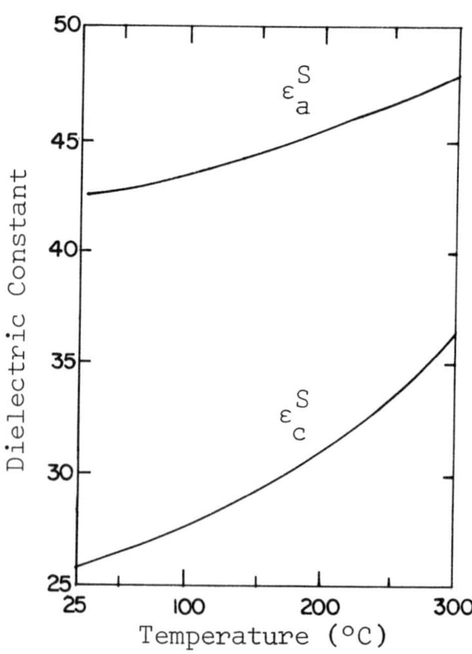


Fig. 23. Dielectric constants of lithium niobate as a function of temperature at 9 GHz [Ohmachi et al.]

As shown in Figure 22, the dielectric constant $\varepsilon_c^T$ increases sharply as the Curie temperature is approached. A typical plot of the dielectric constant versus temperature, in the region of the Curie temperature, is shown in Figure 24. These results of Bergman et al. agree with those of Smolenskii et al., in that a dielectric anomaly is observed only parallel to the 3-fold axis ($\varepsilon_c$); no anomaly in $\varepsilon_a$ is observed. Figure 24 indicates a Curie temperature of about 1150°C for $LiNbO_3$. Bergman et al. , however, show that the Curie temperature varies drastically with melt stoichiometry (see Figure 26).

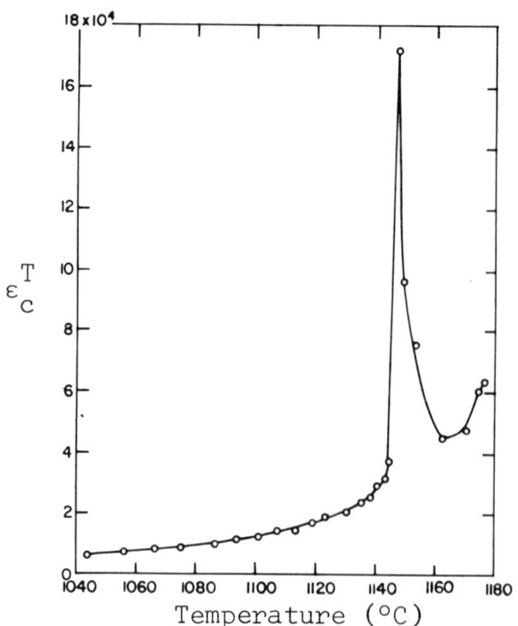


Fig. 24. Typical plot of the dielectric constant of lithium niobate as a function of temperature at 1 kHz [Bergman et al.]

## Dielectric Properties - Loss Tangent

At 130 GHz and 20°C, the dissipation factor is $2.5 \times 10^{-3}$. (Vinagradov et al.)

TABLE 10. LOSS TANGENT OF LITHIUM NIOBATE AT $10^5$ Hz [Nassau et al.].

| T(°C) | tan δ | |
|---|---|---|
| | c-axis | a-axis |
| 400 | 0.001 | 0.0006 |
| 500 | 0.016 | 0.01 |
| 600 | 0.12 | 0.1 |
| 700 | 1.0 | 0.8 |
| 800 | 5 | 8 |
| 900 | 11 | 25 |

See Figure 22 for corresponding dielectric constant data.

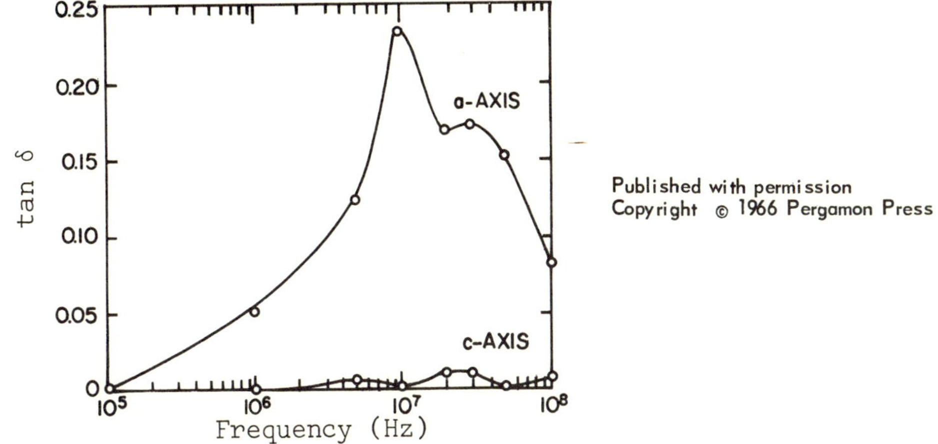




Fig. 25. Loss tangent of lithium niobate as a function of frequency at 25°C [Nassau et al.].

## Dielectric Properties - Electrical Resistivity

Experiments on stoichiometric lithium niobate yield evidence for a considerable amount of ionic conductivity [Bergman]. Nassau et al. quote the following resistivity values for a c-axis slice of $LiNbO_3$:

$5 \times 10^8$ ohm-cm at 400°C

140 ohm-cm at 1200°C

These authors find that the following expression describes the electrical resistivity of lithium niobate over this temperature range:

$$\log \rho = \frac{7150}{T} - 2.823 \qquad (T \text{ in } ^\circ K)$$

# LITHIUM NIOBATE

## Ferroelectric Properties

In 1949, Matthias and Remeika found a ferroelectric effect in lithium niobate; the hysteresis loop observed at 200°C was a perfect rectangle. Based on the inability to observe polarization reversal at room temperature, $LiNbO_3$ has long been viewed as a "frozen" ferroelectric whose direction of polarization is maintained by structural forces too large to be overcome by applied fields, except at elevated temperatures.

Nassau and co-workers, beginning in 1965, have made a very comprehensive investigation of the ferroelectric behavior of lithium niobate. They found that it has a very high Curie temperature (∿1200°C) and that single-domain crystals can be prepared by electric reversal of polarity at surprisingly low electric fields and very high temperatures. The application of fields as low as 1 V/cm for 15 min. at 1200°C converted a multidomain material to a single-domain crystal. This property of domain reversal by an electric field, clearly defined $LiNbO_3$ as a ferroelectric material.

More recently, Camlibel and co-workers have developed a pulsed-field method to measure directly the room temperature, spontaneous polarization of ferroelectric oxides. They found that the ability to reverse the spontaneous polarization is dependent upon the electrode material used. By amploying an aqueous solution of LiCl as the electrode material, electric fields as high as 500 kV/cm could be impressed upon the sample. The availability of such high fields, enabled these investigators to measure the spontaneous polarization in ferroelectrics in which the direction of $P_s$ has been thought previously to be nonreversible. A spontaneous polarization of $P_s = 0.71 \pm 0.02$ C/m$^2$ was reported for $LiNbO_3$.

Differing values for the Curie temperature of lithium niobate have been reported in the literature (Table 11). Differences among these values may be due to variations in stoichiometry; as indicated in Figure 26, the Curie temperature varies drastically with melt stoichiometry. Nassau, in reviewing the research carried out at Bell Telephone Laboratories on $LiNbO_3$, quotes a best value of $T_c = 1195 \pm 15$°C. Thus, among all known ferroelectric materials, lithium niobate has the highest Curie temperature.

TABLE 11. CURIE TEMPERATURE OF LITHIUM NIOBATE

| $T_c$ (°C) | Method | Reference |
|---|---|---|
| 1205±5 | Birefringence; cf. Figure 7 | Warner et al. |
| 1195±15 | Nonlinear optical data; cf. Figure 13 | Miller and Savage |
| 1140 | Dielectric constant | Smolenskii et al. |
| 1210±10 | Single-domain to multidomain transition | Nassau et al. |
| 1177±15 | Heat capacity | Pankratz and King quoted by Bergman et al. |

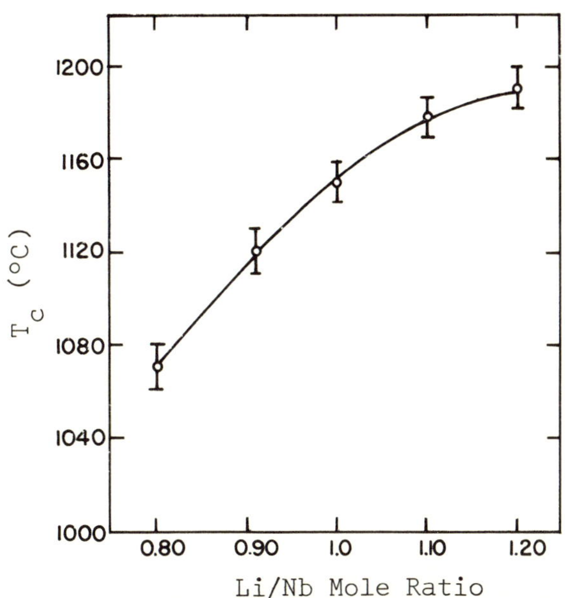


Fig. 26. Curie temperature of $LiNbO_3$ as a function of melt stoichiometry [Bergman et al.].

Thermal Properties

| Melting Point: | 1250°C | Ballman |
|---|---|---|
| | 1253°C | Reisman and Holtzberg |
| | 1260±10°C | Nassau et al. |

Linear Thermal Expansion Coefficient:

$\alpha_a = 16.7 \times 10^{-6}/°C$ (24-800°C) Abrahams et al.

$\alpha_c = \sim 2 \times 10^{-6}/°C$ (24-600°C)

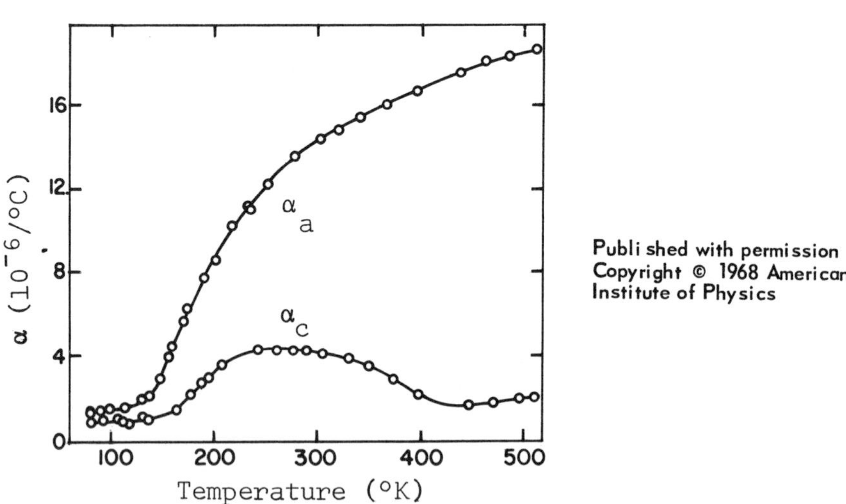




Fig. 27. Linear thermal expansion coefficients of lithium niobate as a function of temperature [Zhdanova et al.]

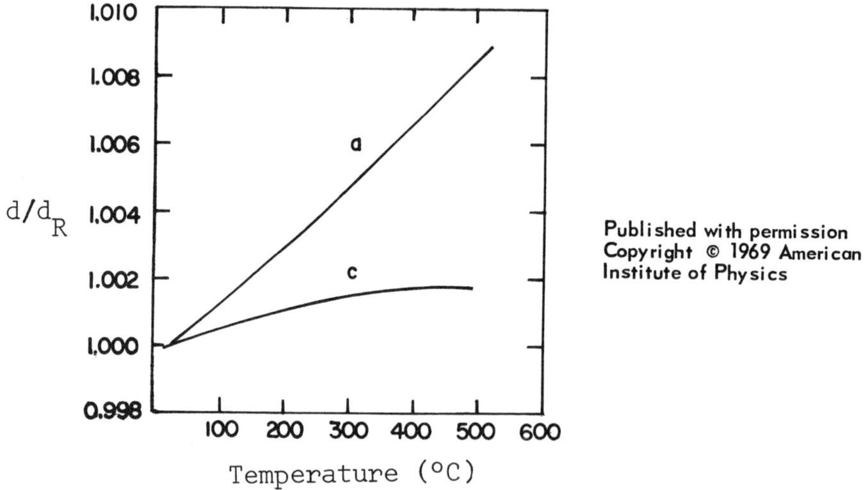




Fig. 28. Relative thermal expansion of lithium niobate ($T_c$ = 1165°C) as a function of temperature.

$$d/d_R = 1 + \alpha\,(T-25) + \beta\,(T-25)^2$$

| | a | c | |
|---|---|---|---|
| $\alpha$ ($10^{-5}$/°C) | 1.54 | 0.75 | |
| $\beta$ ($10^{-9}$/°C$^2$) | 5.3 | -7.7 | [Kim and Smith] |

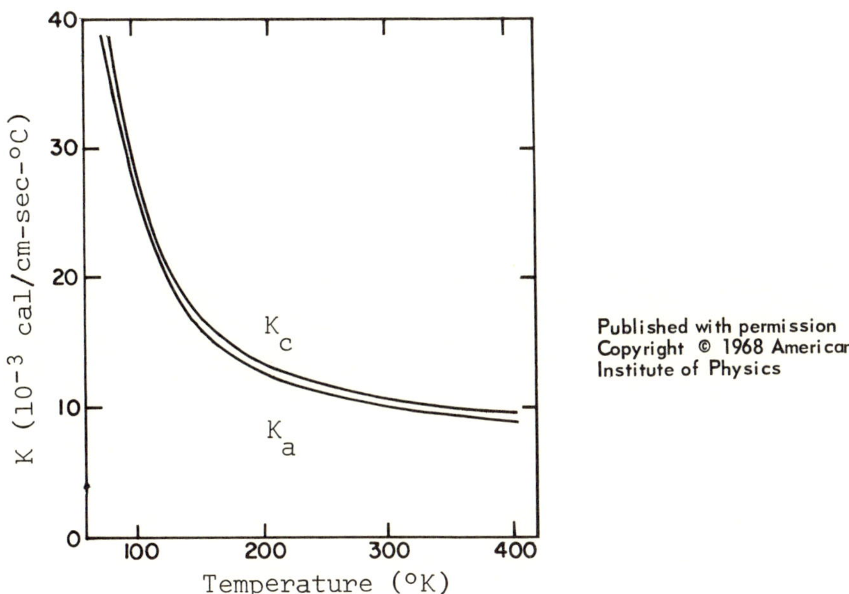


Fig. 29. Thermal conductivity of lithium niobate as a function of temperature. The anisotropy of the thermal conductivity for all temperatures is insignificant ($K_c/K_a \simeq 1.05$) [Zhdanova et al.].

ABRAHAMS, S.C. et al. Ferroelectric Lithium Niobate. Pt. 3: Single Crystal X-Ray Diffraction Study at 24°C. J. OF PHYS. AND CHEM. OF SOLIDS, v. 27, no. 6/7, June-July 1966. p. 997-1012.

ABRAHAMS, S.C. et al. Ferroelectric Lithium Niobate. Pt. 4: Single Crystal Neutron Diffraction Study at 24°C. J. OF PHYS. AND CHEM. OF SOLIDS, v. 27, no. 6/7, June-July 1966. p. 1013-1018.

ABRAHAMS, S.C. et al. Ferroelectric Lithium Niobate. Pt. 5: Polycrystal X-Ray Diffraction Study Between 24°C and 1200°C. J. OF PHYS. AND CHEM. OF SOLIDS, v. 27, no. 6/7, June-July 1966. p. 1019-1026.

ASHKIN, A. et al. Optically-Induced Refractive Index Inhomogeneities in $LiNbO_3$ and $LiTaO_3$. APPLIED PHYS. LETTERS, v. 9, no. 1, July 1966. p. 72-74.

AXE, J.D. and D.F. O'KANE. Infrared Dielectric Dispersion of $LiNbO_3$. APPLIED PHYS. LETTERS, v. 9, no. 1, July 1966. p. 58-60.

BALLMAN, A.A. Growth of Piezoelectric and Ferroelectric Materials by the Czochralski Technique. AMERICAN CERAM. SOC., J., v. 48, no. 2, Feb. 1965. p. 112-113.

BARKER, A.S. and R. LOUDON. Dielectric Properties and Optical Phonons in $LiNbO_3$. PHYS. REV., v. 158, no. 2, June 1967. p. 433-445.

BERGMAN, J.G. et al. Curie Temperature Birefringence and Phase-Matching Temperature Variations in $LiNbO_3$ as a function of Melt Stoichiometry. APPLIED PHYS. LETTERS, v. 12, no. 3, Feb. 1968. p. 92-94.

BERGMANN, G. The Electrical Conductivity of $LiNbO_3$. SOLID STATE COMMUNICATIONS, v. 6, no. 2, Feb. 1968. p. 77-79.

BERNAL, E. et al. Low Frequency Electro-Optic and Dielectric Constants of Lithium Niobate. PHYS. LETTERS, v. 21, no. 3, May 1966. p. 259-260.

BJORKHOLM, J.E. Relative Signs of the Optical Nonlinear Coefficients $d_{31}$ and $d_{22}$ in $LiNbO_3$. APPLIED PHYS. LETTERS, v. 13, no. 1, July 1968. p. 36-37.

BJORKHOLM, J.E. Relative Measurements of the Optical Nonlinearities of KDP, ADP, $LiNbO_3$ and alpha-$HIO_3$. IEEE J. QUANTUM ELECTRONICS, v. QE-4, Nov. 1968. p. 970-972. Also Errata v. QE-5, May 1969. p. 260.

BOSOMWORTH, D.R. The Far Infrared Optical Properties of $LiNbO_3$. APPLIED PHYS. LETTERS, v. 9, no. 9, Nov. 1966. p. 330-331.

BOYD, G.D. et al. $LiNbO_3$: An Efficient Phase Matchable Nonlinear Optical Material. APPLIED PHYS. LETTERS, v. 5, no. 11, Dec. 1964. p. 234-236.

BOYD, G.D. et al. Refractive Index as a Function of Temperature in $LiNbO_3$. J. OF APPLIED PHYS., v. 38, no. 3, Mar. 1967. p. 1941-1943.

BYER, R.L. et al. Growth of High-Quality $LiNbO_3$ Crystals from the Congruent Melt. J. OF APPLIED PHYSICS, v. 41, no. 6, May 1970. p. 2320-2325.

CAMLIBEL, I. Spontaneous Polarization Measurements in Several Ferroelectric Oxides Using a Pulsed-Field Method. J. OF APPLIED PHYSICS, v. 40, no. 4, Mar. 1969. p. 1690-1693.

CARRUTHERS, J.R. et al. Nonstoichiometry and Crystal Growth of Lithium Niobate. J. OF APPLIED PHYSICS, v. 42, no. 5, Apr. 1971. p. 1846-1851.

CHEN, F.S. Optically Induced Change of Refractive Indices in $LiNbO_3$ and $LiTaO_3$. J. OF APPLIED PHYSICS, v. 40, no. 8, July 1969. p. 3389-3396.

CHKALOVA, V.V. et al. Piezoelectric, Elastic and Dielectric Properties of Single Crystal Lithium Niobates and Tantalates. (In Russ.) AKAD. NAUK. IZV. NEORGAN. MAT., v. 3, no. 9, 1967. p. 1715-1716.

DIDOMENICO, M. and S.H. WEMPLE. Oxygen-Octahedra Ferroelectrics. I. Theory of Electro-Optical and Nonlinear Optical Effects. J. OF APPLIED PHYSICS, v. 40, no. 2, Feb. 1969. p. 720-734.

DIXON, R.W. Photoelastic Properties of Selected Materials and Their Relevance for Applications to Acoustic Light Modulators and Scanners. J. OF APPLIED PHYSICS, v. 38, no. 13, Dec. 1967. p. 5149-5153.

DIXON, R.W. and M.G. COHEN. A New Technique for Measuring Magnitudes of Photoelastic Tensors and its Applications to Lithium Niobate. APPLIED PHYS. LETTERS, v. 8, no. 8, Apr. 1966. p. 205-207.

FAY, H. et al. Dependence of Second-Harmonic Phase-Matching Temperature in $LiNbO_3$ Crystals on Melt Composition. APPLIED PHYS. LETTERS, v. 12, no. 3, Feb. 1968. p. 89-92.

GUSEVA, L.M. et al. Investigation of Some of the Optical Characteristics of Ferroelectric Lithium Niobate. ACAD. OF SCI., USSR, BULL., PHYS. SER., v. 31, no. 7, July 1967. p. 1181-1183.

HULME, K.F. et al. Optimum Longitudinal Electrooptic Effect in Oblique-Cut Lithium Niobate Plates. ELECTRONIC LETTERS, v. 5, no. 8, Apr. 1969. p. 171-172.

HULME, K.F. et al. The Signs of the Electro-Optic Coefficients for Lithium Niobate. J. PHYS. C., Ser. 2, v. 2, 1969. p. 855-857.

IRISOVA, N.A. and G.V. KOZLOV. Birefringence of Certain Crystals in the Millimeter Wavelength Range. SOVIET PHYSICS-CRYSTALL., v. 15, no. 5, Mar. 1971. p. 941-942.

ISMAILZADE, I.G. An X-Ray Diffraction Study of Phase Transitions in Lithium Niobate. SOVIET PHYSICS-CRYSTALL., v. 10, no. 3, Nov. 1965. p. 235-237.

ISMAILZADE, I.G. et al. X-Ray Study of Lithium Niobate at High Temperatures. SOVIET PHYSICS-CRYSTALL., v. 13, no. 1, July 1968. p. 25-28.

IWASAKI, H. et al. Dispersion of the Refractive Indices of $LiNbO_3$ Crystals Between 20° and 900°C. JAPAN J. OF APPLIED PHYS., v. 6, no. 9, Sept. 1967. p. 1101-1104.

IWASAKI, H. et al. Temperature and Optical Frequency Dependence of the D.C. Electro-Optic Constants $r_{22}^{T}$ of $LiNbO_3$. JAPAN. J. OF APPLIED PHYS., v. 6, no. 12, Dec. 1967. p. 1419-1422.

KAMINOW, I.P. and W.D. JOHNSTON. Quantitative Determination of Sources of the Electrooptic Effect in $LiNbO_3$ and $LiTaO_3$. PHYS. REV., v. 160, no. 3, Aug. 1967. p. 519-522.

KIM, Y.S. and R.T. SMITH. Thermal Expansion for Lithium Tantalate and Lithium Niobate Single Crystals. J. OF APPLIED PHYS., v. 40, no. 11, Oct. 1969. p. 4637-4641.

KLUDZIN, V.V. Photoelastic Constants of $LiNbO_3$ Crystals. SOVIET PHYSICS, SOLID STATE, v. 13, no. 2, Aug. 1971. p. 540-541.

LAUDISE, R.A. The Search for Nonlinear Optical Materials for Laser Communications. BELL LABS. RECORD, v. 46, no. 1, Jan. 1968. p. 3-7.

LENZO, P.V. et al. Electro-Optic Coefficients in Single-Domain Ferroelectric Lithium Niobate. OPTICAL SOC. OF AMERICA, J., v. 56, no. 5, May 1966. p. 633-635.

LEVINSTEIN, H.J. et al. Reduction of the Susceptibility of Optically Induced Index Inhomogeneities in $LiTaO_3$ and $LiNbO_3$. J. OF APPLIED PHYS., v. 38, no. 8, July 1967. p. 3101-3102.

MALONEY, W.T. et al. Measurement of "Shear" Photoelastic Constants in Lithium Niobate. IEEE PROC., v. 57, no. 7, July 1969. p. 1332-1333.

MATTHIAS, B.T. and J.P. REMEIKA. Ferroelectricity in the Ilmenite Structure. PHYS. REV., v. 76, no. 12, Dec. 1949. p. 1886-1887.

MIDWINTER, J.E. Assessment of Lithium Metaniobate for Linear Optics. APPLIED PHYS. LETTERS, v. 11, no. 4, Aug. 1967. p. 128-130.

MIDWINTER, J.E. Lithium Niobate: Effects of Composition on the Refractive Indices and Optical Second-Harmonic Generation. J. OF APPLIED PHYS., v. 39, no. 7, June 1968. p. 3033-3038.

MILLER, R.C. and A. SAVAGE. Temperature Dependence of the Optical Properties of Ferroelectric $LiNbO_3$ and $LiTaO_3$. APPLIED PHYS. LETTERS, v. 9, no. 4, Aug. 1966. p. 169-171.

MILLER, R.C. et al. Dependence of Second Harmonic Generation Coefficients of $LiNbO_3$ on Melt Composition. J. OF APPLIED PHYS., v. 42, no. 11, Oct. 1971. p.4145-4147.

NASSAU, K. Lithium Niobate - A new Type of Ferroelectric: Growth, Structure and Properties. In Ferroelectricity, Proceedings of the Symposium on Ferroelectricity. General Motors Res. Lab., Warren, Mich., 1966. Ed. WELLER, E.F. Elsevier Pub. Co., N.Y., 1967. p. 259-268.

NASSAU, K. and H.J. LEVINSTEIN. Ferroelectric Behavior of Lithium Niobate. APPLIED PHYS. LETTERS, v. 7, no. 3, Aug. 1965. p. 69-70.

NASSAU, K. et al. Ferroelectric Lithium Niobate. Pt. 1: Growth, Domain Structure, Dislocations and Etching. J. OF PHYS. AND CHEM. OF SOLIDS, v. 27, no. 6/7, June-July 1966. p. 983-988.

NASSAU, K. et al. Ferroelectric Lithium Niobate. Pt. 2: Preparation of Single Domain Crystals. J. OF PHYS. AND CHEM. OF SOLIDS, v. 27, no. 6/7, June-July 1966. p. 989-996.

NIIZEKI, N. et al. Growth Ridges, Etched Hillocks and Crystal Structure of Lithium Niobate. JAPAN. J. OF APPLIED PHYS., v. 6, no. 3, Mar. 1967. p. 318-327.

OHMACHI, Y. et al. Dielectric Properties of $LiNbO_3$ Single Crystals up to 9 GHz. JAPAN. J. OF APPLIED PHYS., v. 6, no. 12, Dec. 1967. p. 1467-1468.

PARFITT, H.T. and D.S. ROBERTSON. Domain Structures in Lithium Niobate Crystals. BRITISH J. OF APPLIED PHYS., v. 18, no. 12, Dec. 1967. p. 1709-1713.

PETERSON, G.E. et al. Electro-Optic Properties of $LiNbO_3$. APPLIED PHYS. LETTERS, v. 5, no. 3, Aug. 1964. p. 62-64.

REINTJES, J. and M.B. SCHULZ. Photoelastic Constants of Selected Ultrasonic Delay-Line Crystals. J. OF APPLIED PHYS., v. 39, no. 11, Oct. 1968. p. 5254-5258.

REISMAN, A. and F. HOLTZBERG. Heterogeneous Equilibria in the Systems $Li_2O$-$Ag_2O$-$Nb_2O_5$ and Oxide-Models. AMERICAN CHEM. SOC., J., v. 80, no. 24, Dec. 1958. p. 6503-6507.

ROITBERG, M.B. et al. Characteristic Features of the Pyroelectric Effect and Electrical Conductivity in Single Crystals of $LiNbO_3$ in the Range 20-250°C. SOVIET PHYSICS-CRYSTALL., v. 14, no. 5, Mar. 1970. p. 814-815.

SHAPIRO, Z.I. et al. Investigation of the $LiTaO_3$ - $LiNbO_3$ System. ACAD OF SCI., USSR, BULL., PHYS. SER., v. 29, no. 6, June 1965. p. 1047-1050.

SMAKULA, P.H. and P.C. CLASPY. The Electro-Optic Effect in $LiNbO_3$ and KTN. AIME METALL. SOC., TRANS., v. 239, no. 3, Mar. 1967. p. 421-424.

SMITH, R.T. and F.S. WELSH. Temperature Dependence of the Elastic, Piezoelectric and Dielectric Constants of Lithium Tantalate and Lithium Niobate. J. OF APPLIED PHYS., v. 42, no. 6, May 1971. p. 2219-2230.

SMOLENSKII, G.A. et al. The Curie Temperature of $LiNbO_3$. PHYS. STATUS SOLIDI, v. 13, no. 2, 1966. p. 309-314.

SPENCER, E.G. et al. Dielectric Materials for Electrooptic, Elastooptic and Ultrasonic Device Applications. IEEE, PROC., v. 55, no. 12, Dec. 1967. p. 2074-2078.

TURNER, E.H. High-Frequency Electro-Optic Coefficients of Lithium Niobate. APPLIED PHYS. LETTERS, v. 8, no. 11, June 1966. p. 303-304.

TURNER, E.H. et al. Dependence of Linear Electro-Optic Effect and Dielectric Constant on Melt Composition in Lithium Niobate. J. OF APPLIED PHYS., v. 41, no. 13, Dec. 1970. p. 5278-5281.

VASILEVSKAYA, A.S. et al. Some Optical Properties of Lithium Niobate Single Crystals. ACAD. OF SCI., USSR, BULL., PHYS. SER., v. 31, no. 7, July 1967. p. 1178-1180.

VEDAM, K. and T.A. DAVIS. Piezo- and Thermo-Optic Behavior of $LiNbO_3$. APPLIED PHYS. LETTERS, v. 12, no. 4, Feb. 1968. p. 138-140.

VINOGRADOV, E.A. et al. Electro-Optic Effect in $LiNbO_3$ in the Millimeter Range. SOVIET PHYS. SOLID STATE, v. 12, no. 3, Sept. 1970. p. 605-607.

WARNER, A.W. et al. Determination of Elastic and Piezoelectric Constants for Crystals in Class (3M). ACOUSTICAL SOC. OF AMERICA, J., v. 42, no. 6, Dec. 1967. p. 1223-1231.

WARNER, J. et al. The Temperature Dependence of Optical Birefringence in Lithium Niobate. PHYS. LETTERS, v. 20, no. 2, Feb. 1966. p. 163-164.

WEMPLE, S.H. and M. DIDOMENICO. Oxygen-Octahedra Ferroelectrics. II. Electro-Optical and Nonlinear-Optical Device Applications. J. OF APPLIED PHYS., v. 40, no. 2, Feb.1969. p. 735-752.

WEMPLE, S.H. et al. Relationship Between Linear and Quadratic Electro-Optic Coefficients in $LiNbO_3$, $LiTaO_3$ and Other Oxygen-Octahedra Ferroelectrics Based on Direct Measurement of Spontaneous Polarization. APPLIED PHYS. LETTERS, v. 12, no. 6, Mar. 1968. p. 209-211.

YAMADA, T. et al. Piezoelectric and Elastic Properties of Lithium Niobate Single Crystals. JAPAN. J. OF APPLIED PHYS., v. 6, no. 2, Feb. 1967. p. 151-155.

ZHDANOVA, V.V. et al. Thermal Properties of Lithium Niobate Crystals. SOVIET PHYS. SOLID STATE, v. 10, no. 6, Dec. 1968. p. 1360-1362.

ZOOK, J.D. et al. Temperature Dependence and Model of the Electro-Optic Effect in $LiNbO_3$. APPLIED PHYS. LETTERS, v. 11, no. 8, Sept. 1967. p. 159-161.

## LITHIUM TANTALATE

### Introduction

Lithium tantalate is a new single crystal material with interesting electrooptic, piezoelectric and ferroelectric properties. Although it has not been as thoroughly investigated as other electrooptic materials, it has important applications in transverse light modulators and other devices.

### Chemical and Physical Properties

| | | |
|---|---|---|
| Chemical Formula | $LiTaO_3$ | |
| Molecular Weight | 235.885 | |
| Density | 7.3 g/cm$^3$<br>7.454 g/cm$^3$ | Ballman<br>Smith |
| Solubility | Insoluble in water and dilute acids | |

### Crystallography

Lithium tantalate was first grown using the flux technique by Matthias and Remeika; a number of investigators have grown $LiTaO_3$ by the Czochralski method. [Ballman and co-workers, Shapiro and co-workers, Ballman et al.] at Bell Telephone Laboratories, have shown that the optical and ferroelectric properties of $LiTaO_3$ are dependent upon melt stoichiometry during the crystal growing process. Lithium tantalate is a ferroelectric crystal with a Curie temperature of about 660°C. It crystallizes into the non-centrosymmetric space group R3c. Above the Curie temperature, there is no apparent change in crystal structure; the crystal remains in the hexagonal (rhombohedral) phase [Shapiro et al., Levinstein et al.].

| | | |
|---|---|---|
| Space Group | | |
| Below $T_c$ | Rhombohedral | R3c or $C_{3v}^6$ |
| Above $T_c$ | Rhombohedral | $R\bar{3}c$ or $D_{3v}^6$ |
| Point Group | | |
| Below $T_c$ | Rhombohedral | 3m or $C_{3v}$ |

TABLE 1. LATTICE CONSTANTS OF $LiTaO_3$ AT ROOM TEMPERATURE

| Hexagonal | | Rhombohedral | | Reference |
|---|---|---|---|---|
| $a_H$(Å) | $c_H$(Å) | $a_{Rh}$(Å) | $\alpha_{Rh}$ | |
| 5.153 | 13.775 | 5.470 | 56° 12' | Shapiro et al. |
| 5.143 | 13.756 | | | Levinstein et al. |
| 5.15428 | 13.78351 | 5.4740 | 56° 10.5' | Abrahams and Bernstein |

The lattice constants of $LiTaO_3$ are dependent on melt stoichiometry, as shown in Figure 1. Abrahams and Bernstein quote the following values of lattice constants corresponding to differing values of the Ta/Li ratio in the melt:

| $a_H$(Å) | $c_H$(Å) | Ta/Li | $T_c$(°C |
|---|---|---|---|
| 5.15359 ± 0.00001 | 13.78070 ± 0.00001 | 0.95 | 625 |
| 5.15428 ± 0.00001 | 13.78351 ± 0.00002 | 1.000 | |

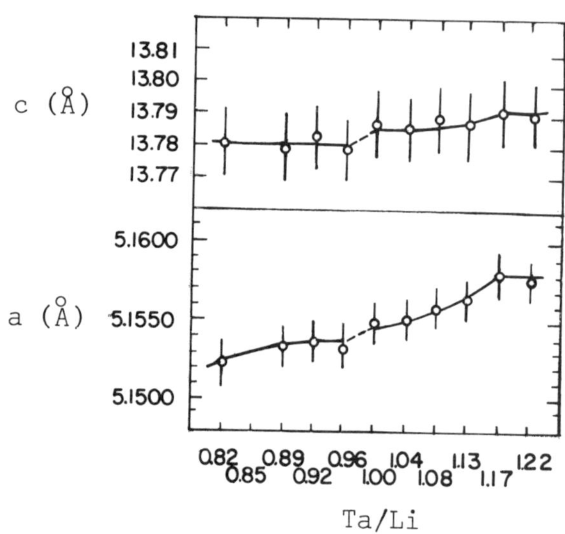


Fig. 1. Lattice constants of lithium tantalate ceramics with various concentration ratio Ta/Li. A plot of the Curie temperature versus the Ta/Li ratio is shown in Figure 11 [Yamada et al.]

The lattice constants of $LiTaO_3$ have been measured to 700°C by Shapiro et al. as shown in Figure 2. Levinstein et al. have reported somewhat different results; they indicated that $a_H$ expands linearly with temperature while $c_H$ remains practically unchanged with temperature to approximately 400°C and then decreases with increasing temperature going through a minimum at 660°C, corresponding to the Curie temperature.

Kim and Smith and Iwasaki et al. have observed a maximum in $c_H$ at 250°C and 450°C, respectively (cf Figure 12).

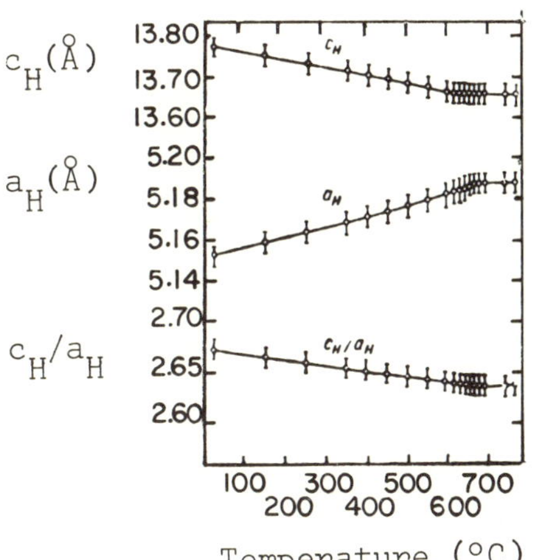


Fig. 2. Lattice constants of $LiTaO_3$ as a function of temperature [Shapiro et al.]



## Optical Properties - Refractive Index

TABLE 2. REFRACTIVE INDEX OF $LiTaO_3$ AT ROOM TEMPERATURE [Bond]

| λ (microns) | $n_o$ | $n_e$ | λ (microns) | $n_o$ | $n_e$ |
|---|---|---|---|---|---|
| 0.45 | 2.2420 | 2.2468 | 2.00 | 2.1066 | 2.1115 |
| 0.50 | 2.2160 | 2.2205 | 2.20 | 2.1009 | 2.1053 |
| 0.60 | 2.1834 | 2.1878 | 2.40 | 2.0951 | 2.0993 |
| 0.70 | 2.1652 | 2.1696 | 2.60 | 2.0891 | 2.0936 |
| 0.80 | 2.1538 | 2.1578 | 2.80 | 2.0825 | 2.0871 |
| 0.90 | 2.1454 | 2.1493 | 3.00 | 2.0755 | 2.0299 |
| 1.00 | 2.1391 | 2.1432 | 3.20 | 2.0680 | 2.0727 |
| 1.20 | 2.1305 | 2.1341 | 3.40 | 2.0601 | 2.0649 |
| 1.40 | 2.1236 | 2.1273 | 3.60 | 2.0513 | 2.0561 |
| 1.60 | 2.1174 | 2.1213 | 3.80 | 2.0424 | 2.0473 |
| 1.80 | 2.1120 | 2.1170 | 4.00 | 2.0335 | 2.0377 |

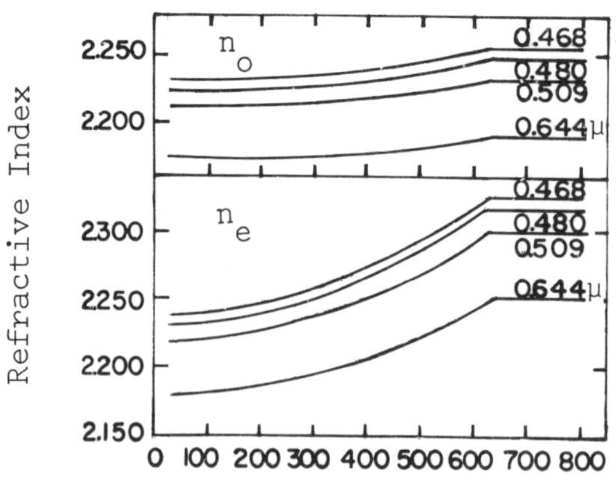


Fig. 3. Refractive index of $LiTaO_3$ as a function of temperature and at four wavelengths [Iwasaki et al.]

## Optical Properties - Birefringence

From a graphical interpolation of the data in Table 2 [Lenzo et al.], lithium tantalate is seen to be optically uniaxial positive with $n_o = 2.175$ and $n_e = 2.180$ at 6328 Å, yielding a birefringence of $n_e - n_o = + 0.005$. The major change in the birefringence with temperature is due to $n_e$, with

$$d(n_e - n_o)/dT = 4.7 \times 10^{-5}/°C$$

at 0.63 micron and 40°C [Denton et al.]. A temperature coefficient less than the above and a zero birefringence can be obtained in $LiTaO_3$ by cooling the sample below room temperature, as suggested by Miller and Savage. These authors also quote a birefringence value of + 0.06 at $T_c$. The birefringence is strongly dependent upon melt stoichiometry, as shown in Figure 4.

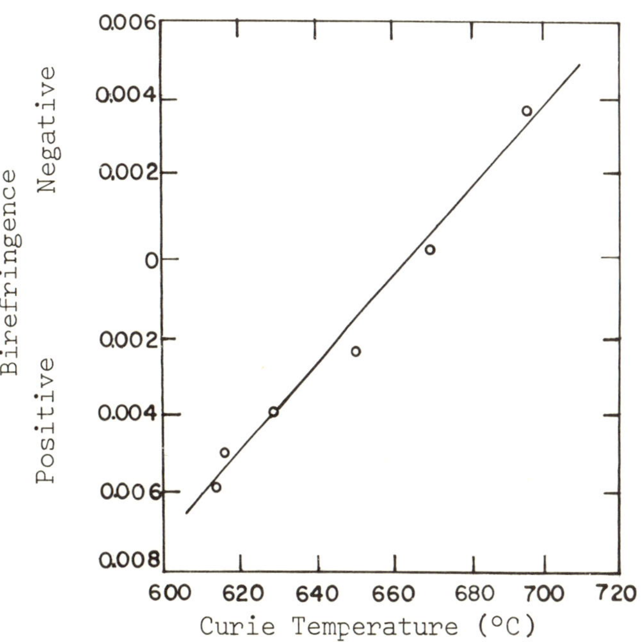


Fig. 4. Birefringence as a function of Curie temperature for single crystal $LiTaO_3$. A plot of the Curie temperature versus melt stoichiometry is shown in Figure 10. [Ballman et al.]



## Optical Properties - Transmission

The optical transmission of $LiTaO_3$ is similar to that of $LiNbO_3$, transmitting from 0.3 micron to 6 microns. However, as pointed out by Loiacono , strong absorption peaks beyond 2 microns limit its usefulness for optical purposes.

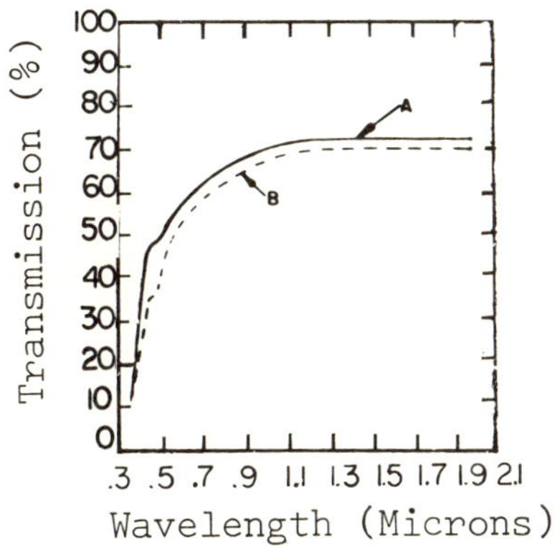


A 10 mm C axis annealed

B 10 mm C axis poled, 125 V/cm.

Fig. 5. Transmission of $LiTaO_3$ as a function of wavelength uncorrected for reflection losses. [Loiacono]

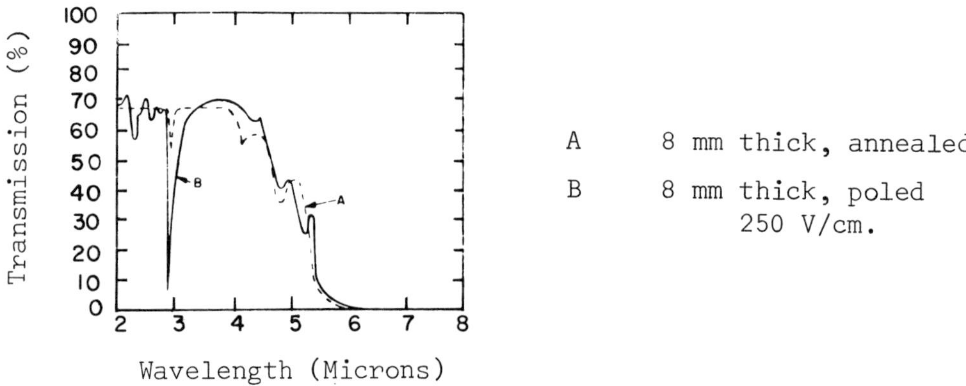


Fig. 6. Transmission of $LiTaO_3$ as a function of wavelength uncorrected for reflection losses. [Loiacono ]

## Optical Properties - Nonlinear Optical Behavior

The following optical nonlinear coefficients, relative to $d_{36}^{2\omega}$ for potassium dihydrogen phosphate (KDP), have been reported Miller and Savage for $LiTaO_3$ using a neodymium laser:

$$d_{22}^{2\omega} = 4.3 \pm 0.5$$

$$d_{31}^{2\omega} = d_{15}^{2\omega} = 2.6 \pm 0.5$$

$$d_{33}^{2\omega} = 40 \pm 5.0$$

Ashkin et al. have observed the presence of an optically induced inhomogeneity in the refractive index of lithium tantalate, caused by power densities as small as $1W/cm^2$ and highly detrimental to the optics of nonlinear devices.

Levinstein et al. have shown that $LiTaO_3$ can be made resistant to such laser-induced inhomogeneities at power levels as high as 500 $W/cm^2$. This is accomplished by annealing $LiTaO_3$ in an electric field of 250 V/cm at a temperature of 700°C for 1/2 hour and then cooling the crystal to room temperature with the field on at a rate of 100°C/hour. Chen has observed local index changes in poled single crystal $LiTaO_3$; $n_e$ was observed to decrease as much as $10^{-3}$ with a focused Ar laser of 20 mW intensity, while the change in $n_o$ was much smaller.

## Electrooptic Properties

A preliminary examination of multidomain $LiTaO_3$ indicated the absence of electrooptic effects , [Peterson et al.]. However, a significant linear, dc and rf electrooptic effect in single-domain crystals grown by the Czochralski technique has been observed and measured by Lenzo et al.

Kaminow and Johnston have derived a simple relationship between an electrooptic coefficient measured at radio frequencies and the corresponding Raman-scattering efficiencies in $LiTaO_3$. They found that the electrooptic effect is due primarily to the lowest-frequency $A_1$-type optic mode and the next lowest E mode, with only a small pure electronic contribution.

Wemple and co-workers have discussed the relationship between the linear and quadratic electrooptic effects in lithium tantalate. Using a directly measured value of the spontaneous polarization ($P_s$ = 0.50 $C/m^2$), these authors show that the linear effect is related fundamentally to a biased quadratic effect associated with each $BO_6$ octahedron in $LiTaO_3$.

TABLE 3. ELECTROOPTIC COEFFICIENTS OF $LiTaO_3$ AT 6328 Å (in $10^{-12}$m/V) [Lenzo et al.]

| | $r_{33}$ | $r_{13}$ | $n_e^3 r_{33} - n_o^3 r_{13}$ | Frequency |
|---|---|---|---|---|
| Clamped | 30.3 | 7.0 | 243 | 50-86 MHz |
| Unclamped | | | 223 | 1000 Hz |

Lenzo et al. have observed two or three sharp acoustic resonances in $LiTaO_3$ between 0.59 MHz and 1.2 MHz; however, they point out that the relatively small difference in the clamped and unclamped electrooptic coefficients should be similar in magnitude as is in fact found to be the case.

The low-frequency (unclamped) electrooptic coefficient given in Table 3 corresponds to a half-wave voltage of 2840 V. The half-wave voltage has been found to be as low as 2700 V [Denton et al.], giving $LiTaO_3$ a slight advantage over $LiNbO_3$ for electrooptic device applications.

## Photoelastic Properties - Elastooptic Coefficients

TABLE 4. ELASTOOPTIC COEFFICIENTS OF LITHIUM TANTALATE AT 6328 Å [Dixon]

| | |
|---|---|
| $p_{11}$ = 0.0804 | $p_{13}$ = 0.094 |
| $p_{12}$ = 0.0804 | $p_{33}$ = 0.150 |
| $p_{44}$ = 0.022 | $p_{41}$ = 0.024 |
| $p_{31}$ = 0.086 | $p_{14}$ = 0.031 |

## Photoelastic Properties - Elastic Constants

TABLE 5. ELASTIC STIFFNESS CONSTANTS OF LITHIUM TANTALATE AT ROOM TEMPERATURE

| | $10^{10}$ $N/m^2$ | | | | | | | Reference |
|---|---|---|---|---|---|---|---|---|
| | $c_{11}$ | $c_{12}$ | $c_{13}$ | $c_{14}$ | $c_{33}$ | $c_{44}$ | $c_{66}$ | |
| Constant Field (E) | 22.98 | 4.41 | 8.11 | -1.04 | 27.81 | 9.68 | 9.29 | Smith |
| | 23.3 | 4.7 | 8.0 | -1.1 | 27.5 | 9.4 | 9.3 | Warner et al. |
| | 22.8 | 3.1 | 7.4 | -1.2 | 27.1 | 9.6 | 9.8 | Yamada et al. |
| Constant Displacement (D) | 23.8 | 2.1 | 7.3 | -2.7 | 28.2 | 11.7 | 10.9 | |
| | 23.9 | 4.1 | 8.0 | -2.2 | 28.4 | 11.3 | 9.9 | Warner et al. |

TABLE 6. ELASTIC COMPLIANCE CONSTANTS OF LITHIUM TANTALATE AT ROOM TEMPERATURE

| | $10^{-12}$ $m^2/N$ | | | | | | | Reference |
|---|---|---|---|---|---|---|---|---|
| | $s_{11}$ | $s_{12}$ | $s_{13}$ | $s_{14}$ | $s_{33}$ | $s_{44}$ | $s_{66}$ | |
| Constant Field (E) | 4.87 | -0.58 | -1.25 | 0.64 | 4.36 | 10.8 | 10.9 | Warner et al. |
| | 4.86 | -0.29 | -1.24 | 0.63 | 4.36 | 10.5 | 10.3 | Yamada et al. |
| Constant Displacement (D) | 4.68 | -0.16 | -1.17 | 1.10 | 4.14 | 9.0 | 9.7 | |
| | 4.76 | -0.50 | -1.20 | 1.02 | 4.19 | 9.3 | 10.5 | Warner et al. |

Smith and Welsh report on the temperature dependence of the elastic and piezoelectric constants from 0°C to 110°C; values for the first are -0.4 to $-6.7\times10^{-4}$/°C for the second, the range is -1.3 to $+1.5\times10^{-4}$/°C

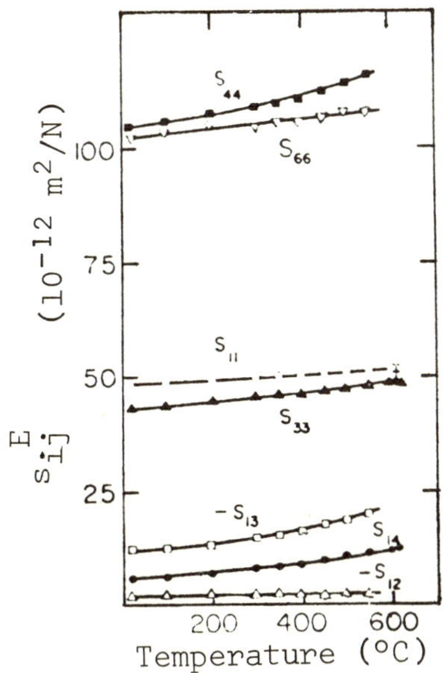


Fig. 6a. Elastic compliance constants of lithium tantalate as a function of temperature [Yamada et al.]

## Piezoelectric Properties

TABLE 7. PIEZOELECTRIC CONSTANTS OF LITHIUM TANTALATE AT ROOM TEMPERATURE

| $d_{ij}$ | $10^{-11}$ C/N | | $e_{ij}$ | $C/m^2$ | | |
|---|---|---|---|---|---|---|
| | a | b | | a | b | c |
| $d_{15}$ | 2.6 | 2.6 | $e_{15}$ | 2.6 | 2.7 | 2.58 |
| $d_{22}$ | 0.7 | 0.85 | $e_{22}$ | 1.6 | 2.0 | 1.59 |
| $d_{31}$ | -0.2 | -0.30 | $e_{31}$ | 0.0 | -0.1 | -0.24 |
| $d_{33}$ | 0.8 | 0.92 | $e_{33}$ | 1.9 | 2.0 | 1.40 |

| $h_{ij}$ | $10^9$ N/C | | $g_{ij}$ | $10^{-2}\ m^2/C$ | |
|---|---|---|---|---|---|
| | a | b | | a | b |
| $h_{15}$ | 7.2 | 7.5 | $g_{15}$ | 5.8 | 5.6 |
| $h_{22}$ | 4.3 | 5.5 | $g_{22}$ | 1.5 | 1.8 |
| $h_{31}$ | 0.0 | -0.3 | $g_{31}$ | -0.6 | -0.77 |
| $h_{33}$ | 5.0 | 5.6 | $g_{33}$ | 2.1 | 2.4 |

a Warner et al.
b Yamada et al.
c Smith

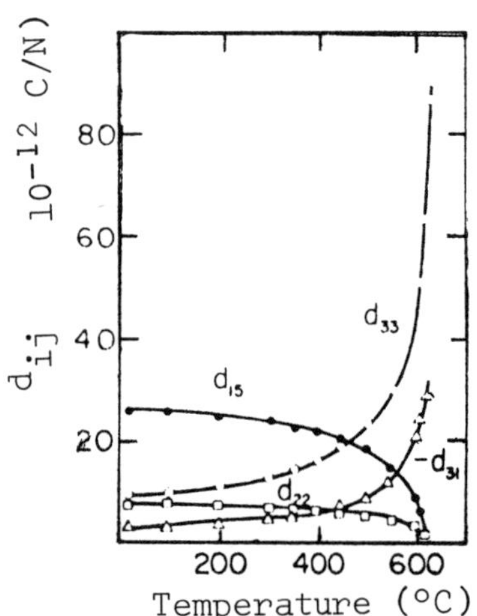


Fig. 6b. Piezoelectric constants of lithium tantalate as a function of temperature [Yamada et al.]

Electromechanical Coupling Coefficients:

| | |
|---|---|
| $k_{33} = 0.30$ | Ballman |
| $k_{33} = 0.20$ | Chkalova et al. |
| $k_{15} = 0.311$ | |

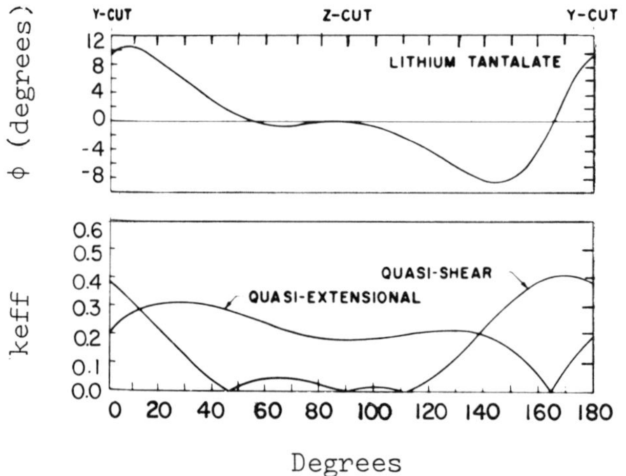




Fig. 7. Effective coupling factors and angle φ between quasi-extensional wave displacement and plate normal for rotated Y-cuts of $LiTaO_3$ [Warner et al.]

Dielectric Properties

Barker et al. have made polarized infrared reflectivity measurements on single crystals at 1 to 500μ both normal and parallel to the c-axis, at 15, 80 and 300°K:

| | 15°K | | 80°K | | 300°K | |
|---|---|---|---|---|---|---|
| | ⊥-c | ‖-c | ⊥-c | ‖-c | ⊥-c | ‖-c |
| $\varepsilon_o$ | 39.2 | 32.5 | 39.4 | 34.1 | 41.4 | 39.8 |
| $\varepsilon_\infty$ | | | | | 4.497 | 4.527 |

TABLE 8. RELATIVE DIELECTRIC CONSTANTS OF LITHIUM TANTALATE AT ROOM TEMPERATURE.

| $\varepsilon_a^S$ | $\varepsilon_c^S$ | $\varepsilon_a^T$ | $\varepsilon_c^T$ | Reference |
|---|---|---|---|---|
| 41 | 43 | 51 | 45 | Warner et al. |
| | 43 | | 47 | Lenzo et al. |
| 41 | 42 | 53 | 44 | Yamada et al. |
| 38.3 | 46.2 | | | Smith |

The temperature dependence of the permittivity has been reported by several investigators, the most recent work is by Smith and Welsh. A typical curve is shown in Fig. 8. Above the Curie point, the variation with temperature is described by a Curie Weiss Law; [Levinstein et al., Razbirin]. The Curie Constant in the paraelectric phase is $1.5x10^5$°K [Yamada et al.].

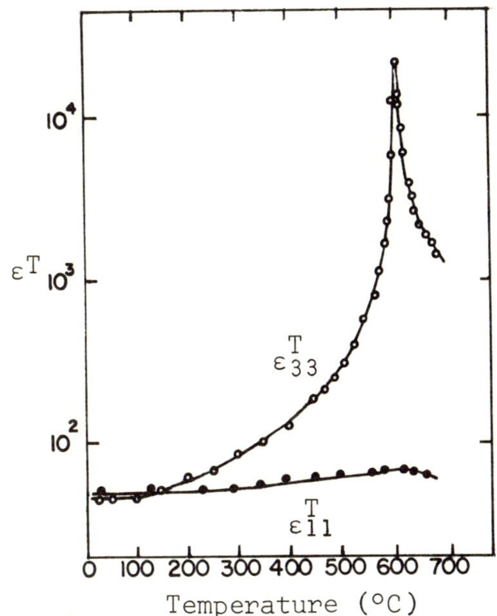


Fig. 8. Dielectric constants of lithium tantalate as a function of temperature measured at 10 kHz [Yamada et al.].

Lithium tantalate has been reported to have very low conductivity and dielectric losses. At room temperature:

| | |
|---|---|
| $\sigma < 10^{-13}$ (ohm-cm)$^{-1}$ | Razbirin, Shapiro et al. |
| $\tan\delta < 0.05$ at 1 kHz | |
| $\tan\delta = 2 \times 10^{-3}$ at 100 MHz | Denton et al. |

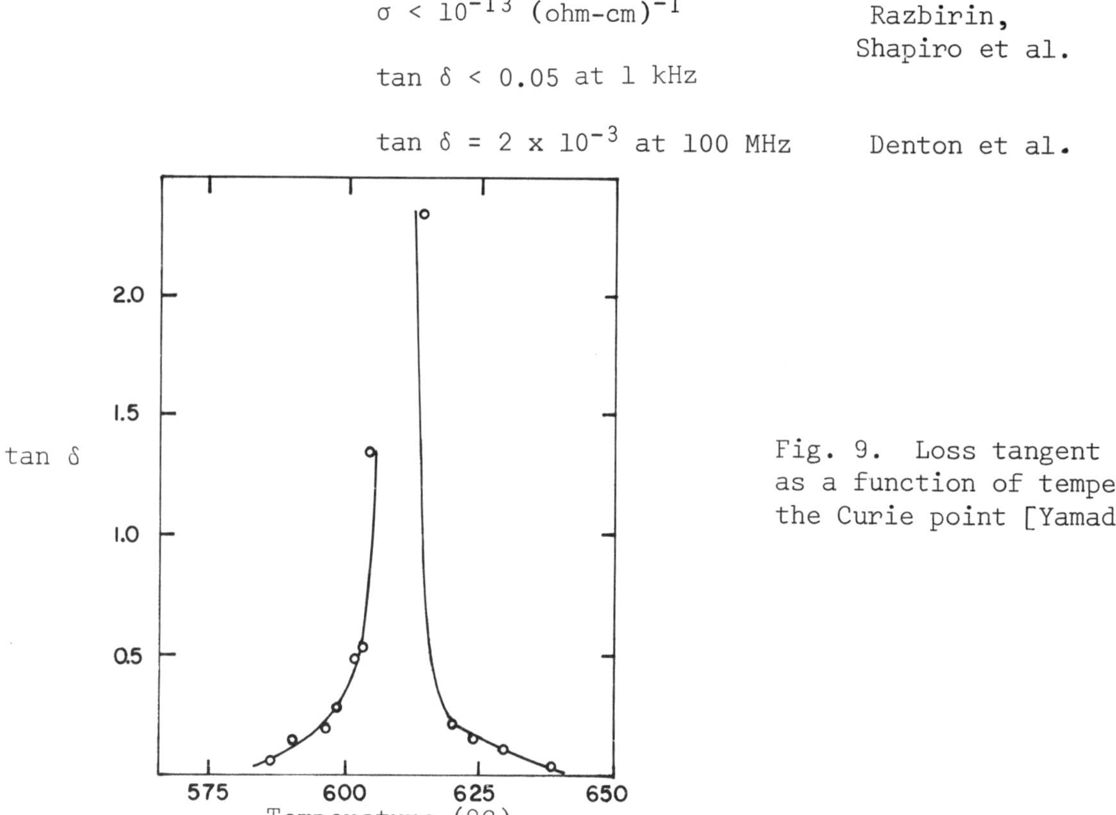


Fig. 9. Loss tangent of $LiTaO_3$ as a function of temperature near the Curie point [Yamada et al.]

## Ferroelectric Properties

Matthias and Remeika in 1949 were the first to report on the ferroelectric behavior of lithium tantalate. Etching experiments by Levinstein et al. in pulled single crystals revealed antiparallel polar domains exist in this material. The maximum domain diameter in the as-grown material is of the order of 5 microns compared to several hundred microns in lithium niobate. Their poling experiments consisted of heating the sample to roughly 700°C and applying a field of 250 V/cm. The sample is then cooled below the Curie temperature with the field applied. These authors also observed that multidomain lithium tantalate has no detectable pyroelectric effect while single domain material has a pyroelectric effect comparable to lithium niobate.

Based on the inability to observe polarization reversal using conventional hysteresis-loop techniques, it has been suggested that $LiTaO_3$ is a "frozen ferroelectric". Camlibel and co-workers , however, have developed a pulsed field method to measure directly the room temperature spontaneous polarization of ferroelectric oxides. The ability to reverse the polarization was found to be dependent on the electrode material used; a spontaneous polarization of $P_s = 0.05 \pm 0.01\ C/m^2$ was reported for $LiTaO_3$.

Differing values reported for the Curie temperature of lithium tantalate, as shown in Table 9, are due to variations in melt stoichiometry during the crystal growth, as indicated in Figures 10 and 11.

Apparently the measurement of Curie temperature becomes a fairly precise determination of relative stoichiometry once a plot such as that shown in Figure 10 is obtained.

The dark circles in Figure 10 represent a series of large crystals (about 50 g) grown to demonstrate that a crystal with a predictable Curie temperature could be grown once a plot of Curie temperature versus composition had been established. The points fall on the plot exceptionally well considering the many variables present in the preparation and growth of this high-temperature material. The single point at the stoichiometric composition but with the Curie temperature well above the curve represents a large crystal grown with 0.05 wt% $MgTiO_3$ added to the melt. The large change in Curie temperature caused by the impurity addition is readily apparent.

TABLE 9. CURIE TEMPERATURE OF $LiTaO_3$

| $T_c$(°C) | Comments | Reference |
|---|---|---|
| 610 ± 10 | 2nd order phase transition indicated from nonlinear optical data | Miller and Savage |
| 660 ± 10 | single crystal | Levinstein et al. |
| 665 ± 10 | ceramic $LiTaO_3$ | Razbirin |
| 618 ± 5 | stoichiometric single crystal | Ballman et al. |
| 615 | single crystal; $TaO_5/Li_2O = 1.1$ | Yamada, et al. |
| 620 ± 10 | single crystal; Ta/Li = 1.1 | Yamada et al. |
| 660 ± 10 | ceramic $LiTaO_3$ | |

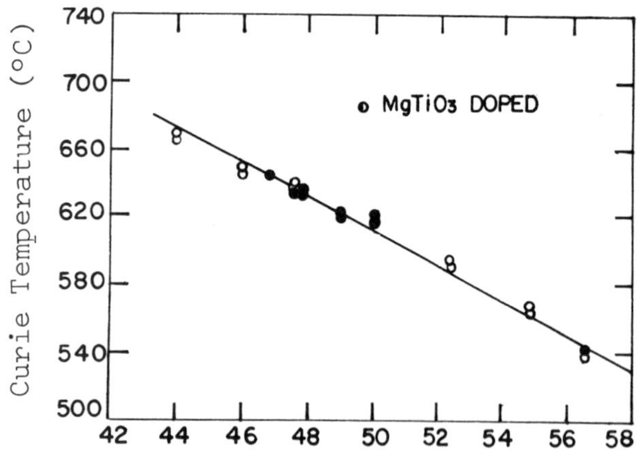


Mole Per Cent Tantalum Oxide in Melt

Fig. 10. Curie temperature of $LiTaO_3$ as a function of melt stoichiometry [Ballman et al.]



Fig. 11. Curie points of lithium tantalate ceramics with various concentration ratio Ta/Li. Open circles indicate the samples sintered at 1400°C, and filled in circles those sintered at 1500°C [Yamada et al.]

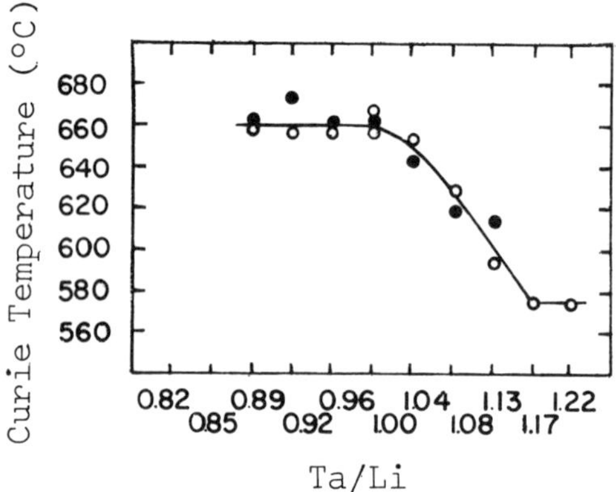


## Thermal Properties

| | | |
|---|---|---|
| Melting Point | 1560°C | Fedulov et al. |
| | 1650°C | Ballman |
| Thermal Expansion ($> T_c$) | $\alpha_c = 5.7 \times 10^{-6}/°C$ | Iwasaki et al. |
| | $\alpha_a = 21 \times 10^{-6}/°C$ | |
| Debye Temperature | 452°K | Abrahams and Bernstein |
| Thermal Conductivity | 0.45 W/cm °K at 20°K (max.)<br>0.05 W/cm °K at 300°K | Oliver and Young |

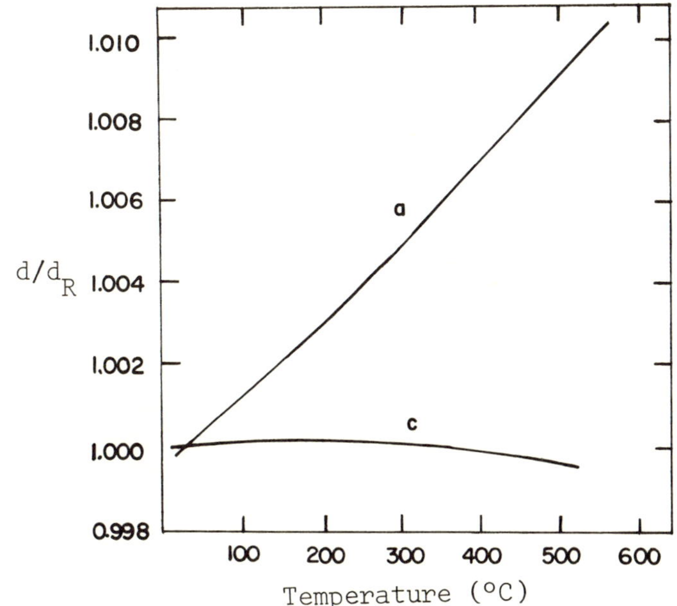


Fig. 12. Relative thermal expansion of lithium tantalate ($T_c$ = 601°C) as a function of temperature.



$$d/d_R = 1 + \alpha\ (T-25) + \beta\ (T-25)^2$$

| | a | c | |
|---|---|---|---|
| $\alpha$ ($10^{-5}$/°C) | 1.62 | 0.22 | |
| $\beta$ ($10^{-9}$/°C$^2$) | 5.9 | -5.9 | Kim and Smith |

## LITHIUM TANTALATE


ABRAHAMS, S.C. and J.L. BERNSTEIN. Ferroelectric Lithium Tantalate. I. Single Crystal X-Ray Diffraction Study at 24°C. J. OF PHYS. AND CHEM. OF SOLIDS, v. 28, no. 9, Sept. 1967. p. 1685-1692.

ASHKIN, A. et al. Optically-Induced Refractive Index Inhomogeneities in $LiNbO_3$ and $LiTaO_3$. APPLIED PHYS. LETTERS, v. 9, no. 1, July 1966. p. 72-74.

BALLMAN, A.A. Growth of Piezoelectric and Ferroelectric Materials by the Czochralski Technique. AMERICAN CERAM. SOC., J., v. 48, no. 2, Feb. 1965. p. 112-113.

BALLMAN, A.A. et al. Curie Temperature and Birefringence Variation in Ferroelectric Lithium Metatantalate as a Function of Melt Stoichiometry. AMERICAN CERAM. SOC., J., v. 50, no. 12, Dec. 1967. p. 657-659.

BARKER, A.S., Jr. et al. Infrared Study of the Lattice Vibrations in $LiTaO_3$. PHYS. REV. B., Ser. 3, v. 2, no. 10, 15 Nov. 1970. p. 4233-4239.

BOND, W.L. Measurement of the Refractive Indices of Several Crystals. J. OF APPLIED PHYS., v. 36, no. 5, May 1965. p. 1674-1677.

CAMLIBEL, I. Spontaneous Polarization Measurements in Several Ferroelectric Oxides Using a Pulsed-Field Method. J. OF APPLIED PHYS., v. 40, no. 4, Mar. 1969. p. 1690-1693.

CHEN, F.S. Optically Induced Change of Refractive Indices in $LiNbO_3$ and $LiTaO_3$. J. OF APPLIED PHYS., v. 40, no. 8, July 1969. p. 3389-3396.

CHKALOVA, V.V. et al. Piezoelectric, Elastic, and Dielectric Properties of Single Crystal Lithium Niobates and Tantalates. AKAD. NAUK SSSR. IZV. NEORGAN. MAT., v. 3, no. 9, 1967. p. 1715-1716.

DENTON, R.T. et al. Lithium Tantalate Light Modulators. J. OF APPLIED PHYS., v. 38, no. 4, Mar. 15, 1967. p. 1611-1617.

DIXON, R.W. Photoelastic Properties of Selected Materials and Their Relevance for Applications to Acoustic Light Modulators and Scanners. J. OF APPLIED PHYS., v. 38, no. 13, Dec. 1967. p. 5149-5153.

FEDULOV, S.A. et al. The Growth of Crystals of $LiNbO_3$, $LiTaO_3$ and $NaNbO_3$ by the Czochralski Method. SOVIET PHYS.-CRYST., v. 10, no. 2, Sept./Oct. 1965. p. 218-221.

IWASAKI, H. et al. Thermal Expansion of $LiTaO_3$ Single Crystal. JAPAN J. OF APPLIED PHYS., v. 6, 1967. p. 1338.

IWASAKI, H. et al. Refractive Indices of $LiTaO_3$ at High Temperatures. JAPAN. J. OF APPLIED PHYS., v. 7, no. 2, Feb. 1968. p. 185-186.

IWASAKI, H. et al. Pyroelectricity and Spontaneous Polarization in $LiTaO_3$. JAPAN. J. OF APPLIED PHYS., v. 6, no. 11, Nov. 1967. p. 1336

KAMINOW, I.P. and W.D. JOHNSTON. Quantitative Determination of Sources of the Electrooptic Effect in $LiNbO_3$ and $LiTaO_3$. PHYS. REV., v. 160, no. 3, Aug. 1967. p. 519-522.

KIM, Y.S. and R.T. SMITH. Thermal Expansion of Lithium Tantalate and Lithium Niobate Single Crystals. J. OF APPLIED PHYS., v. 40, no. 11, Oct. 1969. p. 4637-4641.

LENZO, P.V. et al. Electrooptic Coefficients and Elastic-Wave Propagation in Single-Domain Ferroelectric Lithium Tantalate. APPLIED PHYS. LETTERS, v. 8, no. 4, Feb. 1966. p. 81-82.

LEVINSTEIN, H.J. et al. Domain Structure and Curie Temperature of Single-Crystal Lithium Tantalate. J. OF APPLIED PHYS., v. 37, no. 12, Nov. 1966. p. 4585-4586.

LEVINSTEIN, H.J. et al. Reduction of the Susceptibility to Optically Induced Index Inhomogeneities in $LiTaO_3$ and $LiNbO_3$. J. OF APPLIED PHYS., v. 38, no. 8, July 1967. p. 3101-3102.

LOIACONO, G.M. Optical Transmission Variations in Lithium Tantalate. APPLIED OPTICS, v. 7, no. 3, Mar. 1968. p. 555-556.

MATTHIAS, B.T. and J.P. REMEIKA. Ferroelectricity in the Ilmenite Structure. PHYS. REV., v. 76, no. 12, Dec. 1949. p. 1886-1887.

MILLER, R.C. and A. SAVAGE. Temperature Dependence of the Optical Properties of Ferroelectric $LiNbO_3$ and $LiTaO_3$. APPLIED PHYS. LETTERS, v. 9, no. 4, Aug. 1966. p. 169-171.

GENERAL ELEC. CO., SCHENECTADY, N.Y. RES. AND DEV. CENTER. Microwave Memory Acoustic Crystals. AFML-TR-68-260. By: OLIVER, D.W. and J.D. YOUNG. Contract no. F33615-67-C-1399. Sept. 1968. 66 p.

PETERSON, G.E. et al. Electro-Optic Properties of $LiNbO_3$. APPLIED PHYS. LETTERS, v. 5, no. 3, Aug. 1964. p. 62-64.

RAZBIRIN, B.S. Curie Temperature of Ferroelectric $LiTaO_3$. SOVIET PHYS.-SOLID STATE, v. 6, no. 1, July 1964. p. 254-255.

SHAPIRO, Z.I. et al. Investigation of the $LiTaO_3$ - $LiNbO_3$ System. ACAD. OF SCI., USSR, BULL., PHYS. SER., v. 29, no. 6, June 1965. p. 1047-1050.

SHAPIRO, Z.I. et al. Phase Transitions in $LiNbO_3$ and $LiTaO_3$. SOVIET PHYS.-CRYST., v. 10, no. 6, May-June 1966. p. 725-728.

SMITH, R.T. Elastic, Piezoelectric, and Dielectric Properties of Lithium Tantalate. APPLIED PHYS. LETTERS, v. 11, no. 5, Sept. 1967. p. 146-148.

SMITH, R.T. and F.S. WELSH. Temperature Dependence of the Elastic, Piezoelectric, and Dielectric Constants of Lithium Tantalate and Lithium Niobate. J. OF APPLIED PHYS., v. 42, no. 6, May 1971. p. 2219-2230.

WARNER, A.W. et al. Determination of Elastic and Piezoelectric Constants for Crystals in Class (3M). ACOUSTICAL SOC. OF AMERICA, J., v. 42, no. 6, Dec. 1967. p. 1223-1231.

WEMPLE, S.H. et al. Relationship Between Linear and Quadratic Electro-Optic Coefficients in $LiNbO_3$, $LiTaO_3$, and Other Oxygen-Octahedra Ferroelectrics Based on Direct Measurement of Spontaneous Polarization. APPLIED PHYS. LETTERS, v. 12, no. 6, Mar. 1968. p. 209-211.

YAMADA, T. et al. Curie Point and Lattice Constants of Lithium Tantalate. JAPAN. J. OF APPLIED PHYS., v. 7, no. 3, Mar. 1968. p. 298-299.

YAMADA, T. et al. Dielectric Properties of Lithium Tantalate. JAPAN. J. OF APPLIED PHYS., v. 7, no. 3, Mar. 1968. p. 292.

YAMADA, T. et al. Piezoelectric and Elastic Properties of $LiTaO_3$: Temperature Characteristics. JAPAN. J. OF APPLIED PHYS., v. 8, no. 9, Sept. 1969. p. 1127-1132.

## Introduction

Potassium dideuterium phosphate (abbreviated KDDP or KD*P), an isomorph of potassium dihydrogen phosphate (KDP), has been fairly well explored as an electrooptic material and has found a number of applications as an electrooptic modulator crystal in various devices. The material is also piezoelectric as well as ferroelectric in nature.

Deuterium is an isotopic form of hydrogen, also called heavy hydrogen. Measurements have been made on partially deuterated and fully deuterated samples of KDDP and remarkable property changes occur with the degree of deuteration. The crystal is transparent in the range from 0.2 to 2.0 microns.

## Chemical and Physical Properties

| | | |
|---|---|---|
| Chemical Formula | $KD_2PO_4$ | |
| Molecular Weight | 138.11 | |
| Density (20°C) | 2.344 g/cm$^3$ | Shuvalov and Mnatsakanyan |

KDDP Crystals are very soluble and hygroscopic [Union Carbide]. The density of $KD_2PO_4$ varies with temperature, as shown in Figure 1.

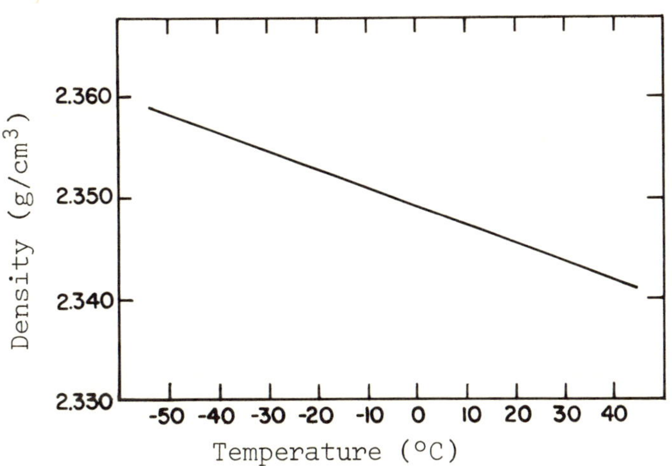


Fig. 1. Density of KDDP as a function of temperature [Jona].

## POTASSIUM DIDEUTERIUM PHOSPHATE (KDDP)

### Crystallography

Above 213°K, potassium dideuterium phosphate is paraelectric and belongs to the tetragonal group $\bar{4}2m$. It undergoes a ferroelectric phase transition at 221°K The tetragonal crystal structure of KDDP is shown in Figure 2 and the deuteron positions in Figure 3.

Crystal Structure

| | | |
|---|---|---|
| Above $T_c$ | Tetragonal | $\bar{4}2m$ |
| Below $T_c$ | Orthorhombic | mm2 |

Lattice Constants

| T(°K) | $a_o$(Å) | $c_o$(Å) | Reference | |
|---|---|---|---|---|
| 222 | 7.4586±0.0007 | 6.9559±0.0008 | Cook | |
| 298 | 7.4697±0.0003 | 6.9766±0.0005 | Cook | Sliker and Burlage |

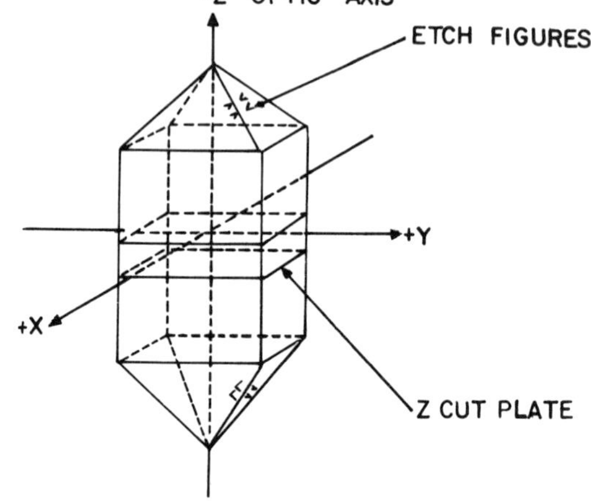


Fig. 2. A KDDP crystal with crystallographic axes labeled.

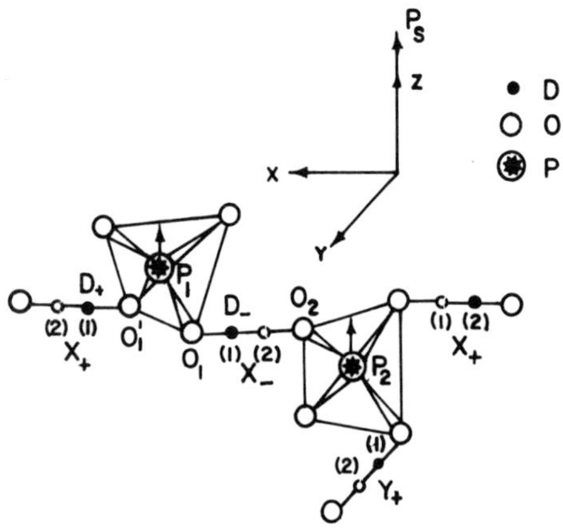


Fig. 3. Deuteron positions in $KD_2PO_4$. Below $T_c$, reversal of polarization is accompanied by a hydrogen shift from position (1) to position (2), with reversal of P-atom displacement along the c axis [Bjorkstam].

## Optical Properties - Refractive Index

TABLE 1. REFRACTIVE INDICES OF KDDP AT ROOM TEMPERATURE

| Wavelength (Å) | $n_o$ | $n_e$ | Reference |
|---|---|---|---|
| 4047 | 1.5189 | 1.4776 | a |
| 4078 | 1.5185 | 1.4772 | a |
| 4358 | 1.5155 | 1.4747 | a |
| 4358 | 1.5160 | 1.4752 | b |
| 4916 | 1.5111 | 1.4710 | a |
| 5461 | 1.5079 | 1.4683 | a |
| 5461 | 1.5077 | 1.4682 | b |
| 5779 | 1.5063 | 1.4670 | a |
| 5893 | 1.5057 | 1.4677 | b |
| 6234 | 1.5044 | 1.4656 | a |
| 6907 | 1.5022 | 1.4639 | a |
| 10000 | 1.47 | 1.44 | c |

a. Phillips
b. Yamazaki and Ogawa    90% deuterated
c. Harshaw

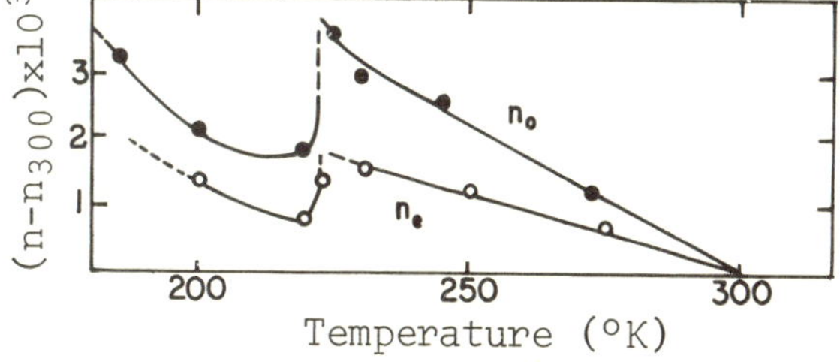


Fig. 4. Change in the refractive indices of 90% deuterated KDDP as a function of temperature at 5893 Å [Yamazaki and Ogawa].

## Optical Properties - Birefringence

Zwicker and Scherrer have measured the change in birefringence as a function of temperature (80° to 300°K) at 5461 Å for deuterated KDP and found an approximate linear relationship, except for the region near the Curie point.

$\partial(n_o-n_e)/\partial T$ at 5461 Å:

| | |
|---|---|
| $-(0.73\pm0.15)\times10^{-5}$ | Phillips |
| $-0.58\times10^{-5}$ | Zwicker and Scherrer |

## Optical Properties - Transmission

Potassium dideuterium phosphate is transparent in the range 0.2 to 2.0 microns, as illustrated in Figures 5 and 6. Sliker and Burlage show that the $KD_2PO_4$ infrared absorption edge is shifted toward longer wavelengths by a factor of 1.35±0.02 from the $KH_2PO_4$ edge.

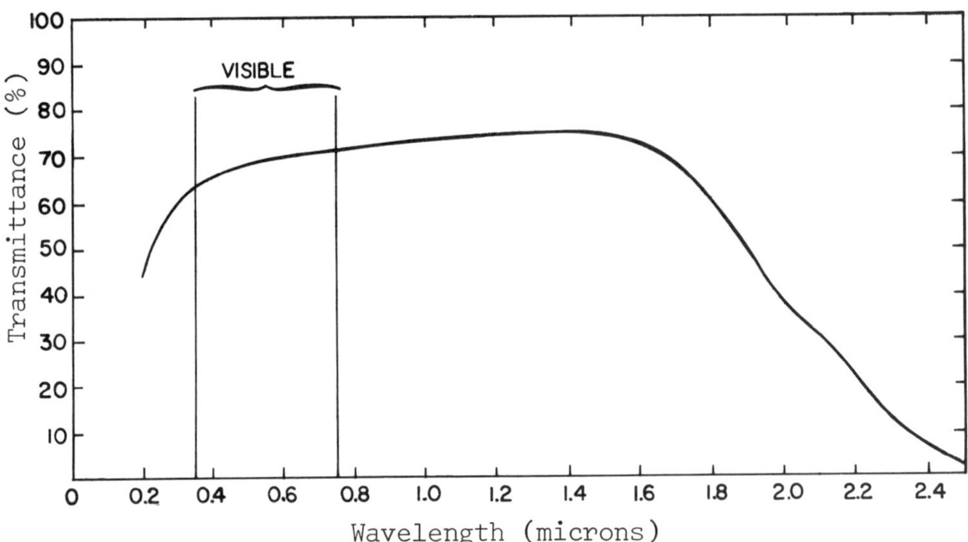


Fig. 5. Transmission as a function of wavelength for KDDP sample 0.022 inches thick. Transmission is uncorrected for reflection losses and is impaired by a small amount by imperfect optical polish [Reitz et al.].

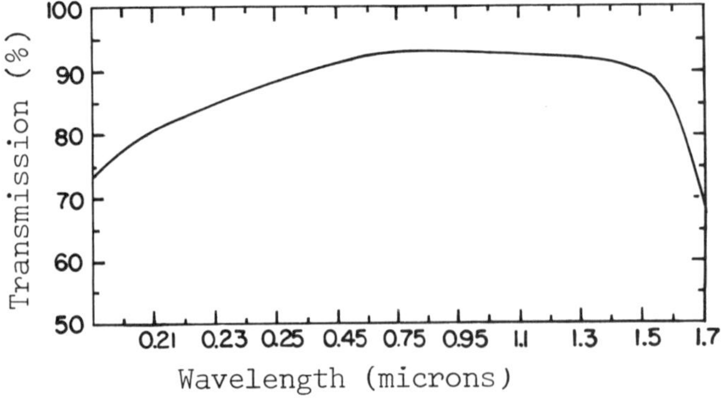


Fig. 6. Transmission as a function of wavelength for KDDP bare polished crystal 0.07 in thick uncorrected for reflection losses [Baird Atomic].

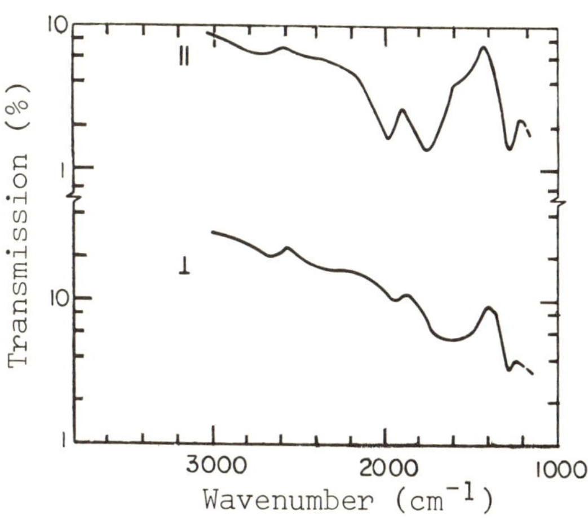


Fig. 7. Transmission as a function of wavenumber for single crystal KDDP at 300°K for light polarized perpendicular and parallel to the z-axis. Broad absorption peaks at 1960 and 1700 $cm^{-1}$ are associated with the hydrogen bonds; small absorptions at 2670 and 2360 $cm^{-1}$ are due to incomplete deuteration [Hill and Ichiki].



Optical Properties - Non-linear Optical Behavior

Miller et al. reported the following second harmonic generation coefficients for KDDP, relative to $d_{36}^{2\omega}$ for KDP: $d_{36}^{2\omega}$ = 0.75 ± 0.02, 0.92 ± 0.004 and $d_{14}^{2\omega}$ = 0.76 ± 0.04, 0.91 ± 0.03 at 0.6943 micron and 1.06 microns, respectively.

In 1962, Bass and coworkers at the University of Michigan reported on the observation of a dc polarization phenomenon accompanying the passage of an intense ruby laser beam through a KDDP crystal. This nonlinear effect permitted them to predict the following optical rectification coefficient:

$$d_{36}^{o} = | X_{zxy}^{o} + X_{zyx} | = 28 \times 10^{-8} \text{esu}$$

$$| X_{xyz}^{o} + X_{xzy}^{o} | = | X_{yxz}^{o} + X_{yzx}^{o} | = \text{no data available}$$

Electrooptic Properties

Potassium dideuterium phosphate exhibits a linear electrooptic effect and has one of the largest known electrooptic coefficients at room temperature (Table 2). It is found that the low-frequency or unclamped electrooptic coefficient $r_{63}^{T}$ in the deuterated material is more than twice that in the undeuterated material. Measurements at lower temperatures indicates a fiftyfold increase in $r_{63}^{T}$ for $KD_2PO_4$ as its Curie point is approached (Figure 8). Vasilevskaya found that the ratio of electrooptic coefficients for clamped and free crystals ($r_{63}^{S}/r_{63}^{T} \approx 0.95$) was fairly constant over the wavelength range 4000 to 7000 Å.

Perfilova and Sonin studied the quadratic electrooptic properties of $KD_2PO_4$ crystals at room temperature and at 5400 Å. Eden described the general modulating performance of $KD_2PO_4$ in Texas Instruments' modulators.

## POTASSIUM DIDEUTERIUM PHOSPHATE (KDDP)

Zwicker and Scherrer observed that the dc (unclamped) electrooptic coefficient, based on the electric field, exhibited a Curie-Weiss behavior as a function of temperature, as illustrated in Figure 8. They found, however, that the electrooptic coefficient based on the dielectric polarization was practically temperature independent. Sliker and Burlage find that the ratio $r_{63}/(\varepsilon_{33}^T - \varepsilon_o)$ is essentially temperature independent and is 0.061 ± 0.003 $m^2/C$ for KDDP. Here $\varepsilon_{33}$ is the permittivity at constant stress and $\varepsilon_o$ is the permittivity of free space.

TABLE 2. ELECTROOPTIC COEFFICIENTS OF KDDP AT ROOM TEMPERATURE.

| $r_{63}^T$ | $r_{63}^S$ | $r_{41}^T$ | $V_{\lambda/2}$ | Wavelength | Sample | Reference |
|---|---|---|---|---|---|---|
| ($10^{-12}$m/V) | | | | | | |
| 20 | | | | 5460 Å | | Zwicker & Scherre |
| 26.4±0.7 | | | | | | |
| | | 8.8±0.4 | | 5460 Å | 83-92% deuterated | Ott & Sliker |
| 25.6 | 24 | | 3.08 kV | 5000 Å | 90% deuterated | Christmas & Wildey |
| 19.7 | 21.0 | | | 5325 Å | | Vasilevskaya & Sonin |

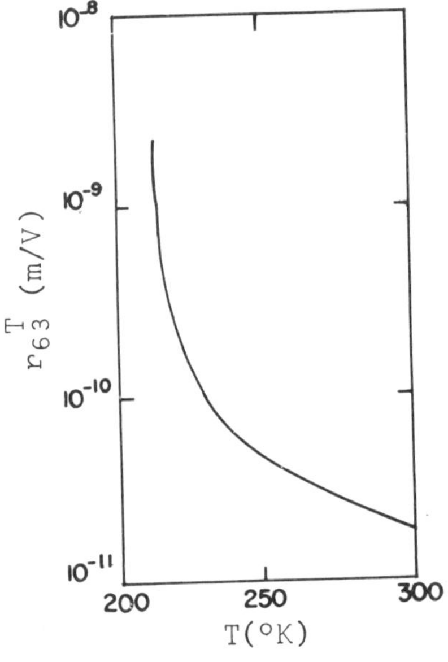


Fig. 8. Electrooptic coefficient of KDDP as a function of temperature [Zwicker and Scherrer].

Vasilevskaya and Sonin also report a piezooptic constant, $\pi_{66}$ of 0.35 x $10^{12}$ $cm^2$/dyne and an elastooptic constant $p_{66}$ of 0.025. These measurements are also made at 5325 Å

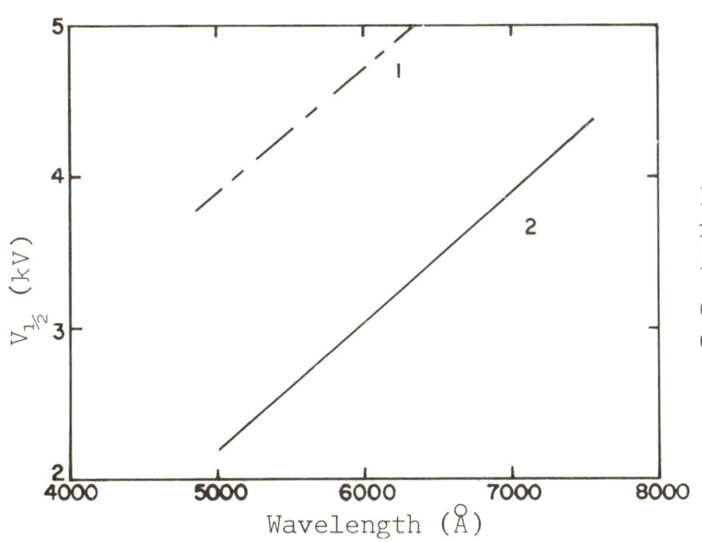


Fig. 9. Half-wave voltage of KDDP at room temperature as a function of wavelength.

Curve 1: >1MHz Pyle
Curve 2: dc Sliker and Burlage

Elastic Constants

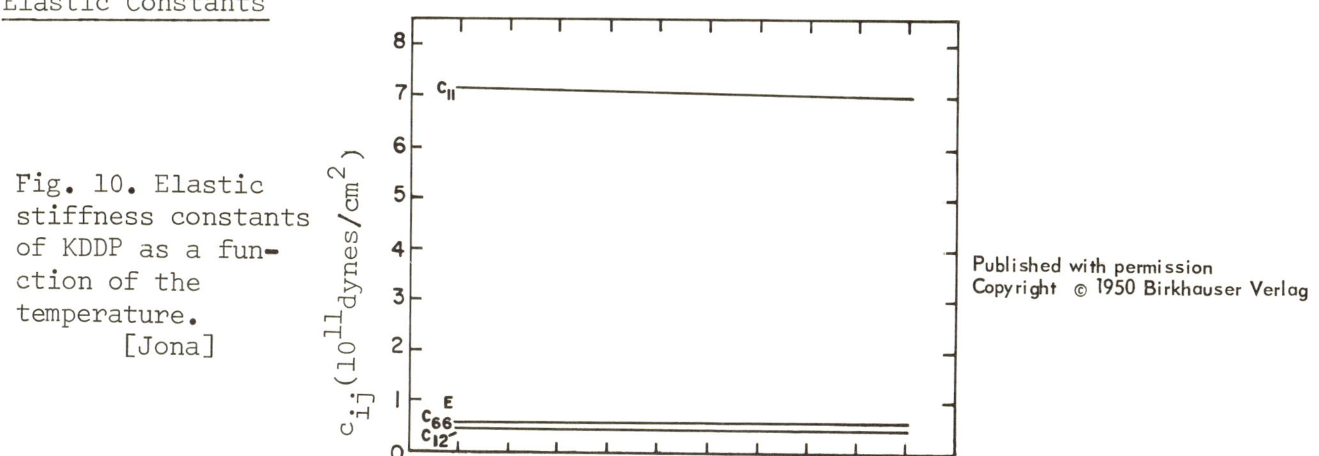


Fig. 10. Elastic stiffness constants of KDDP as a function of the temperature. [Jona]



TABLE 3. ELASTIC CONSTANTS OF KDDP AT 20°C

| Elastic Stiffness Constants | | | | | | Elastic Compliance | | |
|---|---|---|---|---|---|---|---|---|
| $c_{11}$ | $c_{12}$ | $c_{13}$ | $c_{33}$ | $c_{44}$ | $c_{66}$ ($10^{11}$ dynes/cm$^2$) | $s_{66}$ | $s_{44}$ ($10^{-13}$ cm$^2$/dyne) | |
| 7.17 | -0.63 | 1.60 | 5.68 | 1.27 | 0.64 | | | Boyer & Vacher |
| 7.04 | 0.46 | | | | 0.607 | 165 | | Jona |
| | | | | | 0.62 | | | Vasilevskaya & Sonin |
| | | | | | 0.77 (99% deuterated) | | | Skalyo et al. |
| | | | | | | 161 | 77.7 | Landolt-Börnstein |
| | | | | | | 164.7 | | Brown et al. |
| | | | | | | 165±3 | | Sliker & Burlage |

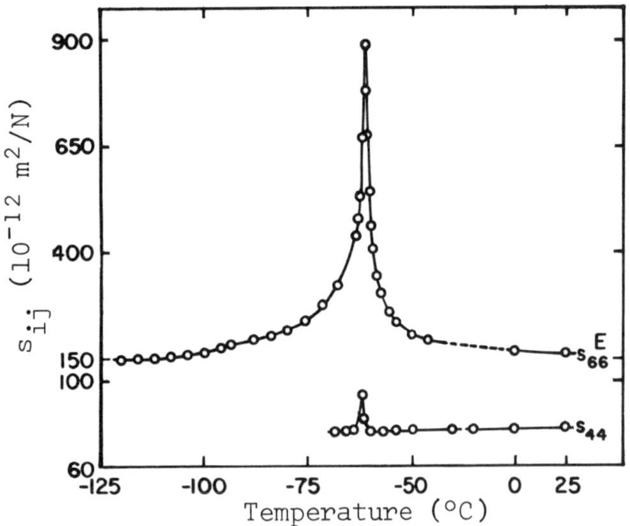


Fig. 11. Elastic compliance constants of KDDP as a function of temperature [Shuvalov and Mnatsakanyan].

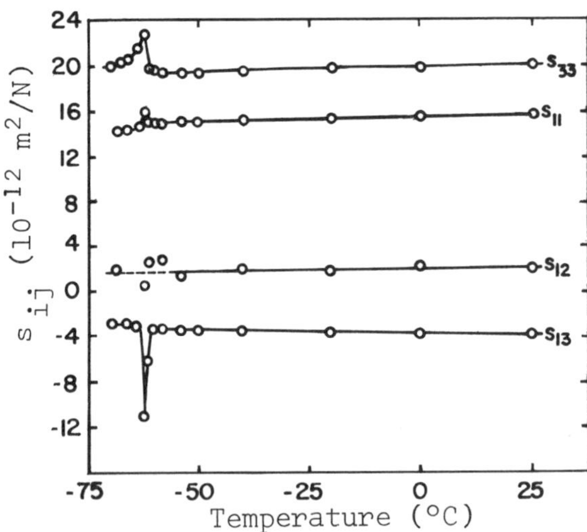


Fig. 12. Elastic compliance constants of KDDP as a function of temperature [Shuvalov and Mnatsakanyan].



## Piezoelectric Properties

TABLE 4. PIEZOELECTRIC CONSTANTS OF KDDP AT ROOM TEMPERATURE

| $d_{36}$ $(10^{-12}$ C/N) | $d_{14}$ $(10^{-12}$ C/N) | $g_{36}$ $(m^2/C)$ | Reference |
|---|---|---|---|
| 52 | | | Brown et al. |
| | | | Landolt-Börnstein |
| 58±2 | | 0.13±0.01 | Sliker and Burlage |
| 52.7±1.7 | 3.3±0.3 | | Shuvalov et al. |

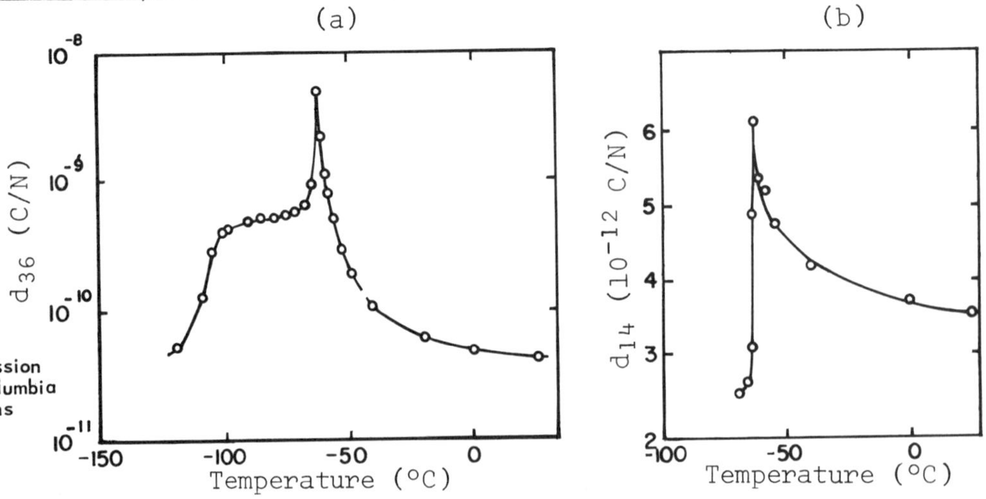

Fig. 13. Piezoelectric constants of KDDP as a function of temperature. (a) $d_{36}$ (b) $d_{14}$ [Shuvalov et al.].

Electromechanical Coupling Coefficient:

| | | |
|---|---|---|
| $k_{36}$ | 0.22±0.01 | Sliker and Burlage |

## Dielectric Properties - Dielectric Constant

The wide range of dielectric constant values reported in the literature for KDDP at room temperature and low frequencies probably can be accounted for by unreported differences in the degree of deuteration. Recently reported values of the dielectric constants of highly deuterated KDDP crystals are given in Table 5.

The results of Shuvalov et al. for the low-frequency dielectric constants of KDDP as a function of temperature are shown in Figure 14. These results agree with the earlier measurements of Sliker and Burlage and Mayer and Bjorkstam. Sliker and Burlage found that the dielectric constant $\varepsilon_c$ follows the Curie-Weiss law, $\varepsilon_c = C/(T-T_c)$, where C = 3760° ± 60°K and $T_c$ = 222° ±1°K. Kamysheva et al. report that at fields to 5 kV.cm and at -110 to -40°C, single crystal platelets show a shift in $T_c$ with field of 0.5°C-cm/kV.

TABLE 5. LOW-FREQUENCY DIELECTRIC CONSTANTS OF KDDP AT ROOM TEMPERATURE

| $\varepsilon_c^T$ | $\varepsilon_a^T$ | f(Hz) | T(°C) | Reference |
|---|---|---|---|---|
| 50±2 | | 1000 | 24 | Sliker and Burlage |
| 51±2 | 57±2 | 400 | 20 | Shuvalov et al. |

$\varepsilon_c = \varepsilon_{33}/\varepsilon_o$; field parallel to the c-axis

$\varepsilon_a = \varepsilon_{11}/\varepsilon_o$; field perpendicular to the c-axis

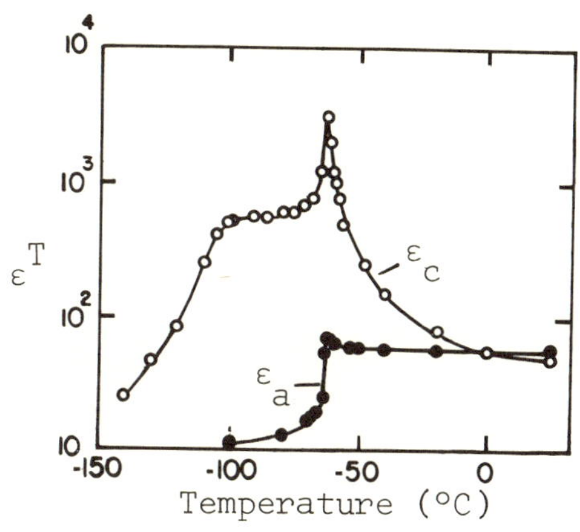


Fig. 14. Dielectric constants of $KD_2PO_4$ as a function of temperature measured at 400 Hz at a field strength of 10 V/cm [Shuvalov et al.].

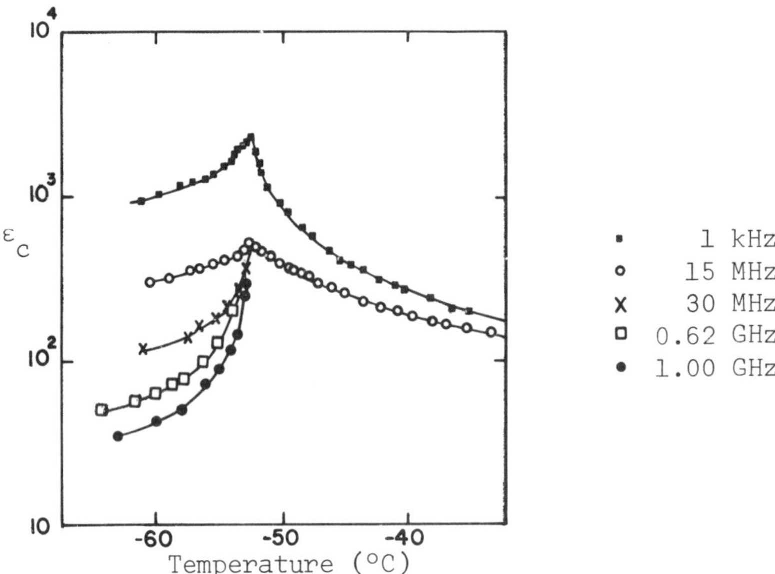


Fig. 15. Dielectric constant as a function of temperature for $KD_2PO_4$ at several frequencies. Main crystal resonances are clamped out between 1 kHz and 15 MHz, as shown, but for $T<T_c$ (-52°C) there is a considerable anomalous response up to 1 GHz [Hill and Ichiki].

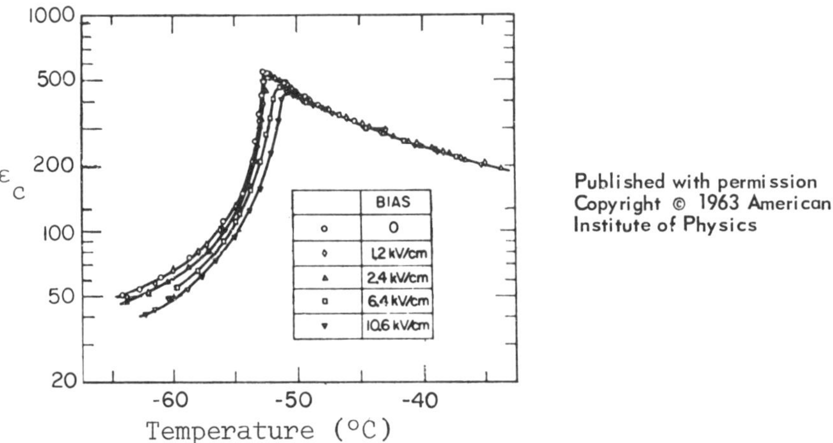




Fig. 16. Dielectric constant as a function of temperature for $KD_2PO_4$ at 620 MHz and various dc bias fields [Hill and Ichiki].

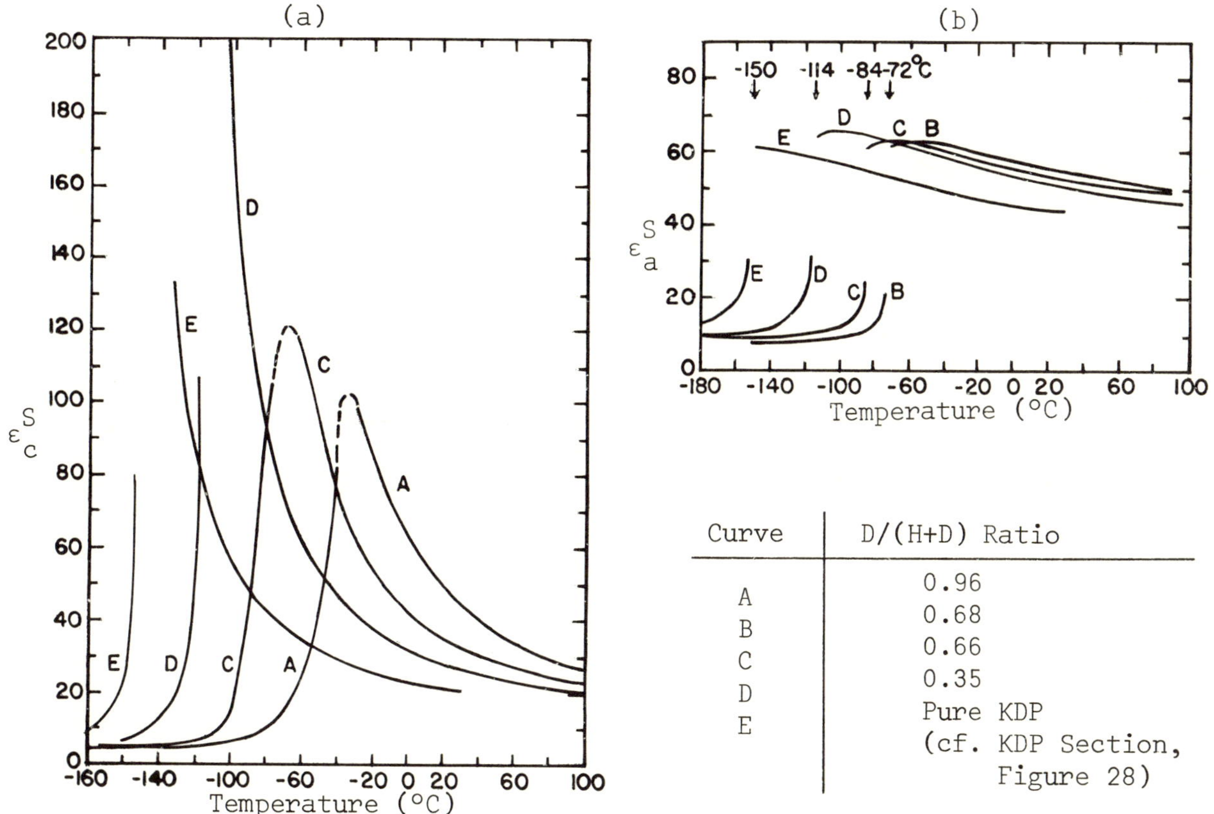


| Curve | D/(H+D) Ratio |
|---|---|
| A | 0.96 |
| B | 0.68 |
| C | 0.66 |
| D | 0.35 |
| E | Pure KDP (cf. KDP Section, Figure 28) |

Fig. 17. Dielectric constant as a function of temperature for partially deuterated KDDP at 9.2 GHz.

(a) $\varepsilon_c^S$ (b) $\varepsilon_a^S$ [Kaminow]

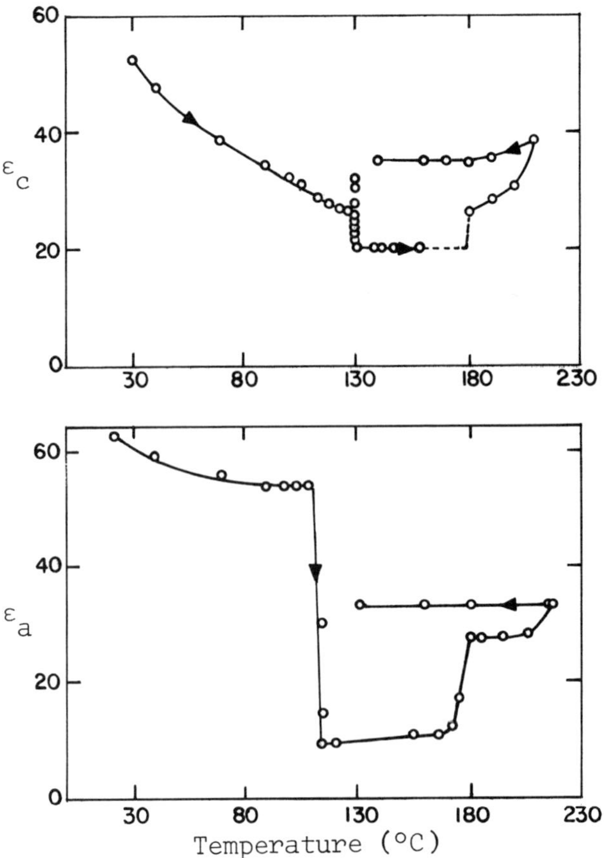


Fig. 18. Dielectric constant of $KD_2PO_4$(~85%D) as a function of temperature at 10 MHz. A phase transition is observed near 180°C [Grinberg et al.].



## Dielectric Properties - Loss Tangent

Stites et al. have measured the dissipation factor at 1kHz of eight samples of KDDP (>90% deuteration) of varying thicknesses and areas and found $\tan\delta_c = 0.42 \pm 0.28$.

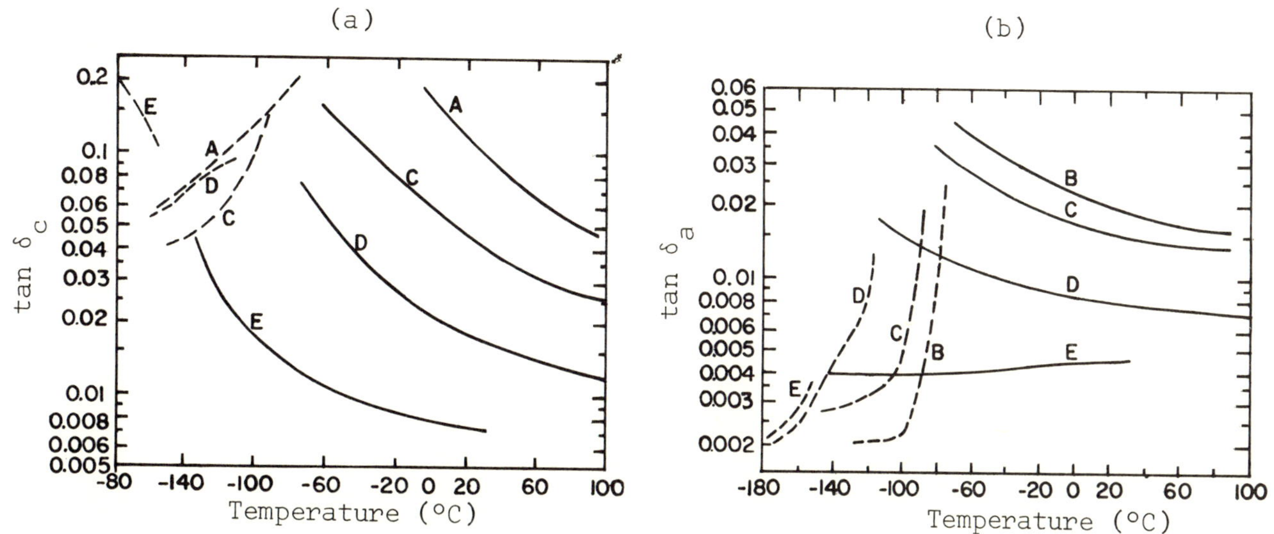




Fig. 19. Loss tangent as a function of temperature for partially deuterated KDDP at 9.2 GHz.

(a) tan $\delta_c$ (b) tan $\delta_a$ [Kaminow]

| Curve | D/(H+D) Ratio |
|---|---|
| A | 0.96 |
| B | 0.68 |
| C | 0.66 |
| D | 0.35 |
| E | Pure KDP (cf. KDP Section, Figure 33) |

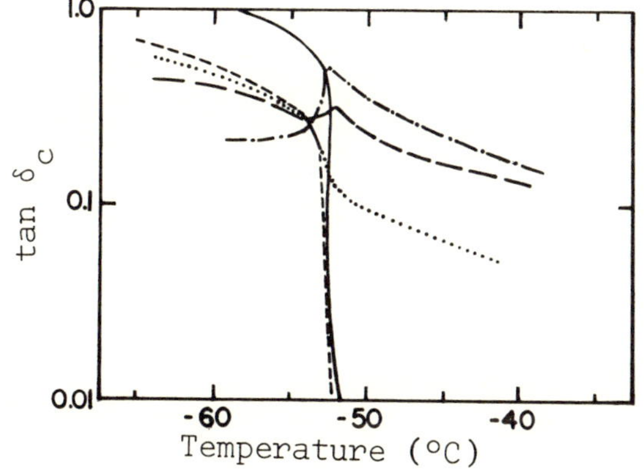


Fig. 20. Loss tangent as a function of temperature for $KD_2PO_4$ at several frequencies [Hill and Ichiki].

--- 15 MHz
—— 30 MHz
······· 0.62 GHz
— — 1.0 GHz
—·— 2.0 GHz

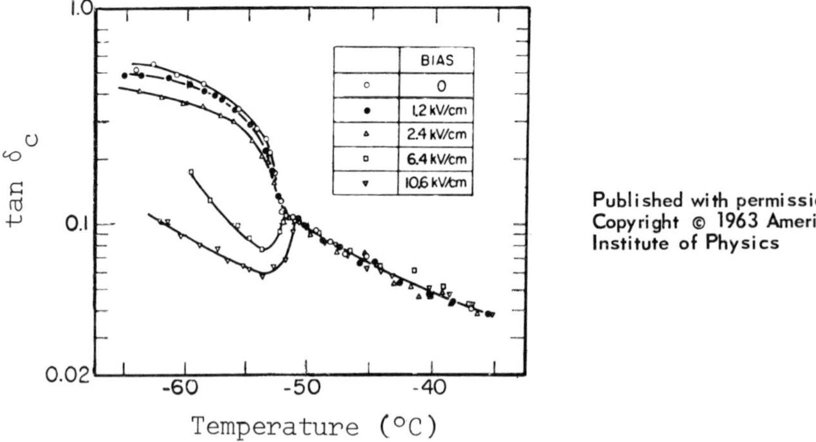


Fig. 21. Loss tangent as a function of temperature for $KD_2PO_4$ at 620 MHz for several values of dc bias field. The anomalous loss for $T<T_c$ (-52°C) is seen to be decreased at large values of bias field [Hill and Ichiki].

Dielectric Properties - Dielectric Strength

TABLE 6. DIELECTRIC STRENGTH OF SEVERAL SAMPLES OF KDDP (>90% DEUTERATED) AT 25°C [Stites et al.]

| Sample Thickness (inch) | Breakdown Voltage (V) | Dielectric Strength (V/mil) |
|---|---|---|
| 0.0092 | 4200 | 456.5 |
| 0.0590 | 10500 | 177.9 |
| 0.0590 | 9800 | 166.1 |
| 0.0560 | 10400 | 185.7 |
| 0.0580 | 10200 | 175.8 |
| 0.0580 | 11000 | 189.6 |
| 0.0580 | 5700 | 98.2 |
| 0.0580 | 6200 | 106.8 |
| 0.0580 | 10100 | 174.1 |
| 0.0175 | 5900 | 337.1 |
| 0.0175 | 6200 | 354.2 |
| 0.0180 | 6800 | 377.7 |
| | | Average = 233.3 |

Dielectric Properties - Electrical Resistivity

Stites et al. have reported a surface resistivity of $1.3 \times 10^{11}$ ohm/sq for a KDDP sample 0.147 cm in thickness and 8.2 $cm^2$ in area.

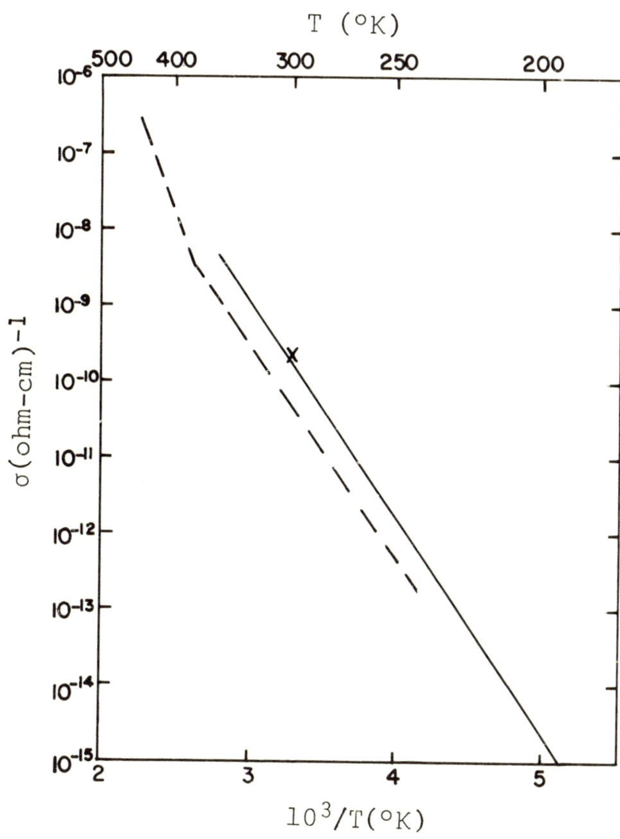


Fig. 22. Electrical conductivity of KDDP as a function of reciprocal temperature.

----- 90% deuterated; perpendicular to the c-axis [O'Keeffe and Perrino]

X >90% deuterated [Stites et al.]

_____ [Schmidt and Uehling]

## POTASSIUM DIDEUTERIUM PHOSPHATE (KDDP)

### Ferroelectric Properties

Bantle (in 1942) was the first to report on the ferroelectric tetragonal form of $KD_2PO_4$. In 1944, Zwicker and Scherrer reported the d-c electrooptic properties of this crystal and related these properties to the ferroelectric behavior of the crystal. They observed that the electrooptic coefficient based on electric fields, was proportional to the dielectric constant, exhibiting a Curie-Weiss behavior as a function of temperature.

The Curie temperature often quoted in the early literature for KDDP is $T_c$=213°K. However, by taking extreme precautions to avoid hydrogen contamination, Sliker and Burlage have observed a Curie temperature of 222±1°K for KDDP. These authors conclude that the lower value previously reported was due to incomplete deuteration. The Curie temperature as a function of the degree of deuteration is shown in Figure 23. From these results [Kaminow ], it appears that completely deuterated KDDP may exhibit a Curie temperature up to 229°K.

Hill and Ichiki studied the high-frequency behavior of $KD_2PO_4$ and found a large domain-associated loss in the ferroelectric region. This loss persisted up to 2 GHz, indicating piezoelectric coupling to very small domains. These authors found that the Curie temperature could be shifted to higher temperatures by a dc bias field, as shown in Figure 24. They also expressed the opinion that the dielectric behavior of $KD_2PO_4$ at high frequencies is not truly characteristic of a second-order transition. Reese and May conducted calorimetric experiments on $KD_2PO_4$ and found the phase transition to be of the first-order type and that a classical, mean-field theory gives a rather satisfactory account of the thermodynamic properties, except within a few degrees of the transition.

Känzig, and Jona and Shirane, in their basic review studies of KDDP ferroelectrics, reported on the importance of the establishment of long-range order of the hydrogen-bond network in the ferroelectric transition of these materials. Recent work by Skalyo et al. on 99% deuterated, single crystals, indicates a Curie temperature of 221°K.

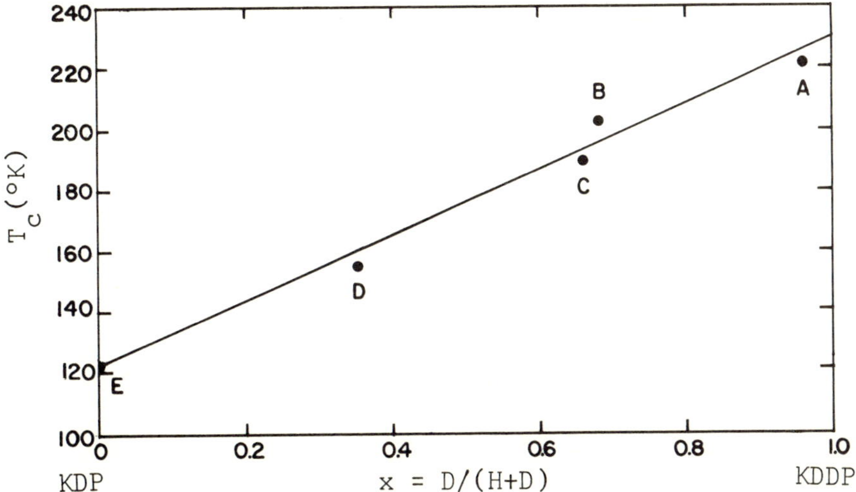


Fig. 23. Curie temperature of KDDP as a function of the degree of deuteration. Deuterium concentrations were determined by mass-spectrometer analysis with an experimental error of about 10%. Curie temperatures were obtained by extrapolating $1/\varepsilon_c$. The solid curve is the linear plot

$$T_c = (123+106x)°K.$$

| Sample | x | $T_c$(°K) | $T_t$(°K)* |
|---|---|---|---|
| A | 0.96 | 222±2 | |
| B | 0.68 | 203±2 | 201 |
| C | 0.66 | | 189 |
| D | 0.35 | 155±2 | 159 |
| E | Pure KDP | 123 | 123 |

*Phase transition observed in $\varepsilon_a$ (Figure 17). [Kaminow]

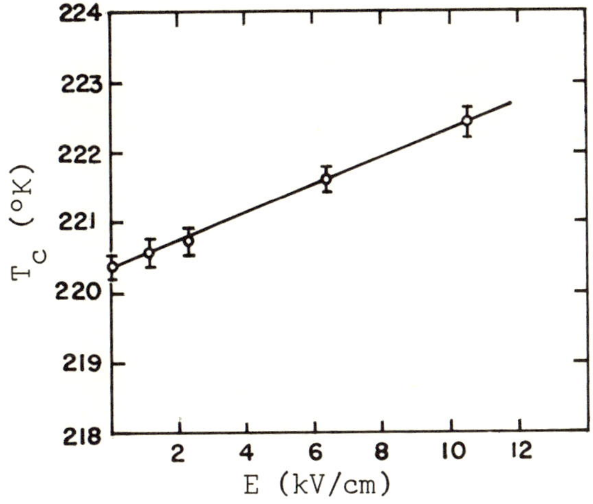


Fig. 24. Curie temperature of KDDP as a function of dc bias field. Data obtained from the dielectric constant results shown in Figure 16. [Hill and Ichiki].

Oettel has made an extensive and detailed study of the ferroelectric domains and domain motion in $KD_2PO_4$ crystals. He found that the domains can be observed in these crystals without the aid of polarizers. The domain walls are visible presumably because of a slight discontinuity in the refractive index at the domain wall caused by strain. Preliminary calculations based on measurement of the threshold angle for total internal reflection indicate that the change in the refractive index at the domain wall is of the order of 0.6%. Domains are of the order of 2 x $10^{-4}$ cm wide in 1 millimeter thick plates and increase in width with plate thickness as $t^{\frac{1}{2}}$; however, a single plate may have domains of various widths ranging over a factor of 2 or 3. Domain widths are the same for $KH_2PO_4$ and $KD_2PO_4$ crystals of the same thickness. Saturation polarization is achieved when a crystal is completely clear of domain structure. $KD_2PO_4$ crystals require many minutes to completely switch to the opposite polarization for an applied field of 3kV/cm. In $KD_2PO_4$ below 135°K, all domains become "pinned," or frozen, and no domain motion is observed over long periods of time even with applied fields as high as 5kV/cm.

## Thermal Properties

| | | |
|---|---|---|
| Melting Point: | 260±1°C | O'Keeffe and Perrino |
| Maximum Safe Operating Temperature: | 100°C | Union Carbide |

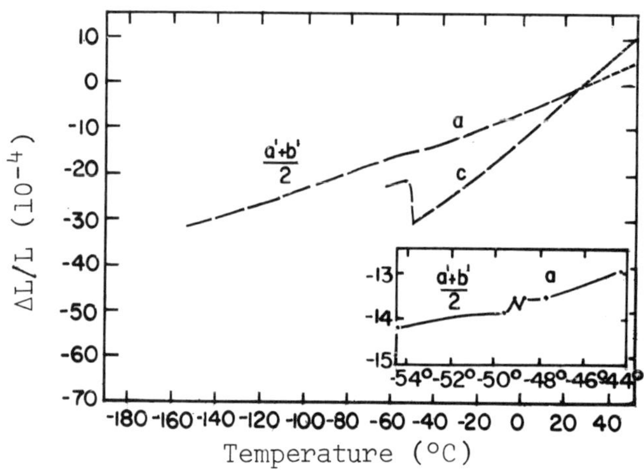


Fig. 25. Thermal expansion of $KD_2PO_4$ as a function of temperature. The inset shows an expanded plot of the expansion along the a-axis near the phase transition [Cook].

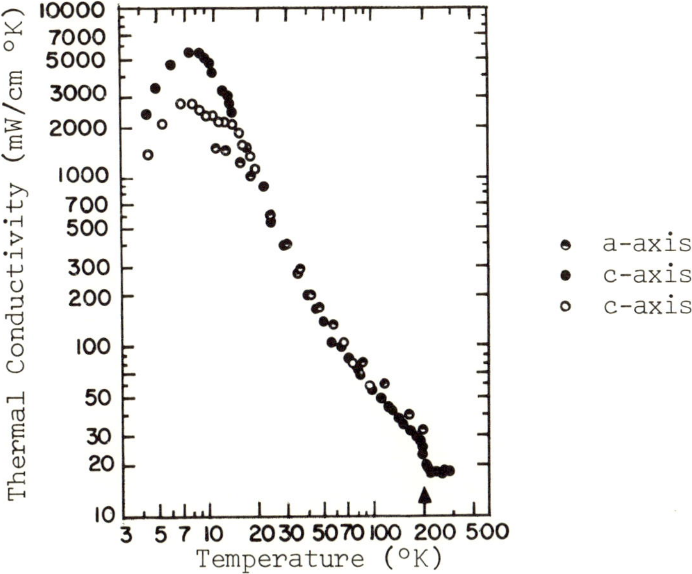


Fig. 26. Thermal conductivity of KDDP as a function of temperature. The arrow indicates the Curie temperature of 213°K [Suemune].

POTASSIUM DIDEUTERIUM PHOSPHATE (KDDP)

BAIRD-ATOMIC, INC., Cambridge, Mass. Electro-optic Light Modulators: KDP, ADP, KDDP, Publication No. EP-5, June 1967. 40 p.

BASS, M. et al. Optical Rectification. PHYS. REV. LETTERS, v. 9, no. 11, 1962. p. 446-448.

BICHARD, V.M. et al. 1.06 micron Absorption Coefficients of Deuterated KDP with 70-100% Deuteration. ELECTRONICS LETTERS, v. 8, no. 6, Mar. 1972, p. 147-148.

BJORKSTAM, J.L. Deuteron-Nuclear-Magnetic-Resonance Study of the Ferroelectric Phase Transition in Deuterated Triglycine Sulfate and $KD_2PO_4$. PHYS. REV., v. 153, no. 2, Jan. 1967. p. 599-605.

BOYER, L. and R. VACHER. Measurement of Elastic Constants of Potassium Dihydrogen Phosphate by Brillouin Diffusion. (In Fr.) PHYS. STAT. SOLIDI, v. 6a, no. 2, Aug. 1971, p. K105-K108.

BROWN, C.S. et al. Piezo-electric Materials--A Review of Progress. IEE., Proc., v. 109B, no. 43, Jan. 1962. p. 99-114.

CHRISTMAS, T.M. and C.G. WILDEY. Pulse Measurement of $r_{63}$ in KD*P. ELECTRONICS LETTERS, v. 6, no. 6, Mar. 1970. p. 152-153.

COOK, W.R. Jr. Thermal Expansion of Crystals with $KH_2PO_4$ Structure. J. OF APPL. PHYS., v. 38, no. 4, Mar. 1967. p. 1637-1642.

TEXAS INSTRUMENTS, INC. Solid State Techniques for Modulation and Demodulation of Optical Waves. By EDEN, D.D. Final Tech. Report ECOM-03250-F, Sept. 1966. 283 pp. AD 489-390.

GRINBERG, J. et al. Isotope Effect in the High Temperature Phase Transition of $KH_2PO_4$. SOLID STATE COMM., v. 5, no. 11, Nov. 1967. p. 863-865.

HARSHAW CHEMICAL CO. CRYSTAL-SOLID STATE DEPT. Harshaw Condensed Catalog of Radiation Detectors, Nuclear Products and Crystal Products. Publ. No. D-2251, Cleveland, Ohio, 1968. 81 pp.

HILL, R.M. and S.K. ICHIKI. High-Frequency Behavior of Hydrogen Bonded Ferroelectrics: Triglycine Sulphate and $KD_2PO_4$. PHYS. REV., v. 132, no. 4, Nov. 1963. p. 1603-1608.

HILL, R.M. and S.K. ICHIKI. Infrared Absorption by Hydrogen Bonds in Single Crystals: $KH_2PO_4$, $KDDPO_4$ and $KH_2AsO_4$. J. OF CHEM. PHYS., v. 48, no. 2, Jan. 1968. p. 838-842.

JONA, F. Elasticity of Piezoelectric and Ferroelectric Crystals (In Ger.). HELV. PHYS. ACTA, v. 23, Dec. 1950. p. 795-844.

JONA, F. and G. SHIRANE. FERROELECTRIC CRYSTALS, 1962. N.Y., MacMillan Co.

KAMINOW, I.P. Microwave Dielectric Properties of $NH_4H_2PO_4$, $KH_2AsO_4$ and Partially Deuterated $KH_2PO_4$. PHYS. REV., v. 138, no. 5A, May 1965. p. A1539-A1543.

KÄNZIG, W. Ferroelectrics and Antiferroelectrics. SOLID STATE PHYS., v. 4, 1957. N.Y., Academic Press. p. 1-197.

POTASSIUM DIDEUTERIUM PHOSPHATE (KDDP)

KAMYSHEVA, L.N. et al. The Effect of A Strong Electric Field on the Dielectric Properties of $KD_2PO_4$. SOVIET PHYS. CRYSTALL. v. 14, no. 1, July 1969. p. 141-142.

LANDOLT-BÖRNSTEIN. NUMERICAL DATA AND FUNCTIONAL RELATIONSHIPS IN SCIENCE AND TECHNOLOGY, New Series, v. 1, Elastic, Piezoelectric, Piezooptic and Electro-Optic Constants of Crystals. Group III: Crystal and Solid State Physics, Ed: HELLWEGE, K.-H. and A.M. HELLWEGE. Berlin, Ger. Springer Verlag, 1966.

MAYER, R.J. and J.L. BJORKSTAM. Dielectric Properties of $KD_2PO_4$. J. OF PHYS. AND CHEM. OF SOLIDS, v. 23, June 1962. p. 619-620.

MILLER, R.C. et al. Quantitative Studies of Optical Harmonic Generation in CdS, $BaTiO_3$ and $KH_2PO_4$ Crystals. PHYS. REV. LETTERS, v. 11, no. 4, Aug. 1963. p. 146-149.

UNIV. OF WASHINGTON, Seattle, Washington. Ferroelectric Domains and Domain Motion in $KH_2PO_4$ and $KD_2PO_4$. by: OETTEL, R.E., M.S. in Electrical Engineering, 1965. 44 p.

O'KEEFFE, M. and C.T. PERRINO. Isotope Effects in the Conductivity of $KH_2PO_4$ and $KD_2PO_4$. J. OF PHYS. AND CHEM. OF SOLIDS, v. 28, no. 6, June 1967. p. 1086-1088.

OTT, J.H. and T.R. SLIKER. Linear Electro-Optic Effects in $KH_2PO_4$ and its Isomorphs. OPTICAL SOC. OF AMERICA, J., v. 54, no. 12, Dec. 1964. p. 1442-1444.

PERFILOVA, V.E. and A.D. SONIN. The Quadratic Electro-Optical Effect in KDP Group Crystals. ACAD. OF SCI., USSR, BULL. PHYS. SER., v. 31, no. 7, July 1967. p. 1154-1157.

PHILLIPS, R.A. Temperature Variation of the Index of Refraction of ADP, KDP and Deuterated KDP. OPTICAL SOC. OF AMERICA, J., v. 56, May 1966. p. 629-632.

PYLE, J.R. Laser Modulation Using Linear Electro-Optic Crystals. Tech. Note PAD 125, Dec. 1966. 35 pp. NASA N67-27128.

REESE, W. and L.F. MAY. Studies of Phase Transition in Order-Disorder Ferro-electrics. II. Calorimetric Investigations of $KH_2PO_4$. PHYS. REV., v. 167, no. 2, Mar. 1968. p. 504-510.

REITZ, E.A. et al. (Autonetics) Electro-Optic Projector Study. Tech. Report No. RADC-TR-66-394, Dec. 1966. AD 646 232.

SCHMIDT, V.H. and E.A. UEHLING. Random Motion of Deuterons in $KD_2PO_4$. PHYS. REV., v. 126, no. 2, Apr. 1962. p. 447-457.

SHUVALOV, L.A. and A.V. MNATSAKANYON. The Elastic Properties of $KD_2PO_4$ Crystals Over a Wide Temperature Range. SOVIET PHYS., CRYSTALL., v. 11, no. 2, Sept. 1966. p. 210-212.

POTASSIUM DIDEUTERIUM PHOSPHATE (KDDP)

SHUVALOV, L.A. et al. Ferroelectric Anomalies of the Dielectric and Piezoelectric Properties of $RbH_2PO_4$ and $KD_2PO_4$ Crystals. ACAD. OF SCI., USSR, Phys. Ser., v. 31, no. 11, Nov. 1967. p. 1963-1966.

SKALYO, J. JR. et al. Ferroelectric-Mode Motion in $KD_2PO_4$. PHYS. REV., B, Ser. 3, v. 1, no. 1, Jan. 1970. p. 278-286.

SLIKER, T.R. and S.R. BURLAGE. Some Dielectric and Optical Properties of $KD_2PO_4$. J. OF APPL. PHYS., v. 34, no. 7, July 1963. p. 1837-1840.

AUTONETICS, ANAHEIM, CALIF. Electro-Optic Projection Study. Final Rept., Dec. 1963-Dec. 1964, by STITES, R.S. et al. Contract No. AF 30(602)-3263. Apr. 1965. 137 p. AD 617 087.

SUEMUNE, Y. Thermal Conductivity of the $KH_2PO_4$ Type. PHYS. SOC. OF JAPAN, J., v. 21, no. 4, Apr. 1966. p. 802.

UNION CARBIDE CORP., Linde/Electronics, Div., San Diego, California. Linde Crystals. 4 pp.

VASILEVSKAYA, A.S. The Electro-Optical Properties of Crystals of KDP Type. SOVIET PHYS.-CRYSTAL., v. 11, no. 5, Mar. 1967. p. 644-647.

VASILEVSKAYA, A.S. and A.S. SONIN. The Relationship of Structure to Electrooptic and Elastooptic Properties in Crystals of the KDP Group. SOVIET PHYS.-CRYSTALL., v. 14, no. 4, Jan. 1970. p. 611-613.

YAMAZAKI, M. and T. OGAWA. Temperature Dependences of the Refractive Indices of $NH_4H_2PO_4$, $KH_2PO_4$ and Partially Deuterated $KH_2PO_4$. OPTICAL SOC. OF AMERICA, J., v. 56, no. 10, Oct. 1966. p. 1407-1408.

ZWICKER, B. and P. SCHERRER. Electro-Optical Properties of the Ferroelectric Crystals $KH_2PO_4$. (In Ger.) HELV. PHYS. ACTA, v. 17, no. 5, Sept. 1944. p. 346-373.

# POTASSIUM DIHYDROGEN ARSENATE (KDA)

## Introduction

Potassium dihydrogen arsenate (KDA) is an isomorph of potassium dihydrogen phosphate (KDP). It is piezoelectric and ferroelectric in nature. In a recent (1967) survey of readily available electrooptic materials [Bicknell et al.], KDA was described to be the best for use in low drive-power, wide-bandwidth modulators.

## Chemical and Physical Properties

| | |
|---|---|
| Chemical Formula | $KH_2AsO_4$ |
| Molecular Weight | 180.05 |
| Density | 2.867 $g/cm^3$ |
| Solubility in Water (6°C) | 19 g/100 g water |

## Crystallography

Above 95°K, KDA is paraelectric and belongs to the tetragonal group $\bar{4}2m$. It undergoes a ferroelectric phase transition at 95°K.

| | |
|---|---|
| Crystal Symmetry | Tetragonal |
| Point Group | $\bar{4}2m$ or $D_{2d}$ |
| Lattice Constants [Cook] | |

| T(°C) | $a_o$(Å) | $c_o$(Å) |
|---|---|---|
| 25 | 7.6295 ± 0.0006 | 7.1605 ± 0.0008 |
| -178 ($T_c$) | 7.6034 ± 0.0009 | 7.1006 ± 0.0011 |
| 18 | 7.627 | 7.162 [Donnay] |

POTASSIUM DIHYDROGEN ARSENATE (KDA)

## Optical Properties

TABLE 1. REFRACTIVE INDEX OF POTASSIUM DIHYDROGEN ARSENATE [Ott and Sliker]

| Wavelength (Å) | $n_o$ | $n_e$ |
|---|---|---|
| 4861 | 1.5762 | 1.5252 |
| 5460 | 1.5707 | 1.5206 |
| 5893 | 1.5674 | 1.5179 |
| 6563 | 1.5632 | 1.5146 |

Thermal Birefringence Coefficient [Yap and Bicknell]

$$\partial (n_e - n_o) / \partial T = 2.6 \times 10^{-5}/°C$$

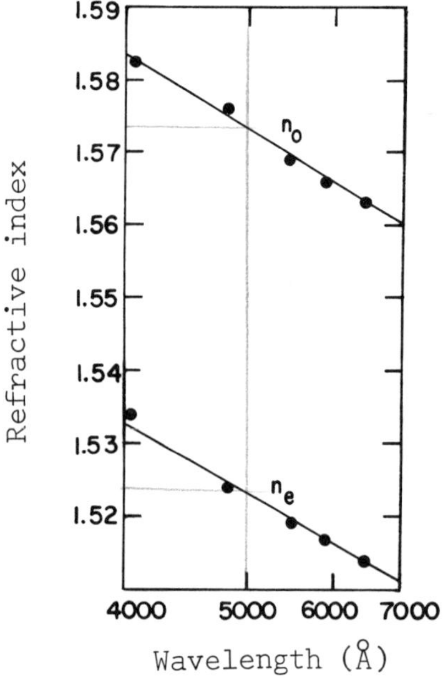


Fig. 1. Refractive index of KDA as a function of wavelength [Adhav]

Fig. 2. Transmission of a 2 mm thick polished sample of KDA as a function of wavelength [Adhav]

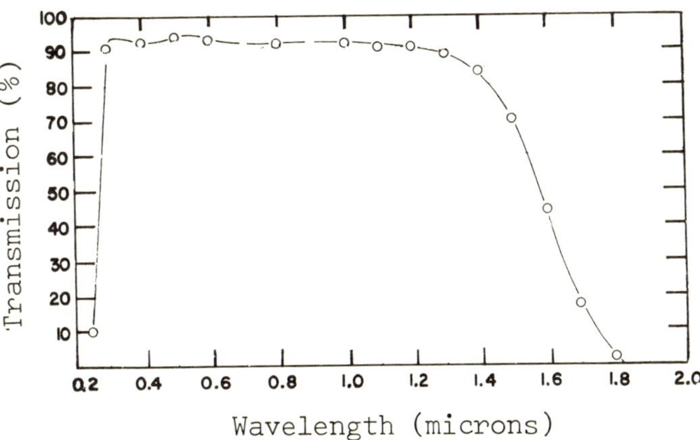


Miller has reported the following nonlinear coefficients for second harmonic generation in KDA, relative to $d_{36}^{2\omega}$ in $KH_2PO_4$ (KDP):

$$d_{36}^{2\omega} / d_{36}^{2\omega}\ (\mathrm{KDP}) = 1.06 \pm 0.06$$

at 1.06 microns

$$d_{14}^{2\omega} / d_{36}^{2\omega}\ (\mathrm{KDP}) = 1.12 \pm 0.05$$

## Electrooptic Properties

TABLE 2. ELECTROOPTIC PROPERTIES OF KDA

| Wavelength (Å) | $r_{63}^{T}$ ($10^{-12}$ m/V) | $r_{41}^{T}$ ($10^{-12}$ m/V) | $V_{1/2}$ (kV) | Reference |
|---|---|---|---|---|
| 5460 | 10.9 ± 0.1 | 12.5 ± 0.4 | 6.43 ± 0.06 | Ott and Sliker |
| 5500 | 10.9 | | 6.5 | Adhav |

Vasilevskaya and Sonin report the following values from measurements made at 5350 Å and 20°C: $r_{63}^{T} = 8.7 \times 10^{-12}$ m/V, $r_{63}^{S} = 8.3 \times 10^{-12}$ m/V. The elastooptic constant, $p_{66}$ is calculated to be 0.020.

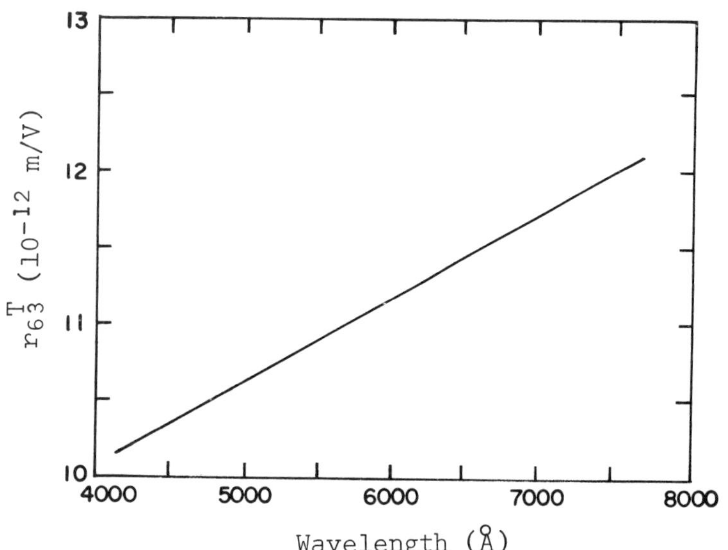


Fig. 3. Electrooptic coefficient $r^T_{63}$ for KDA as a function of wavelength [Adhav]

Ott and Sliker found that $r^T_{41}$ increases roughly 3% as the wavelength decreased from 5460 to 4360 Å. However, this result is tentative since the 3% increase corresponds approximately to the uncertainty in $r^T_{41}$ at a given wavelength.

Fig. 4. Half-wave voltage of KDA as a function of wavelength.

1. Ott and Sliker

2. Adhav

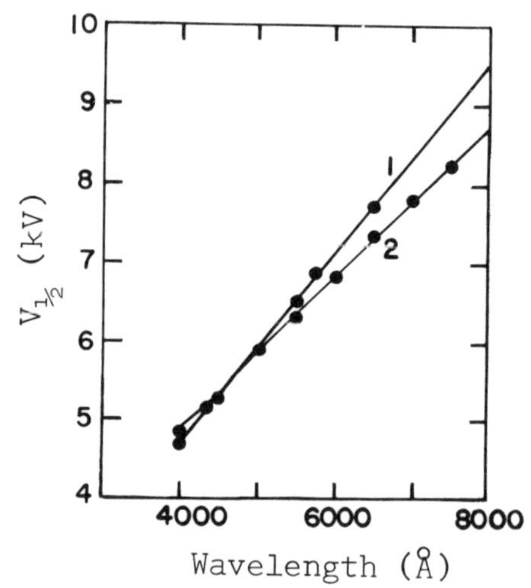

## POTASSIUM DIHYDROGEN ARSENATE (KDA)

Somewhat larger values of $V_{\frac{1}{2}}$ than shown in Figure 4 (6.1, 9.6 and 14.3 kV at 4000, 5400, and 7250 Å, respectively) are reported by Vasilevskaya et al.

### Photoelastic Properties

Vasilevskaya et al. report values of the piezooptic coefficient $\pi_{66}$ for KDA ranging from 2.5 to 3 x $10^{-12}$ $m^2/N$ in the wavelength range 4500 to 7000 Å.

TABLE 3. ELASTIC CONSTANTS OF KDA

| $c_{ij}$ ($10^{10}$ $N/m^2$) | | | | | | | |
|---|---|---|---|---|---|---|---|
| $c_{11}$ | $c_{33}$ | $c_{12}$ | $c_{13}$ | $c_{44}$ | $c_{66}$ | T(°C) | Reference |
| 5.3 | 3.7 | -0.6 | -0.2 | 1.2 | 0.7 | 25 | a |
| 6.482 | 4.824 | 0.077 | 1.358 | 1.075 | 0.663 | 20 | b |
| | | | | | 0.74 | | c |
| $(1/c_{ij})(dc_{ij}/dT)$ ($10^{-6}$/°C) | | | | | | | |
| -525 | -490 | -1200 | -180 | -280 | -530 | 0 | b |
| $s_{ij}$ ($10^{-12}$ $m^2/N$) | | | | | | | |
| $s_{11}$ | $s_{33}$ | $s_{12}$ | $s_{13}$ | $s_{44}$ | $s_{66}$ | | |
| 19 | 27 | 2 | 1 | 86 | 152 | 25 | a |
| 17.83 | 23.84 | 2.6 | -6.6 | 91.7 | 147.4 | 25 | d |
| | | | | 88.6 | 149 | | e |
| $(1/s_{ij})(ds_{ij}/dT)$ ($10^{-6}$/°C) | | | | | | | |
| 395 | 370 | 2100 | -670 | 280 | 510 | 20-80 | d |

(a) Officer, as cited by Landolt-Börnstein
(b) Haussühl
(c) Vasilevskaya et al.
(d) Adhav
(e) Mason

## Piezoelectric Properties

TABLE 4. PIEZOELECTRIC CONSTANTS OF KDA

| $10^{-12}$ C/N | | T(°C) | Reference |
|---|---|---|---|
| $d_{14}$ | $d_{36}$ | | |
| 26.6 | 22.4 | 0 | Niemiec |
| 25 | 22 | 25 | Van Dyke and Gordon as cited by Landolt-Börnstein |
| | 23 | | Vasilevskaya et al. |
| 23.5 | 22 | | Mason |
| 17.1 | 19.7 | 25 | Adhav |
| 15 | 17 | 80 | |

Other piezoelectric constants of KDA at 25°C are as follows [Van Dyke and Gordon] as cited by Landolt-Börnstein:

$$e_{14} = 0.29 \quad C/m^2$$
$$e_{36} = 0.14$$
$$g_{14} = 8.58 \times 10^{-2} \; m^2/C$$
$$g_{36} = 10.11$$
$$h_{14} = 9.9 \times 10^8 \; N/C$$
$$h_{36} = 6.9$$

Electromechanical coupling coefficients:

$k_{14} = 0.095$ Sliker

$k_{36} = 0.13$

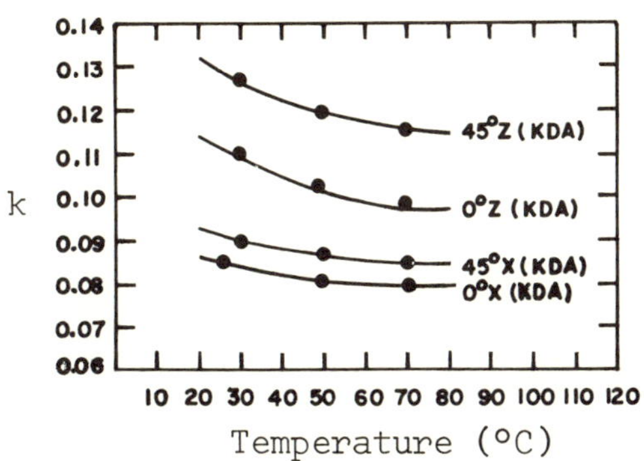


Fig. 5. Electromechanical coupling coefficients for KDA as a function of temperature [Adhav]

## Dielectric Properties

Dielectric constant [Sliker]

$\varepsilon_c^T = 21.0 \qquad \varepsilon_a^T = 53.7$

$\varepsilon_c^S = 20.6 \qquad \varepsilon_a^S = 53.0$

Here, $\varepsilon_c = \varepsilon_{33}/\varepsilon_o$ and $\varepsilon_a = \varepsilon_{11}/\varepsilon_o$, where $\varepsilon_o$ is the permittivity of free space.

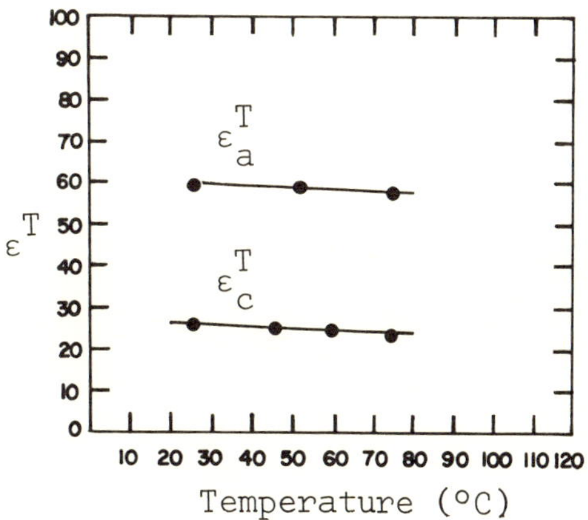


Fig. 6. Dielectric constant of KDA at 1 kHz as a function of temperature [Adhav]

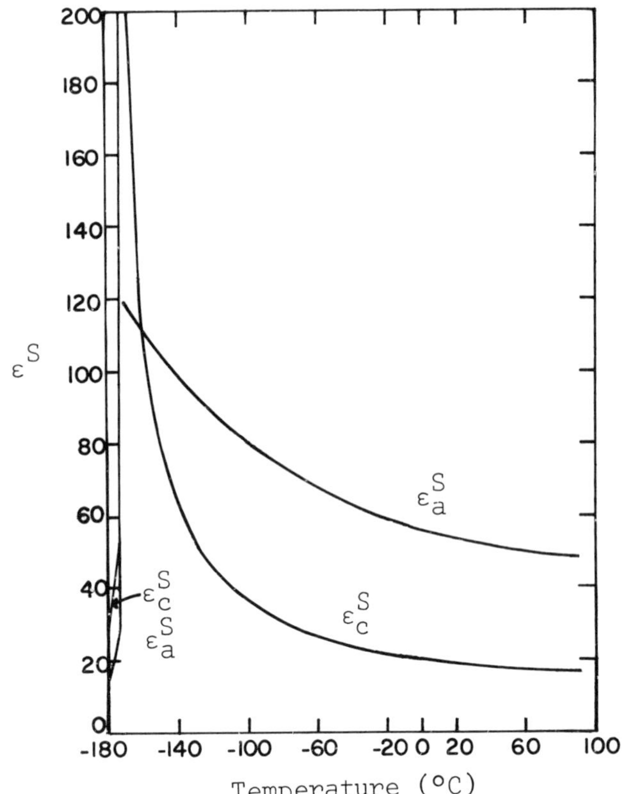


Fig. 7. Dielectric constant of KDA at 9.2 GHz as a function of temperature [Kaminow]



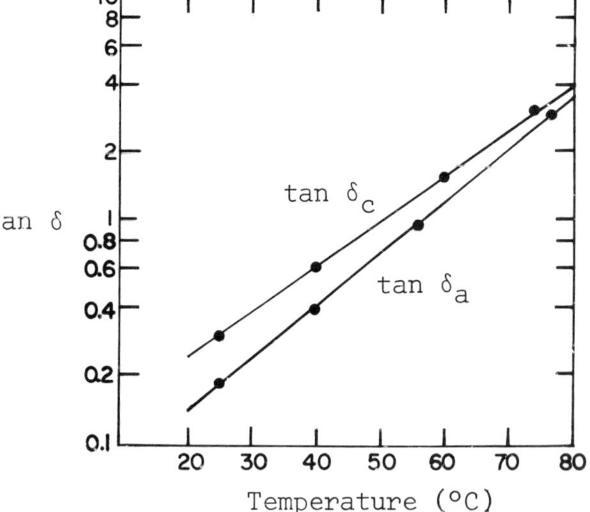


Fig. 8. Loss tangent of KDA at 1 kHz as a function of temperature [Adhav]

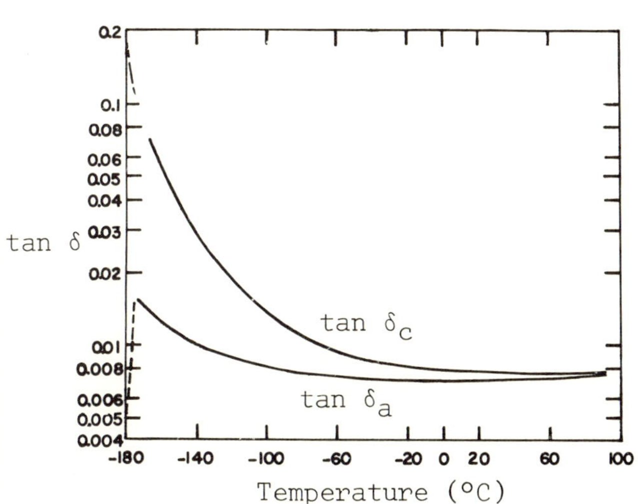


Fig. 9. Loss tangent of KDA at 9.2 GHz as a function of temperature [Kaminow]



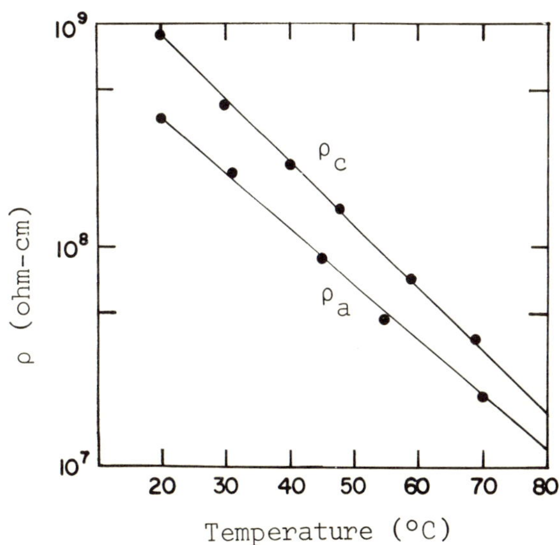


Fig. 10. Electrical resistivity of KDA as a function of temperature [Adhav]

## Ferroelectric Properties

Potassium dihydrogen arsenate belongs to the KDP-type ferroelectric class of materials.

| | |
|---|---|
| $T_c$ = 95.57 ± 0.05 °K | Stephenson and Zettlemoyer |
| $dT_c/dP$ = -3.3°K/kbar | Frenzel et al. from measurements to 1.2 kbar |

## Thermal Properties

| | | |
|---|---|---|
| Melting Point | 288°C | |
| Safe Operating Temperature | 100°C | Sliker |
| Thermal Conductivity | 8.1 mW/cm°K | Yap and Bicknell |

TABLE 5. THERMAL COEFFICIENT OF EXPANSION OF KDA

| Temperature Range (°C) | Expansion Coefficient ($10^{-6}$/°C) | | Reference |
|---|---|---|---|
| | along a-axis | along c-axis | |
| -50 to 50 | 24.2 | 47.1 | Cook |
| 27 to 147 | 24.9 | 50.0 | Deshpande and Khan |
| 20 | 23.4 | 48.5 | Haussühl |
| 0 to 100 | 21.5 | 46.5 | Adhav |

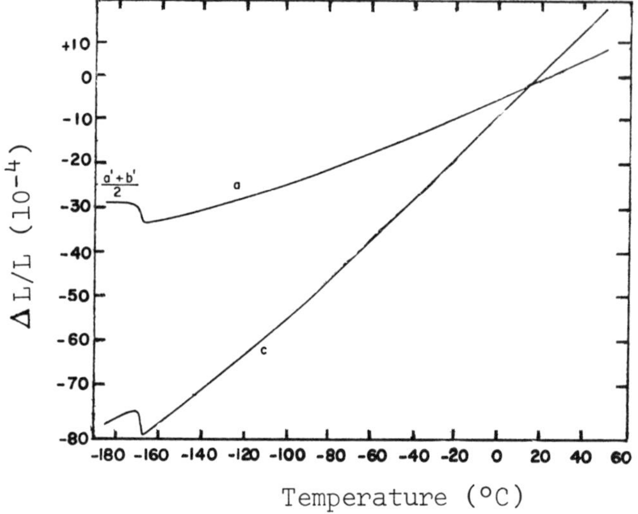


Fig. 11. Thermal expansion of KDA as a function of temperature [Cook]

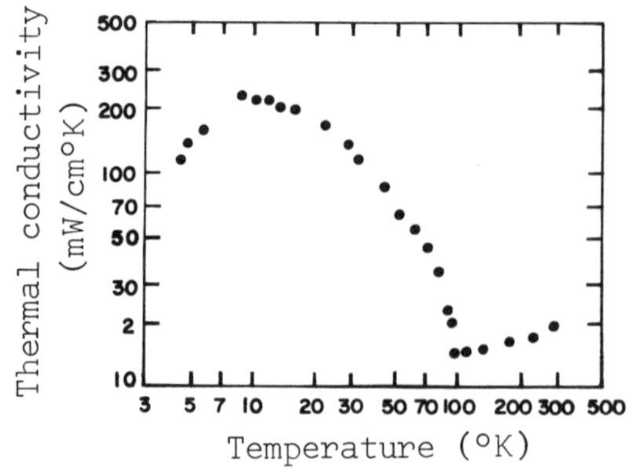


Fig. 12. Thermal conductivity of KDA as a function of temperature; the arrow indicates the Curie temperature [Suemune]

POTASSIUM DIHYDROGEN ARSENATE (KDA)


ADHAV, R.S. Some Physical Properties of Single Crystals of Normal and Deuterated Potassium Dihydrogen Arsenate. I. Piezoelectric and Elastic Properties. J. OF APPL. PHYS., v. 39, no. 9, Aug. 1968. p. 4091-4094.

ADHAV, R.S. Some Physical Properties of Single Crystals of Normal and Deuterated Potassium Dihydrogen Arsenate. II. Electro-Optic and Dielectric Properties. J. OF APPL. PHYS., v. 39, no. 9, Aug. 1968. p. 4095-4098.

ADHAV, R.S. Linear Electro-Optic Effects in Tetragonal Phosphates and Arsenates. OPTICAL SOC. OF AMERICA, J., v. 59, no. 4, Apr. 1969. p. 414-418.

SYLVANIA ELECTRONIC SYSTEMS. WALTHAM, MASS. Wideband Optical Modulation Techniques. by: BICKNELL, W.E. et al. Final Rept. F-5154-1. Contract No. AF33(615)-3118. Mar. 1967. 186 p. AD 810 458.

COOK, W.R., JR. Thermal Expansion of Crystals with $KH_2PO_4$ Structure. J. OF APPL. PHYS., v. 38, no. 4, Mar. 1967. p. 1637-1642.

DESHPANDE, V.T. and A.A. KHAN. X-Ray Determination of the Thermal Expansion of Potassium Dihydrogen Arsenate. ACTA CRYST., v. 18, pt. 5, 1965. p. 977-978.

DONNAY, J.D.H. (Ed.) Crystal Data. Determinative Tables. 2nd Ed. AMERICAN CRYSTALLOGRAPHIC ASSN. Apr. 1963. ACA Monograph no. 5.

FRENZEL, C. et al. The Influence of Hydrostatic Pressure on the Phase Transition of Ferroelectric Crystals of the $KH_2PO_4$-Type. PHYSICA STATUS SOLIDI, a, v. 2, no. 2, June 1970, p. 273-279.

HAUSSÜHL, S. Elastic and Thermoelastic Properties of $KH_2PO_4$, $KH_2AsO_4$, $NH_4PO_4$ and $RbH_2PO_4$. Z. FUER KRYSTALLOGRAPHIE, v. 120, 1964, p. 401-414.

KAMINOW, I.P. Microwave Dielectric Properties of $NH_4H_2PO_4$, $KH_2AsO_4$ and Partially Deuterated $KH_2PO_4$. PHYS. REV., v. 138, no. 5A, May 1965. p. A1539-A1543.

LANDOLT-BÖRNSTEIN. NUMERICAL DATA AND FUNCTIONAL RELATIONSHIPS IN SCIENCE AND TECHNOLOGY, New Series, v. 1, Elastic, Piezoelectric, Piezooptic and Electrooptic Constants of Crystals. Group III: Crystal and Solid State Physics, ed. by: HELLWEGE, K.-H and A.M. HELLWEGE. Berlin, Ger. Springer Verlag, 1966.

MASON, W.P. Elasto-Electric Constants of $KH_2PO_4$ Type Crystal. Table III. PHYSICAL ACOUSTICS, v. 1, Pt. A. Academic Press, N.Y. 1964. p. 181.

MILLER, R.C. Optical Second Harmonic Generation in Piezoelectric Crystals. APPLIED PHYS. LETTERS, v.5, no. 1, July 1964. p. 17-19.

NIEMIEC, T. A Method of Measurement of Small Periodic Displacements and Its Application to Determining the Piezoelectric Constants of Potassium Dihydrogen Arsenate. PHYS. REV., v. 75, no. 1, 1949. p. 215-216.

OTT, J.H. and T.R. SLIKER. Linear Electro-Optic Effects in $KH_2PO_4$ and its Isomorphs. OPTICAL SOC. OF AMERICA, J., v. 54, no. 12, Dec. 1964. p. 1442-1444.

CLEVITE CORP. Reference Data on Linear Electro-Optic Effects, by: SLIKER, T.R. Eng. Memo. 64-10. May 15, 1964. 9 p.

## POTASSIUM DIHYDROGEN ARSENATE (KDA)

STEPHENSON, C.C. and A.C. ZETTLEMOYER. The Heat Capacity of $KH_2AsO_4$ from 15 to 300°K. The Anomaly at the Curie Temperature. AMERICAN CHEM. SOC., J., v. 66, no. 9, 1944. p. 1402-1405.

SUEMUNE, Y. Thermal Conductivity of the $KH_2PO_4$ Type. PHYS. SOC. OF JAPAN, J., v. 21, no. 4, Apr. 1966. p. 802.

VASILEVSKAYA, A.S. and A.S. SONIN. The Relationship of Structure to Electrooptic and Elastooptic Properties in Crystals of the KDP Group. SOVIET PHYS. CRYST., v. 14, no. 4, Jan. 1970. p. 611-613.

VASILEVSKAYA, A.S. et al. Electrooptical and Elastooptical Properties of Alkali Metal Dihydrogen Arsenates. SOVIET PHYS. SOLID STATE, v. 9, no. 4, Oct. 1967. p. 986-987.

SLYVANIA ELECTRONIC SYSTEMS. Solid State Techniques for Modulating Optical Waves. Final Report, by: YAP, B.K. and W.E. BICKNELL. Tech. Report ECOM-01283-F, Oct. 1966. AD 803 162.

# POTASSIUM DIHYDROGEN PHOSPHATE (KDP)

## Introduction

Potassium dihydrogen phosphate is one of the most widely known and explored electrooptic crystals. Crystals are grown at room temperature from a water solution; excellent crystals as large as 5 cm in any dimension can be obtained commercially. Although the crystals are water soluble and fragile, they can be handled, cut, and polished without difficulty. These crystals are transparent throughout the visible and ultra-violet. Applications of KDP-type crystals was pioneered by the Brush Development Co. (now part of Clevite) and Baird Associates. Since the development of the laser in 1960, KDP has been the most widely used electrooptic material; large samples of the required optical quality have been available and the electrooptic effect is quite large.

## Chemical and Physical Properties

| | | |
|---|---|---|
| Chemical Formula | $KH_2PO_4$ | |
| Molecular Weight | 136.09 | |
| Density | 2.338 $g/cm^3$ | Bystrova and Federov, Mason |
| Solubility in water, at 25°C | 33 g/100g water | |
| Solubility in alcohol | Insoluble | |

KDP is very hygroscopic which can present problems of water on the exposed surface causing arcing or breakdown under high voltage.

## POTASSIUM DIHYDROGEN PHOSPHATE (KDP)

### Crystallography

Cook has described the equipment for growing KDP crystals at the Clevite Corp. and some problems associated with the aqueous growth method. The KDP crystals for electrooptic applications must be grown from high purity materials in exacting laboratory conditions. Close control of growth variables is necessary to achieve crystals free of strain and of the highest optical quality.

Above 123°K, potassium dihydrogen phosphate (KDP) belongs to the tetragonal group $\bar{4}2m$ ($V_d$), which lacks a center of inversion, and exhibits a linear electrooptic effect. It undergoes a ferroelectric phase transition at 123°K. In the paraelectric phase (above 123°K), KDP is optically uniaxial with the optic axis along the tetragonal z axis (Figure 1). At $T_c$, the symmetry is lowered to orthorhombic by the appearance of a spontaneous strain $X_y$. The orthorhombic axes, a and b, are in the x,y plane but are at 45° to x and y; c is parallel to z, Reese and May. Figure 2 shows crystal structure of KDP below 123°K.

A characteristic feature of the crystal structure of KDP consists in the existence of short hydrogen bonds oriented almost perpendicularly to the c-axis and connecting two $PO_4$ tetrahedra, Känzig. X-ray diffraction data have firmly established that in the ferroelectric phase, $K^+$, $P^{+5}$ and $O^{-2}$ ions are displaced along the c-axis relative to their symmetric positions in the paraelectric phase; these displacements explain the observed magnitude of the saturated polarization satisfactorily. It has been well established that the ferroelectric phase transition in $KH_2PO_4$ crystals is triggered by a cooperative ordering in the proton arrangements and the spontaneous polarization appears by the finite displacements of K, P, and O ions along the c-axis, Kobayashi.

# POTASSIUM DIHYDROGEN PHOSPHATE (KDP)

Crystal Structure

| | | |
|---|---|---|
| Above $T_c$ | Tetragonal | $\bar{4}2m$ or $D_{2d}$ |
| Below $T_c$ | Orthorhombic | mm2 or $C_{2v}$ |

Lattice Constants

| T(°K) | $a_o$(Å) | $c_o$(Å) | Reference |
|---|---|---|---|
| 123 | 7.4256±0.0005 | 6.9296±0.0009 | Cook |
| 298 | 7.4529±0.0002 | 6.9751±0.0006 | Cook, Sliker and Burlage |
| | 7.4528±0.0004 | 6.9683±0.0004 | Sirdeshmukh and Deshpande |

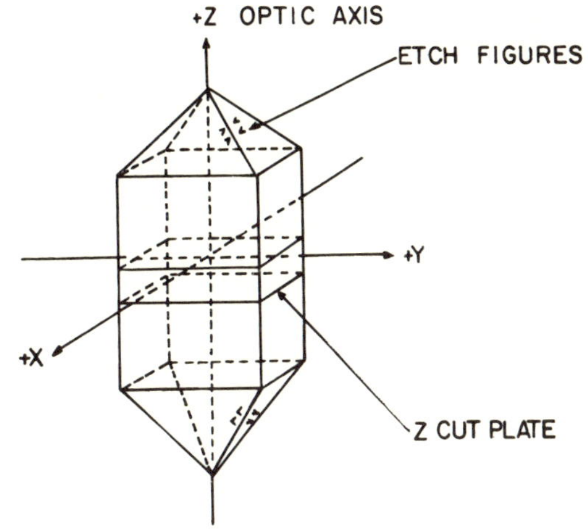


Fig. 1. A KDP tetragonal crystal with crystallograhic axes labelled.

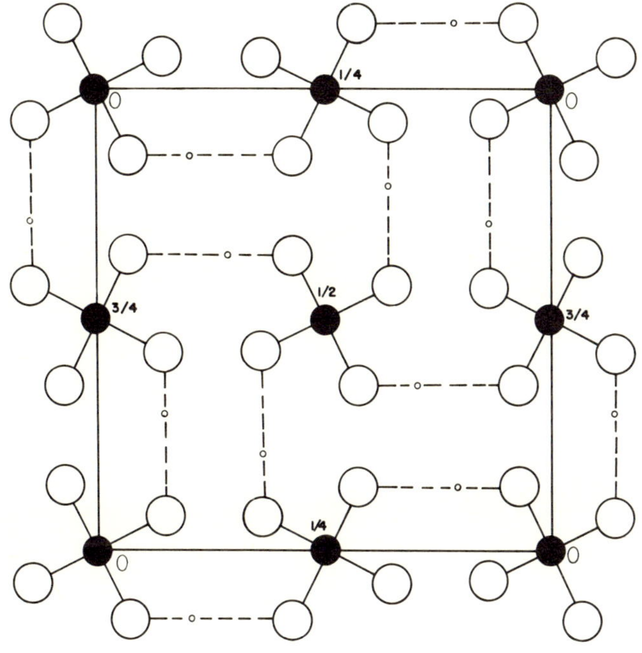


Fig. 2. The structure of $KH_2PO_4$ below 123°K, projected in the a-b plane. The heights of the P atoms are marked on the projection as fractions of the unit cell heights. The K atoms are displaced by about 1/2 from the P atoms [Cook].

● Potassium & Phosphorus

○ Oxygen

∘ Hydrogen (in low temperature structure)

POTASSIUM DIHYDROGEN PHOSPHATE (KDP)

## Optical Properties - Refractive Index

Zernicke has made the most extensive and accurate refractive index measurements of KDP from 2138 to 15290 Å. His data are summarized in Table 1 and Figure 3. Various refractive index measurements made by other investigators are tabulated in Table 2.

TABLE 1. REFRACTIVE INDICES OF KDP AT 24.8°C [Zernicke].

| Wavelength (μ) | Index in air | | Absolute index corrected to vacuum | |
|---|---|---|---|---|
| | Ordinary ray | Extraordinary ray | Ordinary ray | Extraordinary ray |
| 0.2000 | 1.621996 | 1.563315 | 1.622630 | 1.563913 |
| 0.3000 | 1.545084 | 1.497691 | 1.545570 | 1.498153 |
| 0.4000 | 1.524035 | 1.479814 | 1.524481 | 1.480244 |
| 0.5000 | 1.514498 | 1.472068 | 1.514928 | 1.472486 |
| 0.6000 | 1.508851 | 1.467856 | 1.509274 | 1.468267 |
| 0.7000 | 1.504817 | 1.465193 | 1.505235 | 1.465601 |
| 0.8000 | 1.501508 | 1.463303 | 1.501924 | 1.463708 |
| 0.9000 | 1.498514 | 1.461830 | 1.498930 | 1.462234 |
| 1.0000 | 1.495628 | 1.460590 | 1.496044 | 1.460993 |
| 1.1000 | 1.492730 | 1.459481 | 1.493147 | 1.459884 |
| 1.2000 | 1.489751 | 1.458443 | 1.490169 | 1.458845 |
| 1.3000 | 1.486645 | 1.457436 | 1.487064 | 1.457838 |
| 1.4000 | 1.483381 | 1.456437 | 1.483803 | 1.456838 |
| 1.5000 | 1.479938 | 1.455427 | 1.480363 | 1.455829 |
| 1.6000 | 1.476302 | 1.454395 | 1.476729 | 1.454797 |
| 1.7000 | 1.472459 | 1.453333 | 1.472890 | 1.453735 |
| 1.8000 | 1.468400 | 1.452234 | 1.468834 | 1.452636 |
| 1.9000 | 1.464118 | 1.451093 | 1.464555 | 1.451495 |
| 2.0000 | 1.459603 | 1.449906 | 1.460044 | 1.450308 |

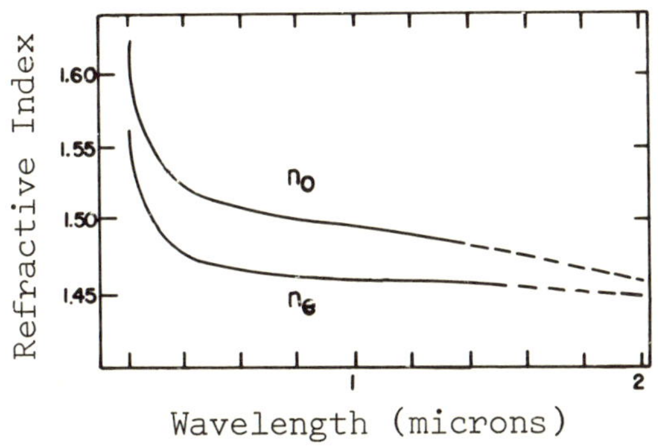


Fig. 3. Refractive indices of KDP as a function of wavelength at 24.8°C. Curves are dashed in those spectral regions where no actual measurements were made [Zernicke].

TABLE 2. REFRACTIVE INDICES OF KDP

| Temperature (°K) | Wavelength (Å) | $n_o$ | $n_e$ | Reference |
|---|---|---|---|---|
| 300 | 4358 | 1.5202 | 1.4770 | Yamazaki and Ogawa |
| --- | 4880 | 1.5158 | 1.4727 | Pyle |
| 300 | 5461 | 1.5120 | 1.4702 | Yamazaki and Ogawa |
| --- | 5500 | 1.5114 | 1.4696 | Pyle |
| 295 | 5560 | 1.5100 | 1.4684 | Carpenter |
| 300 | 5893 | 1.5100 | 1.4682 | Yamazaki and Ogawa |
| --- | 5893 | 1.5095 | 1.4684 | Calucci |
| --- | 6330 | 1.5073 | 1.4668 | Pyle |
| 296 | 6563 | 1.506 | ------ | Union Carbide |

The variation of the refractive indices of KDP with temperature is detailed in Figures 4 and 5 and Tables 3 and 4.

TABLE 3. REFRACTIVE INDICES OF KDP AT VARIOUS TEMPERATURES [Vishnevskii and Stefanskii].

| | λ = 5770 Å | | λ = 5460 Å | |
|---|---|---|---|---|
| T(°C) | $n_o$ | $n_e$ | $n_o$ | $n_e$ |
| -196 | 1.5179 | 1.4747 | 1.5198 | 1.4760 |
| -50 | 1.5128 | 1.4709 | 1.5147 | 1.4723 |
| +20 | 1.5102 | 1.4692 | 1.5120 | 1.4706 |
| +60 | 1.5090 | 1.4681 | 1.5108 | 1.4696 |
| +100 | 1.5072 | 1.4671 | 1.5090 | 1.4686 |
| +150 | 1.5053 | 1.4659 | 1.5070 | 1.4673 |
| +200 | 1.5034 | 1.4647 | 1.5053 | 1.4661 |

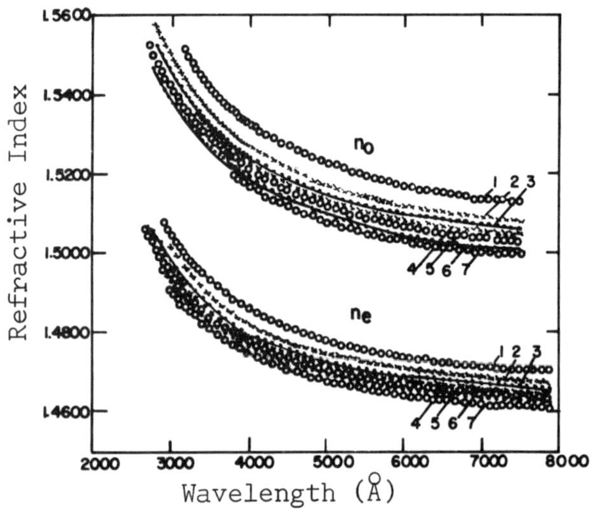


Fig. 4. Refractive indices of KDP as a function of wavelength at various temperatures [Vishnevskii and Stefanskii].

| Curve | T(°C) |
|---|---|
| 1 | -196 |
| 2 | -50 |
| 3 | +20 |
| 4 | +64 |
| 5 | +100 |
| 6 | +150 |
| 7 | +200 |

TABLE 4. CHANGE IN THE REFRACTIVE INDICES OF KDP WITH TEMPERATURE [Phillips].

| Wavelength (Å) | Index at 298°K | | Increase at 201°K from 298°K | | Increase at 154°K from 298°K | |
|---|---|---|---|---|---|---|
| | $n_o$ | $n_e$ | $n_o$ | $n_e$ | $n_o$ | $n_e$ |
| 6907 | 1.5052 | 1.4655 | 0.0033 | 0.0022 | 0.0047 | 0.0034 |
| 6234 | 1.5079 | 1.4672 | 0.0033 | 0.0022 | 0.0048 | 0.0033 |
| 5791 | 1.5099 | 1.4686 | 0.0034 | 0.0023 | 0.0048 | 0.0033 |
| 5461 | 1.5117 | 1.4700 | 0.0034 | 0.0023 | 0.0049 | 0.0033 |
| 4916 | 1.5152 | 1.4727 | 0.0034 | 0.0023 | 0.0049 | 0.0033 |
| 4358 | 1.5200 | 1.4766 | 0.0034 | 0.0023 | 0.0048 | 0.0034 |
| 4078 | 1.5232 | 1.4792 | 0.0034 | 0.0023 | 0.0050 | 0.0034 |
| 4047 | 1.5235 | 1.4795 | 0.0035 | 0.0023 | 0.0051 | 0.0033 |
| 3653 | 1.5292 | 1.4843 | 0.0037 | 0.0024 | ... | ... |

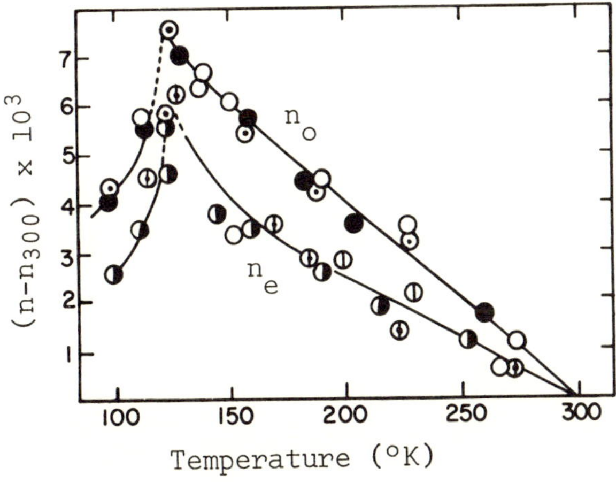


Fig. 5. Change in the refractive indices of KDP as a function of temperature [Yamazaki and Ogawa].

$n_o$ { ⊙ 4358Å, ○ 5461Å, ● 5893Å }

$n_e$ { 4358Å, 5461Å, ◑ 5893Å }

## Optical Properties - Birefringence

The value of the thermal birefringence coefficient for KDP, as measured by Phillips is in good agreement with the earlier results of Zwicker and Scherrer . However, later work by Dowley, using parametric fluorescence at 0.5μ yields a rather higher value of $d(n_e-n_o)/dT = 1.745 \times 10^{-5}/°C$.

$\partial(n_o-n_e)/\partial T$ at 5461 Å:

| | |
|---|---|
| $-(1.1 \pm 0.1) \times 10^{-5}/°C$ | Phillips |
| $-1.20 \times 10^{-5}/°C$ | Zwicker and Scherrer |

The optical behavior of KDP above $T_c$ (to 300°K) and as a single domain below $T_c$ (to 80°K) has been carefully examined at 5461 Å by Zwicker and Scherrer. Above $T_c$, it is uniaxial but shows a large electric field-dependent birefringence which is linear with the polarization (or strain) induced by the field. Below $T_c$, the crystal is biaxial and has a large "spontaneous" birefringence which has the same dependence on the polarization or strain. When Zwicker and Scherrer removed the dc bias field below $T_c$, the linear "spontaneous" birefringence disappeared, presumably due to the formation of equal volumes of oppositely polarized domains such that linear effects would cancel, Hill and Ichiki. The dependence of the spontaneous change in the refractive indices on the spontaneous polarization is shown in Figure 6. Birefringence-polarization relationships have been derived by Cook.

Kaminow has observed strain birefringence, due to heating, in a KDP rod used as a cavity-type microwave intensity modulator. When 1.6 W is dissipated in a cylindrical rod of any length and diameter, a deviation of the phase retardation of ±2.9 rad and a reduction in the degree of intensity modulation by a factor or 1/2 is observed.

Shamburov and Kucherova have investigated anomalous birefringence (non-zero birefringence along the optic axis) in KDP produced by optical inhomogenieties introduced during crystal growth.

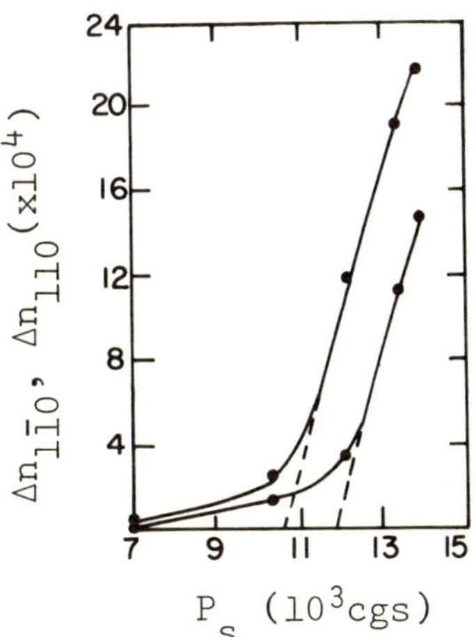


[Blokh and Lutsiv-Shumskii]

Fig. 6. The spontaneous change of the refractive index in two directions as a function of the spontaneous polarization at 6000 Å and just below the Curie point (120°K).

$$\Delta n_{1\bar{1}0} = n_{1\bar{1}0} - n_o$$
$$\Delta n_{110} = n_{110} - n_o$$

The slope of the rectilinear sections of the curves has the value $n_o^3 r_{63}/2$. From this relationship, the electrooptic coefficient is found to be $r_{63} = (19 \pm 2) \times 10^{-8}$ cgs.

## Optical Properties - Transmission

Potassium dihydrogen phosphate is transparent in the range 0.4 to 1.3 microns (Figures 7-9). KDP is quite opaque in the infrared region; the absorption coefficient is approximately 3000 $cm^{-1}$ above 3 microns. Hill and Ichiki have studied the infrared absorption by hydrogen bonds in KDP from 3000 to 1000 $cm^{-1}$ over a temperature range 20-300°K (Figure 10). Three broad absorptions are found at 2670, 2360 and 1720 $cm^{-1}$, plus a sharper absorption at 1850 $cm^{-1}$ appearing below the Curie temperature.

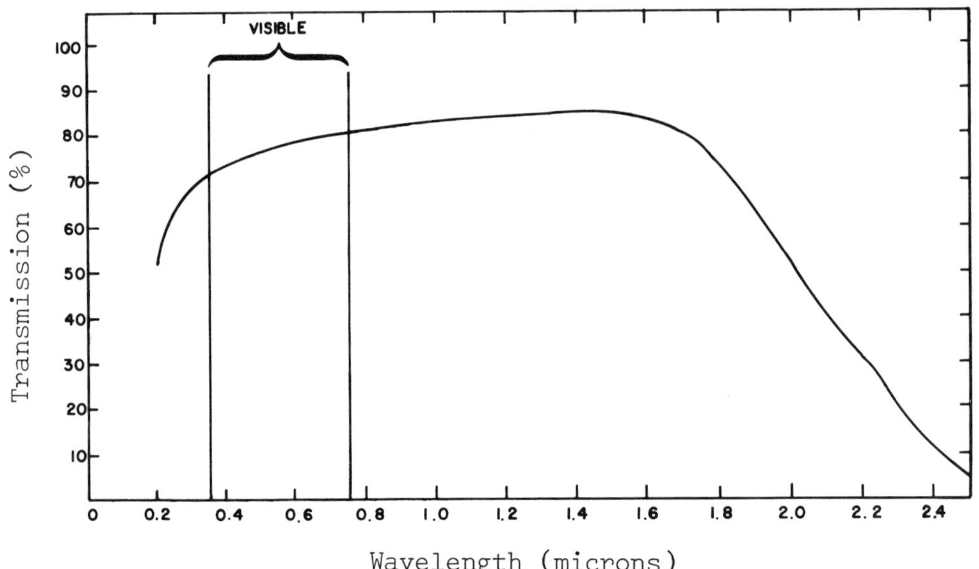


Fig. 7. Transmission as a function of wavelength for KDP sample 0.022 in. thick. Transmission is uncorrected for reflection losses [Reitz et al.].

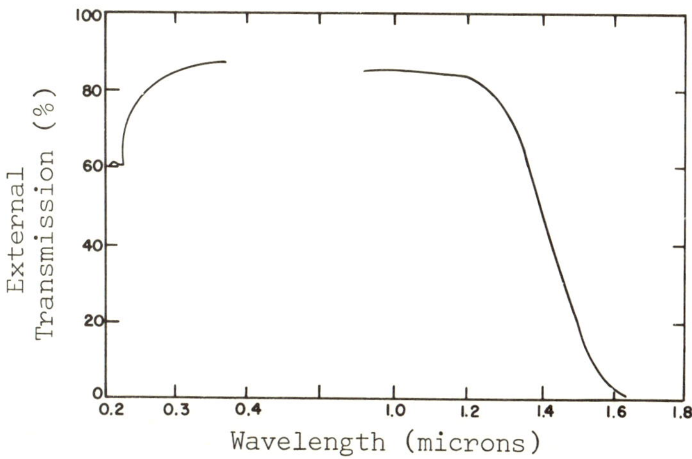


Fig. 8. Transmission as a function of wavelength for KDP samples 0.06 in. thick on the short wavelength portion and 0.5 in. thick on the long wavelength side [Valpey Corp.].

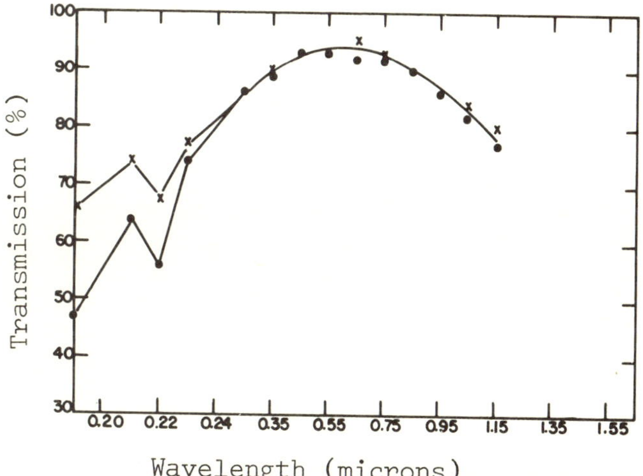


Fig. 9. Transmission as a function of wavelength for a bare polished crystal of KDP 1/2 in. thick (x) and 3/4 in. thick (•). Transmission is uncorrected for reflection losses [Baird Atomic].

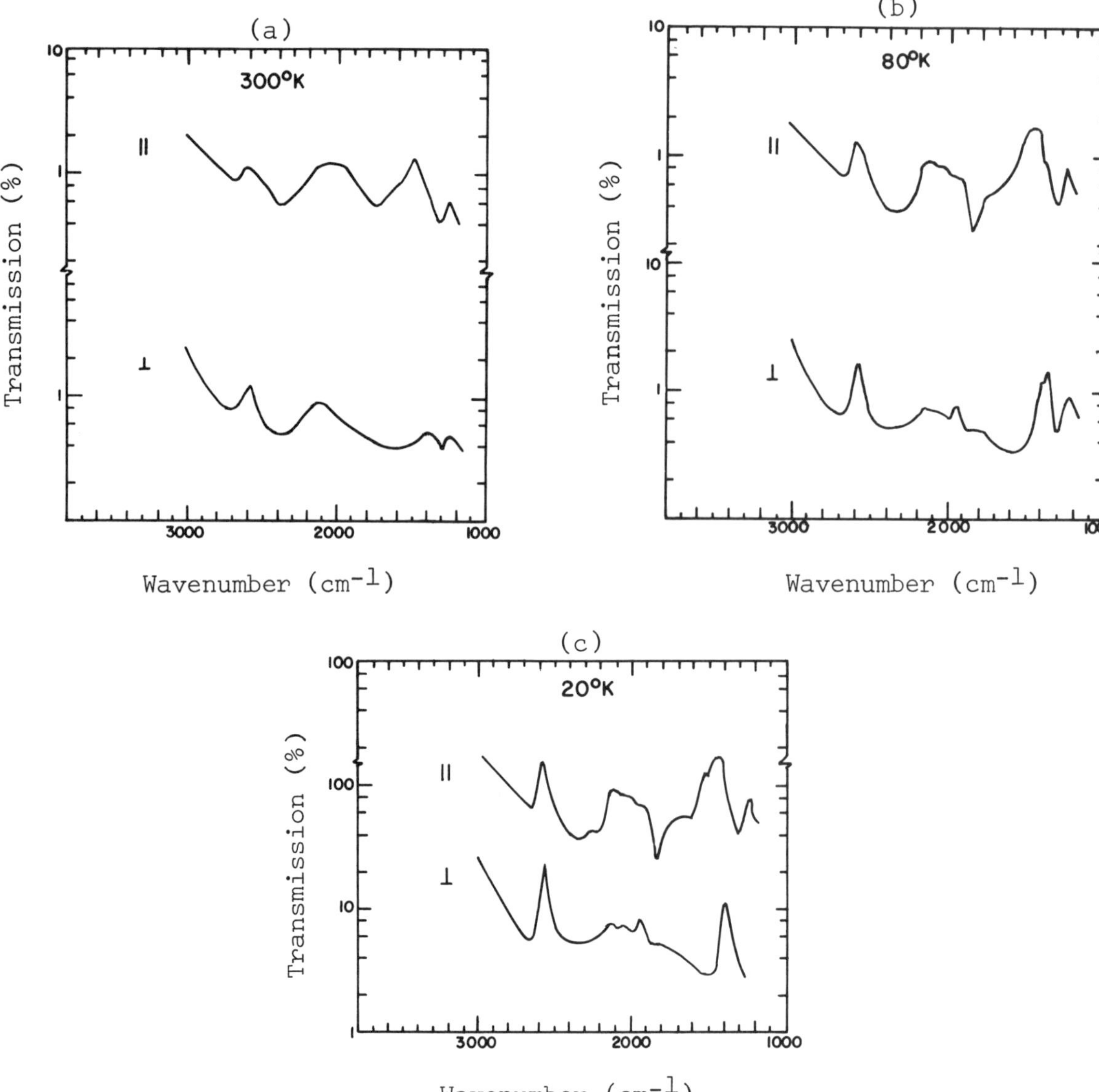


Fig. 10. Transmission as a function of wavenumber for single crystal KDP at (a) 300°K, (b), 80°K, and (c) 20°K for light polarized perpendicular (10μ thick z- cut crystal) and parallel (50μ thick x-cut crystal) to the z-axis [Hill and Ichiki].

## Optical Properties - Nonlinear Optical Behavior

Ward and Franken have reviewed and interpreted the data on nonlinear optical phenomenon in KDP and shown that the second harmonic effect is dominated by energy levels in the ultraviolet ("electronic" levels) whereas the dc and the linear electro-optic effects may have contributions due to processes simultaneously dependent on ultraviolet and infrared ("ionic") levels.

The following nonlinear second order polarization coefficients are non-zero in KDP:

paraelectric phase, above $T_c$: $d_{14}^{2\omega}$, $d_{25}^{2\omega} = d_{14}^{2\omega}$, $d_{36}^{2\omega}$

ferroelectric phase, below $T_c$: $d_{15}^{2\omega}$, $d_{24}^{2\omega}$, $d_{31}^{2\omega}$, $d_{33}^{2\omega}$

Maker et al. have reported that the passage of ruby laser light (6943 Å) through KDP crystals produces second harmonic generation (SHG) blue light of frequency 3470 Å. Ashkin et al. observed nonlinear SHG on a continuous basis in KDP using the 1.1526-micron transition of the helium-neon laser, and found $d_{36}^{2\omega} = (3 \pm 1) \times 10^{-9}$ esu. Since SHG coefficients often are given relative to the coefficient $d_{36}^{2\omega}$ in KDP, the absolute value of this quantity is particularly important. Robinson has pointed out that the above value for $d_{36}^{2\omega}$ is now believed to be too large. Relative measurements indicate that $d_{36}^{2\omega}$ is nearly identical in KDP and $NH_4H_2PO_4$ (ADP); the most recent absolute measurements on ADP give:

[Bjorkholm and Siegman] $1.38 \pm 0.22 \times 10^{-9}$ esu at 6328 Å

Jerphagnon and Kurtz have recently used Maker fringes to determine $d_{36}$ in KDP and ADP; they report the following relation: $d_{36}(\text{ADP}) = (1.20 \pm 0.09) d_{36}(\text{KDP})$ at 1.064μ

Miller et al. report the following relative values for $d_{14}^{2\omega}$ in KDP:

$$d_{14}^{2\omega}/d_{36}^{2\omega} = 0.95 \pm 0.06 \text{ at } 0.694 \text{ microns}$$

$$d_{14}^{2\omega}/d_{36}^{2\omega} \text{ (KDP)} = 1.01 \pm 0.05 \text{ at } 1.06 \text{ microns}$$

Van der Ziel and Bloembergen measured optical SHG from KDP as a function of temperature above and below the ferroelectric transition. Their results are presented in Figure 11. Although signals due to $d_{33}^{2\omega}$ were in the noise level, they determined that $d_{33}^{2\omega}$ at 77°K was less than 0.03 times the value of $d_{31}^{2\omega}$ at 300°K for KDP.

In 1962, a dc polarization accompanying the passage of an intense ruby laser beam through KDP was observed by Bass et al. They measured the optical rectification coefficient

$$d^{o}_{36} = |X^{o}_{zxy} + X^{o}_{zyx}| = |2X^{o}_{zxy}| = 5 \times 10^{-8} \text{ e.s.u.}$$

with an estimated accuracy of a factor of 3. Using the intimate relationship between optical rectification and the linear electrooptic effect Armstrong, the value $|d^{o}_{36}| = 6.0 \times 10^{-8}$ e.s.u. was predicted by Bass et al., using the electrooptic data of Carpenter. A comparison of the temperature dependence of the optical rectification effect with that of the linear electrooptic effect is shown in Figure 12. The temperature dependence of optical rectification in KDP is shown in more detail in Figure 13.

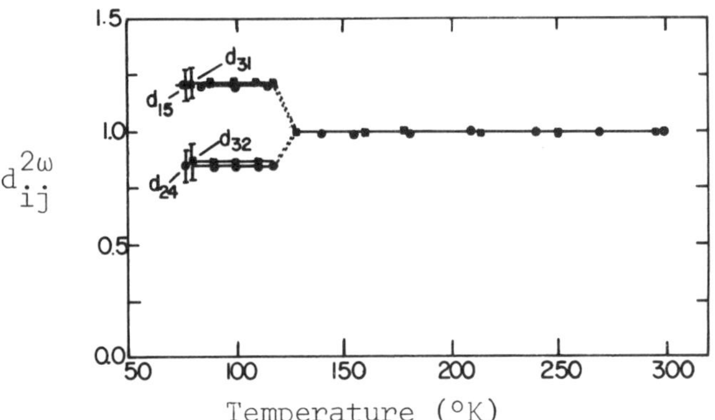


Fig. 11. Relative values of the nonlinear second harmonic generation coefficients in KDP as a function of temperature. The vertical lines at 77°K indicate the root-mean-square deviation from the average [Van der Ziel and Bloembergen].



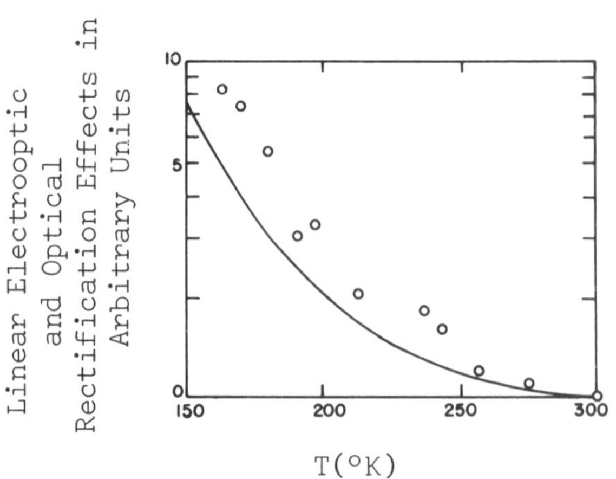


Fig. 12. The optical rectification effect (open circles) and the linear electrooptic effect (solid curve) for KDP as a function of temperature. The electrooptic data are taken from Zwicker and Scherrer, and both sets of data are normalized to 300°K [Bass et al.].

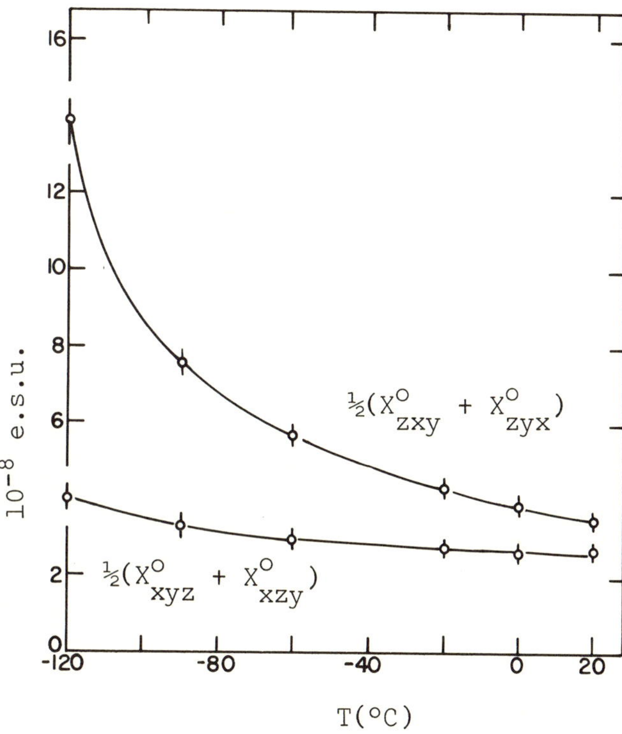


Fig. 13. Optical rectification coefficients in KDP as a function of temperature [Ward and New].

| T(°C) | $\frac{1}{2}(X^{o}_{xyz}+X^{o}_{xzy})$ ($10^{-8}$e.s.u.) | $\frac{1}{2}(X^{o}_{zxy}+X^{o}_{zyx})$ ($10^{-8}$e.s.u.) |
|---|---|---|
| 20 | 2.69 | 3.49 |
| 0 | 2.64 | 3.87 |
| -20 | 2.74 | 4.33 |
| -60 | 3.01 | 5.72 |
| -90 | 3.35 | 7.61 |
| -120 | 4.06 | 13.9 |

## Electrooptic Properties

Carpenter investigated the linear electrooptic effect of potassium dihydrogen phosphate, utilizing conventional light sources. The magnitude of this phenomena was found to be the largest observed with any crystal; the electrooptic coefficient for KDP is more than one and a half times the coefficient for ADP. KDP possesses only two independent electrooptic coefficients: $r_{63}$ and $r_{41} = r_{52}$.

In 1944, Zwicker and Scherrer reported the dc electrooptic properties of $KH_2PO_4$ and $KD_2PO_4$ and related these properties to the ferroelectric behavior of these crystals. They observed that the electrooptic coefficient based on the electric field was proportional to the dielectric constant, exhibiting a Curie-Weiss behavior as a function of temperature. The electrooptic coefficient based on the dielectric polarization was found to be the same temperature-independent constant for both KDP and KDDP. Sonin et al. in 1967, studied the electrooptic properties of KDP single crystals near the Curie point (123°K).

In the paraelectric phase (above 123°K), KDP possesses the non-centrosymmetric point group $\bar{4}2m$ and, consequently, it exhibits both linear and quadratic electrooptic effects at room temperature. The quadratic electrooptic effect can be readily distinguished in the static regime from the linear effect by appropriate orientation of the applied electric field and the direction of propagation of the incident light. Induced birefringence, due only to the quadratic effect, appears in cases when the electric field and the light have the following directions; Perfilova and Sonin

| Direction of the field | Direction of the light |
|---|---|
| <100> | <001> |
| <100> | <010> |
| <110> | <001> |
| <001> | <010> |

The electrooptic coefficients of KDP at room temperature are listed in Table 5. The dc electrooptic coefficient $r_{63}^T$, corresponding to a free crystal, is related to the electrooptic coefficient $r_{63}^S$, measured at high modulating frequencies, corresponding to a "clamped" crystal, by

$$r_{63}^T = r_{63}^S + p_{66}\, d_{36} ,$$

where $p_{66}$ is an elastooptic coefficient and $d_{36}$ is a piezoelectric constant. This relationship is illustrated in Figure 14, where $r_{63}$ is shown as a function of modulatin frequency. From Table 7 and 10, respectively, $p_{66}$ = 0.058 and $d_{36}$ = 23.2 x $10^{-12}$m/V at 20°C. Thus,

$$r_{63}^{T} - r_{63}^{S} = 1.3 \times 10^{-12}\text{m/V} \quad ,$$

in good agreement with the data given in Table 5. These data indicate that $r_{63}^{S}/r_{63}^{T} \simeq 0.9$.

TABLE 5. ELECTROOPTIC COEFFICIENTS OF KDP AT ROOM TEMPERATURE

| Electrooptic Coefficient ($10^{-12}$m/V) | | | λ (Å) | Reference |
|---|---|---|---|---|
| $r_{41}^{T}$ | $r_{63}^{T}$ | $r_{63}^{S}$ | | |
| 8.6 | 10.5 | | 5560 | Carpenter |
| 8.77 | 10.3 | | 5460 | Ott and Sliker |
| | 10.2 | | 5460 | Pursey et al. |
| | | 8.5 | 5461 | Pisarevskii et al. |
| | | 9.5 | | Myers quoting unpublished data of Cook and Jaffe of Clevite Corp. |
| | 10.5 | 9.6 | 5461 | Calculated using half-wave voltage data of Sterzer et al. (Table 6) and refractive index data of Yamasaki and Ogawa (Table 2). |
| | | 8.8 | 6330 | Rosner et al. |

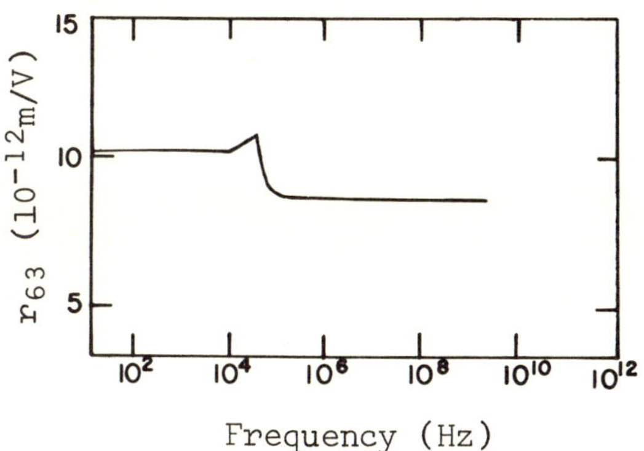


Fig. 14. Electrooptic coefficient $r_{63}$ of KDP as a function of modulating frequency Pisarevskii et al. These data agree with the results of Carpenter.

Blokh, Pursey et al. and Vasilevskaya find that $r_{63}^T$ and the ratio $r_{63}^S/r_{63}^T$ are practically independent of wavelength in the visible region of the spectrum (4000-7000 Å). The dispersion of $r_{63}^T$ over a wider wavelength range is shown in Figure 15.

In the longitudinal configuration, with the incident light parallel to the electric field, values of $r_{63}$ may be calculated from the relation

$$r_{63} = \lambda/2\, V_{\frac{1}{2}}\, n_o^{\;3} \quad ,$$

where $\lambda$ is the wavelength, $V_{\frac{1}{2}}$ is the half-wave voltage and $n_o$ is the ordinary refractive index. Experimental values of $V_{\frac{1}{2}}$ are given in Table 6 and are plotted in Figure 16. The half-wave voltage is linear in the region 4000-7000 Å; in this region, the dispersion of $r_{63}$ is due only to the dispersion of the refractive index $n_o$. The dependence of $V_{\frac{1}{2}}$ on $\lambda$ departs from the linear in the regions of increased absorption (see Figure 8), corresponding to the increase in $r_{63}^T$, illustrated in Figure 15.

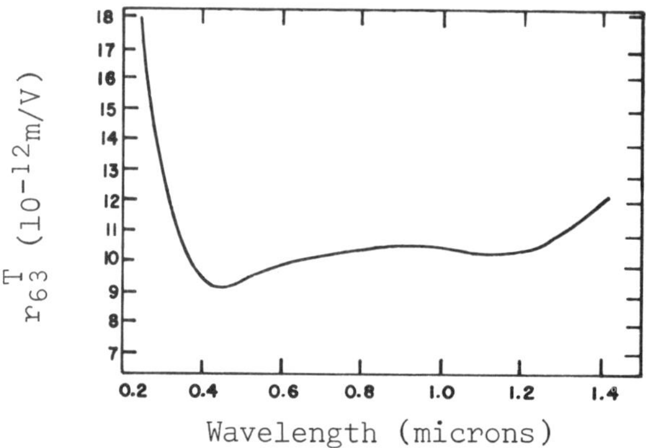


Fig. 15. Electrooptic coefficient of KDP as a function of wavelength [Blokh and Lutsiv-Shumskii].

Blokh et al. have since measured the temperature and wavelength dependence of $r_{41}$ at temperatures of 150 to 300°K and wavelengths of 0.3 to 0.7 microns.

TABLE 6. LONGITUDINAL HALF-WAVE VOLTAGE OF UNCLAMPED KDP AT ROOM TEMPERATURE

| $V_{1/2}$ (kV) | λ (Å) | Reference |
|---|---|---|
| 7.7 | 5560 | Carpenter |
| 7.5* | 5461 | Sterzer et al. |
| 6.04 | 4360 | Ott and Sliker |
| 7.65 | 5460 | |
| 8.17 | 5780 | |
| 7.4 | 4880 | Pyle |
| 8.3 | 5500 | |
| 9.6 | 6330 | |

*Sterzer et al. report a value of $V_{1/2}$ = 8.2 kV for a clamped crystal; they also report transverse half-wave voltage data.

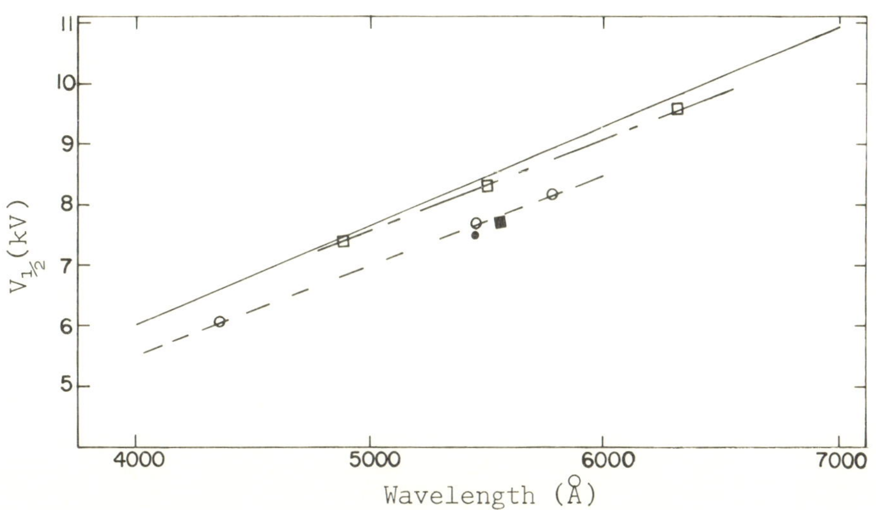


Fig. 16. Longitudinal half-wave voltage of KDP at room temperature as a function of wavelength.

——— Blokh

--□-- Pyle

--o-- Ott and Sliker

● Sterzer et al.

■ Carpenter

## POTASSIUM DIHYDROGEN PHOSPHATE (KDP)

Zwicker and Scherrer observed that the dc (unclamped) electrooptic coefficient, based on the electric field, exhibited a Curie-Weiss behavior as a function of temperature, as illustrated in Figure 17. They found, however, that the electrooptic coefficient based on the dielectric polarization was practically temperature independent. Ott and Sliker find that the ratio $r_{63}/\varepsilon_{33}^{T} - \varepsilon_o)$ is essentially temperature independent and is $0.058 \pm 0.003$ $m^2/C$ for KDP. Here $\varepsilon_{33}^{T}$ is the permittivity at constant stress and $\varepsilon_o$ is the permittivity of free space.

Perfilova and Sonin have reported the following quadratic electrooptic coefficients (in $10^{-13}$ $cm^2/V^2$) of KDP at room temperature and at a wavelength of 5400 Å:

$$n_e^3 R_{33} - n_o^3 R_{13} = 3.1$$
$$n_e^3 R_{31} - n_o^3 R_{11} = 1.35$$
$$n_o^3 (R_{12} - R_{11}) = 0.89$$
$$n_o^3 R_{66} = 0.30$$

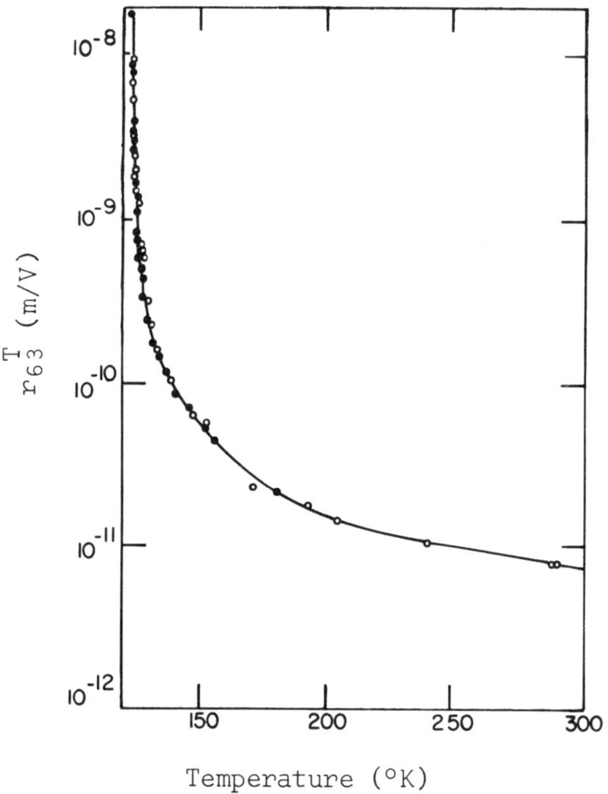


Fig. 17. Electrooptic coefficient of KDP as a function of temperature.

- ● light beam parallel to electric field
- o light beam perpendicular to electric field

[Zwicker and Scherrer]

## Photoelastic Properties - Elastooptic Coefficients

Elastooptic coefficients of KDP, measured at 6328 Å by Dixon, are given in Table 7. Dixon notes that earlier measurements at 5600 Å by West and Makas gave $p_{66}^{E}$ = -0.0685 and $-0.007<\frac{1}{2}(p_{11}-p_{12})<0$. Brody and Cummins employ Brillouin scattering to obtain $p_{66}^{E=0}$ = -0.0345-2.132/(T-117.45°K) resulting from Curie-Weiss anomaly.

TABLE 7. ELASTOOPTIC COEFFICIENTS OF KDP [Dixon].

| $p_{ij}$ | Magnitude |
|---|---|
| $p_{11}$ | 0.251 |
| $p_{12}$ | 0.249 |
| $p_{31}$ | 0.225 |
| $p_{13}$ | 0.246 |
| $p_{33}$ | 0.221 |
| $p_{66}^{E}$ | 0.058 |

## Photoelastic Properties - Piezooptic Coefficients

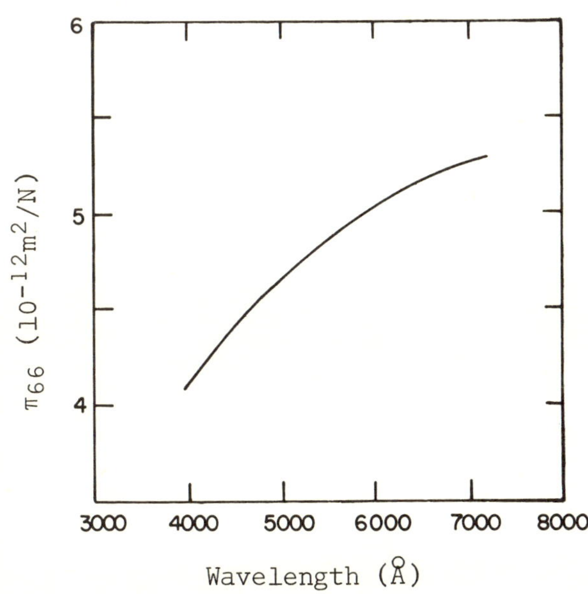


Fig. 18. Piezooptic coefficient $\pi_{66}$ of KDP as a function of wavelength [Vasilevskaya].

At 5600 Å, West and Makas find $\pi_{66}^{E}$ = -11.25 and $-0.3<(\pi_{11}-\pi_{12})<0$ in units of $10^{-12}m^2/N$. Blokh and Lutsiv-Shumskii report on $\pi_{44}$ at 20°C and 6500 Å = $5\times10^{-12}m^2/N$.

## Photoelastic Properties - Elastic Constants

Elastic stiffness ($c_{ij}$) and elastic compliance ($s_{ij}$) constants of KDP, at room temperature as cited by Landolt-Börnstein and Hearmon, are given in Tables 8 and 9, respectively. The temperature dependence of these constants are shown in Figures 19-22. The following temperature coefficients for the elastic stiffness constants have been reported:

| | Gerber | Mason |
|---|---|---|
| $\frac{1}{c_{44}} \frac{dc_{44}}{dT}$ | $-4.1 \times 10^{-4}$ | $-3.9 \times 10^{-4}$ |
| $\frac{1}{c_{66}^D} \frac{dc_{66}^D}{dT}$ | $-6.0 \times 10^{-4}$ | $-7.6 \times 10^{-4}$ |

TABLE 8. ELASTIC STIFFNESS CONSTANTS OF KDP AT 20°C

| $10^{10}$ N/m$^2$ | | | | | | Method | Reference | |
|---|---|---|---|---|---|---|---|---|
| $c_{11}$ | $c_{33}$ | $c_{12}$ | $c_{13}$ | $c_{44}$ | $c_{66}$ | | | |
| 6.91 | 5.56 | -0.600 | 1.22 | 1.29 | 0.600 | Light Diffraction | 1946 | Zwicker |
| 7.08 | 5.84 | -0.383 | 1.55 | 1.28 | 0.633 | Resonant Frequency | 1946 | Zwicker citing Bantle et al. |
| 7.8 | 7.7 | 3.23 | 3.84 | 1.27 | 0.61 | Resonant Frequency | 1950 | Mason |
| 7.14 | 5.62 | -0.49 | 1.29 | 1.27 | 0.624 | Pulse Echo | 1950 | Price and Huntington |
| 7.04 | 5.67 | -0.491 | 1.35 | 1.28 | 0.620 | Average of earlier results, excluding Mason's | 1952 | Hearmon |
| 7.4 | 6.8 | 1.8 | 2.7 | 1.35 | 0.63 | Resonant Frequency | 1953 | Barkla and Finlayson |
| 7.165 | 5.640 | -0.627 | 1.494 | 1.248 | 0.621 | | 1964 | Haussühl |

TABLE 9. ELASTIC COMPLIANCE CONSTANTS OF KDP AT 20°C

| $10^{-12}$ $m^2/N$ | | | | | | Method | Reference | |
|---|---|---|---|---|---|---|---|---|
| $s_{11}$ | $s_{33}$ | $s_{12}$ | $s_{13}$ | $s_{44}$ | $s_{66}$ | | | |
| 15.3 | 19.6 | 2.0 | -3.8 | 77.5 | 168 | Light Diffraction | 1946 | Zwicker |
| 15.2 | 19.5 | 1.8 | -4.5 | 77.9 | 158 | Resonant Frequency | 1946 | Zwicker citing Bantle et al. |
| 18 | 20 | -4 | -7 | 79 | 164 | Resonant Frequency | 1950 | Mason |
| 14.7 | 19.5 | 1.7 | -3.8 | 79 | 159 | Pulse Echo | 1950 | Price and Huntington |
| 15.1 | 19.5 | 1.8 | -4.0 | 78.1 | 162 | Average of earlier results, excluding Mason's | 1952 | Hearmon |

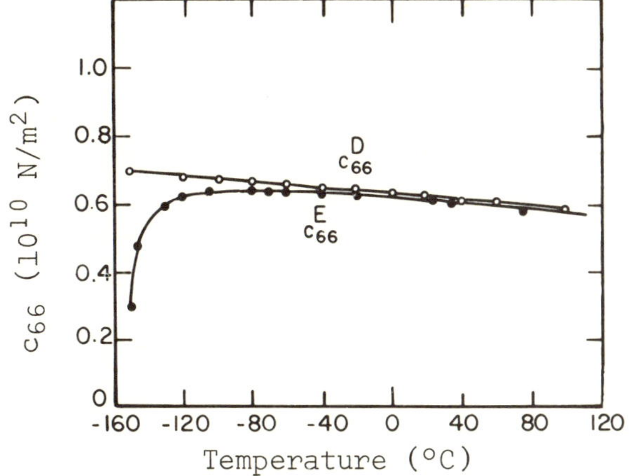


Fig. 19. Elastic stiffness constants $c_{66}^E$ and $c_{66}^D$ $(=c_{66}^P)$ of KDP as a function of temperature [Mason].



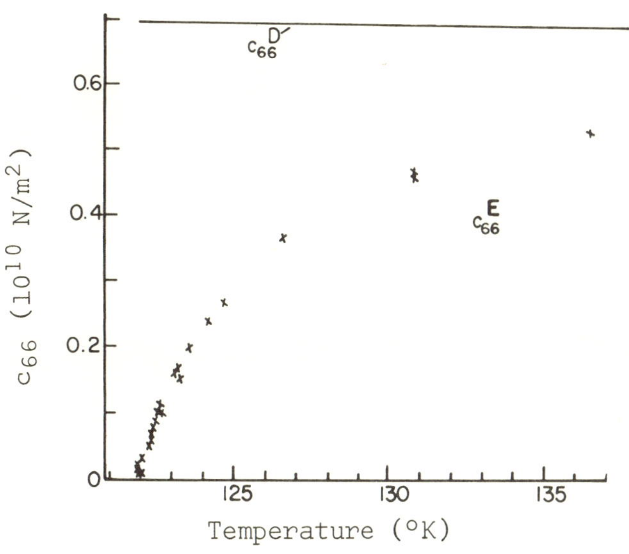


Fig. 20. Elastic stiffness constants as a function of temperature.

—— $c_{66}^D$ Mason

x $c_{66}^E$ Garland and Novotny

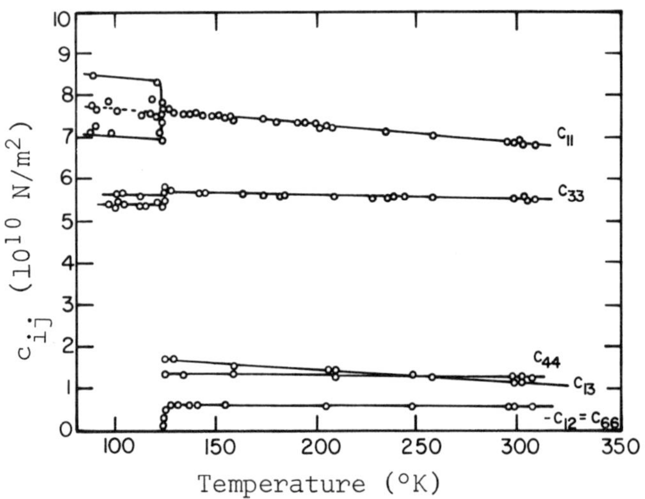


Fig. 21. Elastic stiffness constants of KDP as a function of temperature [Zwicker].



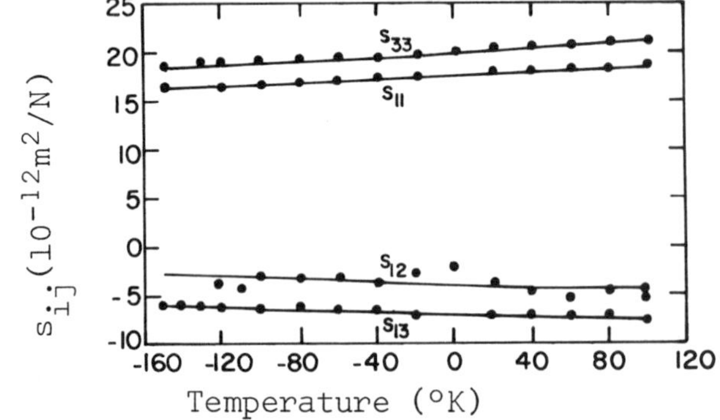


Fig. 22. Elastic compliance constants of KDP as a function of temperature [Mason].



Piezoelectric Properties

The piezoelectric behavior of $KH_2PO_4$ has been reviewed in great depth by Känzig and Jona and Shirane. KDP is piezoelectric above and below its Curie point, Cook. Both the piezoelectric coefficients and the polarization obey a Curie-Weiss law similar to that exhibited by the electrooptic coefficients. KDP possesses only two independent piezoelectric coefficients: $d_{36}$ and $d_{14} = d_{25}$.

Mason has measured the piezoelectric constant $d_{36}$ as a function of temperature (Figure 23). He determined values for all forms of the piezoelectric constants, as listed in Table 10, from the relations

$$e_{36} = d_{36}/s_{36}^{E} \quad ;$$
$$g_{36} = 4\pi d_{36}/\varepsilon_c^{F} \quad ;$$
$$h_{36} = g_{36}\, c_{66}^{D} \quad ,$$

where $\varepsilon_c^F$ is the "free" dielectric constant for KDP cut normal to the z axis.

POTASSIUM DIHYDROGEN PHOSPHATE (KDP)

As shown in Figure 23, $d_{36}$ obeys a Curie-Weiss law of the type

$$d_{36} = (d_{36})_o + \frac{B}{T-T_c} ,$$

where $B = 4.2 \times 10^{-9}$ C/N and $(d_{36})_o \simeq -2.7 \times 10^{-12}$ C/N; Jona and Shirane, using the data of Bantle and coworkers.

Other piezoelectric constants are given in Table 11 and Figure 24. The piezoelectric coupling constant of KDP is shown in Figure 25.

TABLE 10. PIEZOELECTRIC CONSTANTS OF KDP [Mason].

| T(°C) | $d_{36}$ ($10^{-12}$ C/N) | $e_{36}$ ($10^{-1}$ C/m$^2$) | $g_{36}$ ($10^{-2}$ m$^2$/C) | $h_{36}$ ($10^8$ N/C) |
|---|---|---|---|---|
| 100 | 16.8 | 0.970 | 11.04 | 6.48 |
| 80 | 18.0 | 1.057 | 11.16 | 6.63 |
| 60 | 19.7 | 1.167 | 11.58 | 6.96 |
| 40 | 21.1 | 1.267 | 11.64 | 7.05 |
| 20 | 23.2 | 1.420 | 11.82 | 7.32 |
| 0 | 25.4 | 1.577 | 11.88 | 7.50 |
| -20 | 28.6 | 1.800 | 12.12 | 7.74 |
| -40 | 32.9 | 2.08 | 12.30 | 7.98 |
| -60 | 39.7 | 2.52 | 12.45 | 8.19 |
| -80 | 51.0 | 3.25 | 12.60 | 8.37 |
| -100 | 67.3 | 4.27 | 12.66 | 8.43 |
| -120 | 111.3 | 6.93 | 12.90 | 8.79 |
| -130 | 160.0 | 9.67 | 12.96 | 8.91 |
| -140 | 292 | 15.77 | 13.02 | 8.94 |
| -145 | 488 | 23.3 | 13.08 | 9.06 |
| -150 | 1467 | 43.3 | 13.20 | 9.21 |

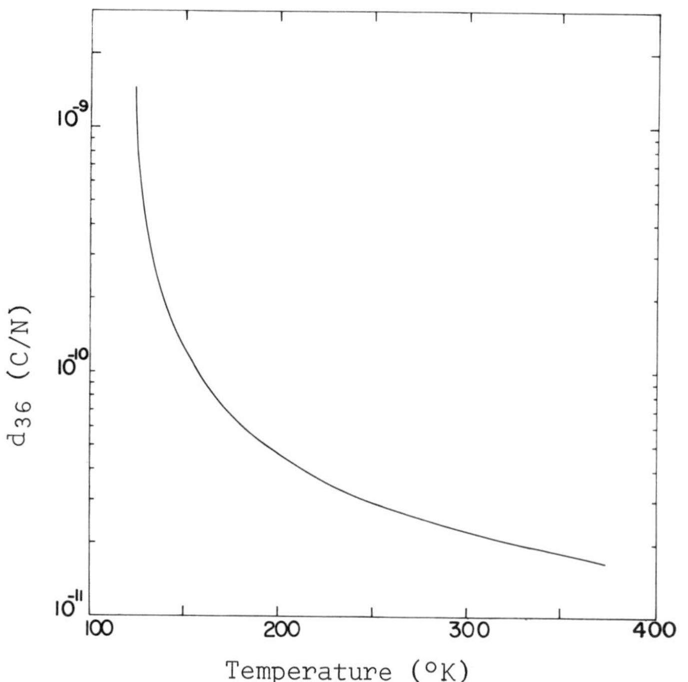


Fig. 23. Piezoelectric constant $d_{36}$ of KDP as a function of temperature [Mason].

TABLE 11. PIEZOELECTRIC CONSTANTS OF KDP

| Piezoelectric Constant | Value | Unit | T(°C) | Reference |
|---|---|---|---|---|
| $d_{14}$ | 1.4<br>1.28 | $10^{-12}$ C/N | 20<br>20 | Mason<br>Landolt-Börnstein |
| $g_{14}$ | 0.33 | $10^{-2}$ $m^2/N$ | 0 | ↓ |
| $h_{14}$ | 0.42 | $10^8$ N/C | 0 | ↓ |

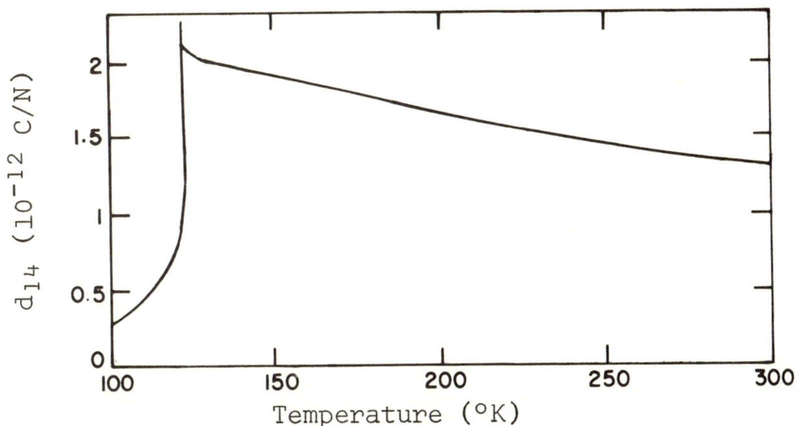


Fig. 24. Piezoelectric constant $d_{14}$ of KDP as a function of temperature [Jona and Shirane, citing the data of Ess].

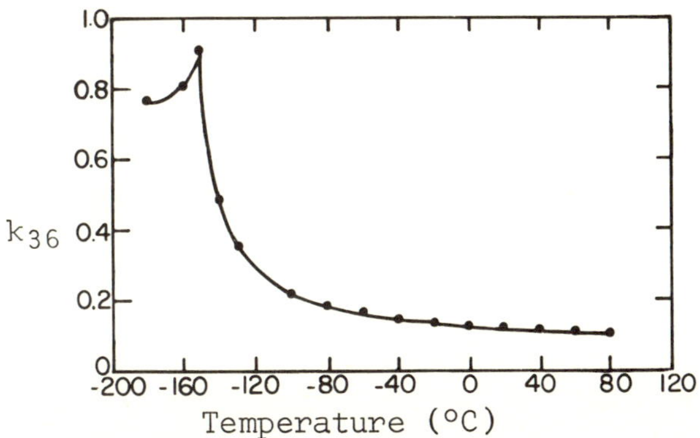


Fig. 25. Piezoelectric coupling constant as a function of temperature [Mason].

$k_{14}$ = 0.008

Sliker

$k_{36}$ = 0.121

POTASSIUM DIHYDROGEN PHOSPHATE (KDP)

## Dielectric Properties - Dielectric Constant

Potassium dihydrogen phosphate is classified as a nonlinear dielectric material. Its dielectric properties have been measured over wide ranges of temperature and frequency. The dielectric properties of KDP may be described by two dielectric constants, one for fields along the c-axis ($\varepsilon_c$) and the other for fields perpendicular to that direction, or along the a-axis ($\varepsilon_a$).* The best representative data for the dielectric constant of KDP at room temperature are those of Von Hippel and Belyaev et al. (Table 12). Single room temperature values reported by various other investigators have fallen in line with these data and are not reported here.

TABLE 12. DIELECTRIC CONSTANT OF KDP AT ROOM TEMPERATURE

| Frequency (Hz) | $\varepsilon_c^{(A)}$ | $\varepsilon_c^{(B)}$ | $\varepsilon_a^{(A)}$ | $\varepsilon_a^{(B)}$ |
|---|---|---|---|---|
| $10^2$ | 21.8 ± 0.5 | 21.4 | 43.7 ± 1.5 | 44.5 |
| $10^3$ | 21.3 ± 0.5 | 20.7 | 43.4 ± 1.5 | 44.3 |
| $10^4$ | 20.8 ± 0.5 | 20.5 | 43.2 ± 1.5 | 44.3 |
| $10^5$ | 20.1 ± 0.5 | 20.3 | 43.0 ± 1.5 | 44.3 |
| $10^6$ | - | 20.2 | - | 44.3 |
| $10^7$ | - | 20.2 | - | 44.3 |
| $10^8$ | - | 20.2 | - | 44.3 |
| $9.8 \times 10^8$ | 20.0 ± 0.5 | - | 42.5 ± 1.5 | - |
| $9.4 \times 10^9$ | 19.7 ± 0.5 | - | 42.3 ± 1.5 | - |
| $3.96 \times 10^{10}$ | 19.6 ± 0.5 | - | 42.0 ± 1.5 | - |

(A) Belyaev et al.
(B) Von Hippel

*$\varepsilon_c = \varepsilon_{33}/\varepsilon_o$, $\varepsilon_o$ is the permittivity of free space.

Peshikov, Burdanina et al. and Tokamaru and Abe have reported changes in the dielectric properties of KDP from gamma or fast neutron irradiation, apparently as a result of structure changes. Dielectric anomalies and phase transitions are also reported at temperatures of 150 to 200°C by Gruenberg et al. and Pereverzeva et al.

Upon cooling, $\varepsilon_c$ increases to a very high value, of the order of $10^5$, at the ferroelectric transition temperature ($T_c$ = 123°K), as shown in Figure 26. In the temperature range of about 50°C above the transition temperature, the dielectric constant $\varepsilon_c$ follows the Curie-Weiss law: $\varepsilon_c = C/(T-T_o)$ with C = 3250° and $T_o$ = -150°C. Thus the Curie-Weiss temperature $T_o$ coincides with the transition temperature $T_c$, Jona and Shirane.

Craig, in studying the ferroelectric transition in KDP, found that ε remains almost constant at 5 x $10^4$ for almost one degree below the transition temperature, and then drops gradually. Above the transition, ε = 3300/(T - 121.062). These high and low temperature curves extrapolate to an intersection point (transition temperature) at 121.127°K. Between 121.127 and 121.177°K, hysteresis in ε was observed, thereby showing the KDP transition to be first rather than second order, according to this investigator. Reese and May reported the crossing of the high- and low-temperature behavior to occur at 122.58°K, approximately 0.1°K below the temperature of the observed heat-capacity anomaly (Figure 27). Using equipment that was not as sensitive as that employed by Craig, these authors observed no hysteresis in the dielectric measurements.

Kaminow and Harding have measured the dielectric constant of KDP at 9.6 GHz, corresponding to a "clamped" crystal (Figure 28). They find that the dielectric constant obeys a Curie-Weiss law: $\varepsilon_c = A/(T-T_c) + B$ with A = 2820°K and B = 4.7. Here, the appropriate Curie temperature is -154.5°C. These results are in agreement with those of Mason.

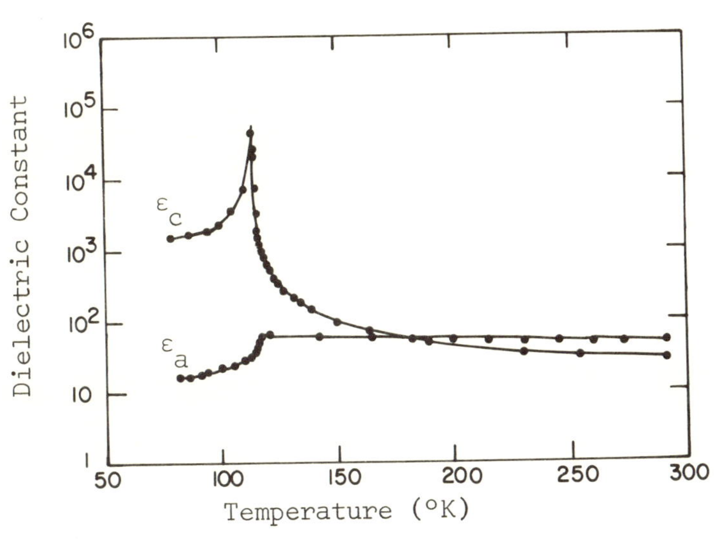


Fig. 26. Dielectric constant of $KH_2PO_4$ as a function of temperature measured at 800 Hz with a signal of 200 V/cm [Busch].

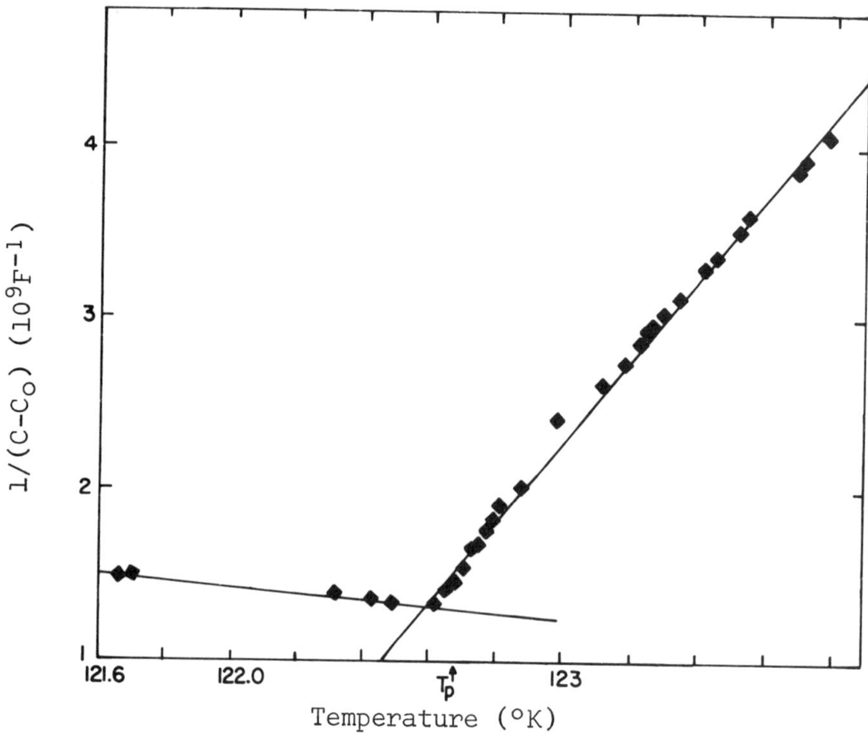


Fig. 27. The results of dielectric-constant measurements at 1 kHz and 30 V/cm are shown in the form of a plot of the reciprocal of the measured capacity minus lead capacitance ($C-C_o$) as a function of temperature. Such a plot displays the Curie-Weiss law behavior above the dielectric maximum. The temperature of the heat-capacity maximum, $T_p$ is shown and occurs about 0.1°K above the maximum in the dielectric constant [Reese and May].



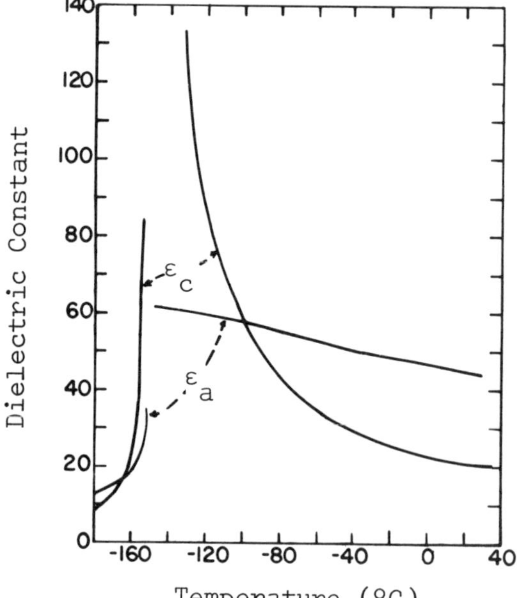


Fig. 28. Dielectric constant of KDP as a function of temperature at 9.2 GHz. [Kaminow and Harding].

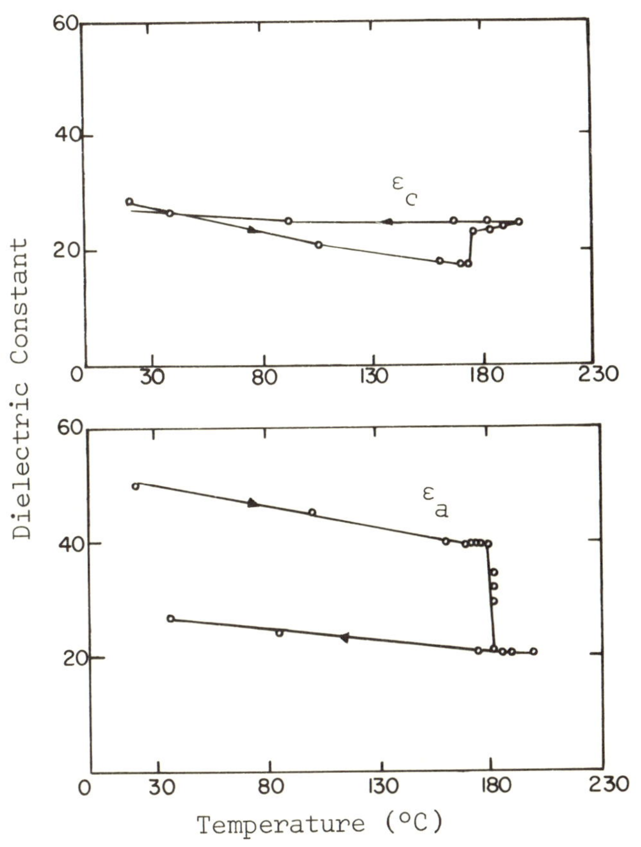


Fig. 29. Dielectric constant of KDP as a function of temperature at 10 MHz. A phase transition is observed at 180° ± 3°C. [Grinberg et al.].



Fig. 30. Dielectric constant of KDP as a function of temperature measured at 1 kHz with a signal of 1 V/cm, for different values of dc biasing field E: 1) E = 0, 2) E = 1.11, 3) E = 2.02, 4) E = 2.83, 5) E = 3.56 kV/cm. The dotted curve indicates a shift of $T_c$ with E of 0.30 to 0.60 cm °C/kV [Kamysheva et al.].



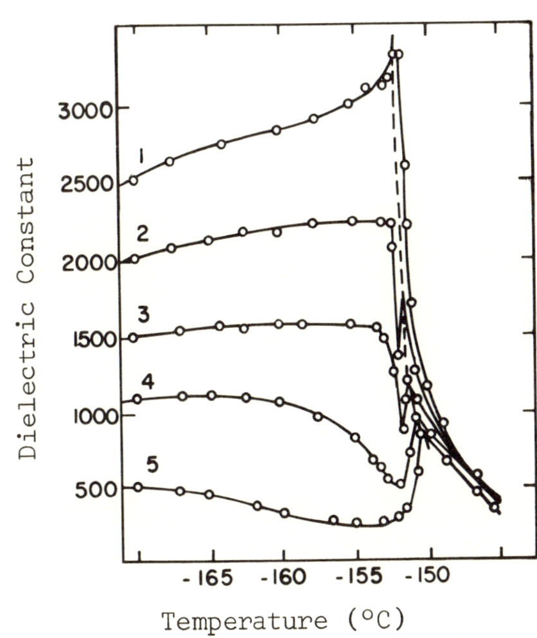

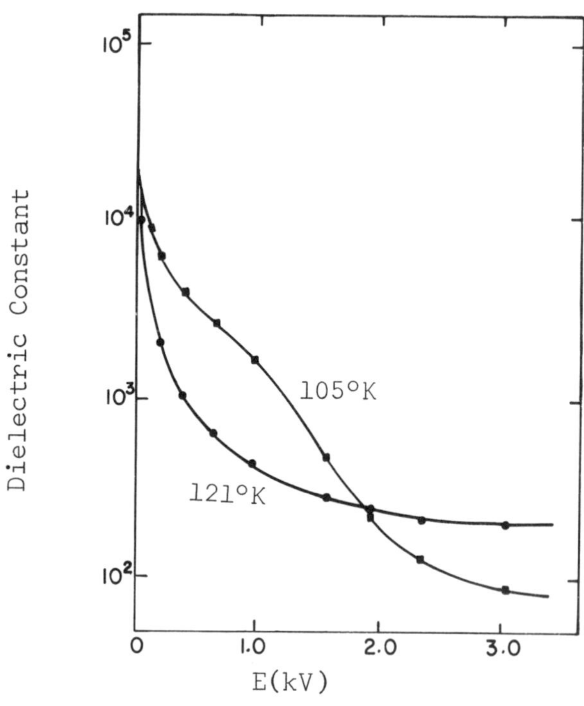


Fig. 31. Dielectric constant of KDP as a function of dc bias field E measured at 1 kHz, and two temperatures [Fouskova et al.].

## Dielectric Properties - Loss Tangent

The best and most representative data for loss tangent of KDP at room temperature are those of Von Hippel and Belyaev et al. (Table 13).

TABLE 13. LOSS TANGENT OF KDP AT ROOM TEMPERATURE

| Frequency (Hz) | $\tan \delta_c^{(A)}$ | $\tan \delta_c^{(B)}$ | $\tan \delta_a^{(A)}$ | $\tan \delta_a^{(B)}$ |
|---|---|---|---|---|
| $10^2$ | 0.06 | 0.0170 | None Reported | 0.0098 |
| $10^3$ | 0.008 | 0.0024 | | 0.0015 |
| $10^4$ | 0.002 | <0.0020 | | <0.0005 |
| $10^5$ | 0.0006 | <0.0005 | | <0.0005 |
| $10^6$ | - | <0.0005 | | <0.0005 |
| $10^7$ | - | <0.0005 | | <0.0005 |
| $10^8$ | - | <0.0005 | | <0.0005 |
| $9.8 \times 10^8$ | 0.0005 | - | | - |
| $9.4 \times 10^9$ | 0.0008 | - | | - |
| $3.96 \times 10^{10}$ | 0.003 | - | | - |

(A) Belyaev et al.
(B) Von Hippel

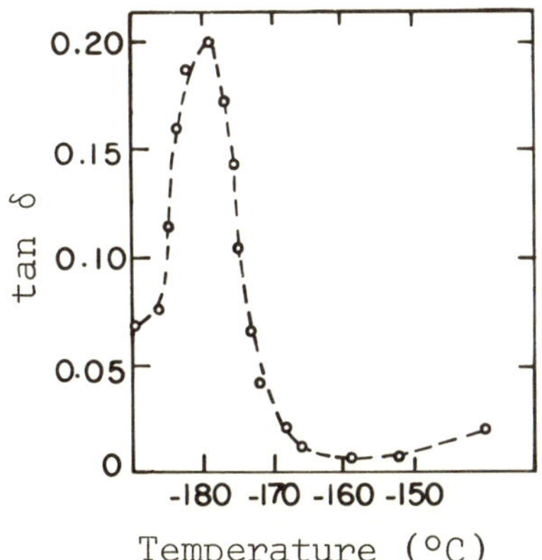


Fig. 32. Loss tangent of KDP as a function of temperature measured at 1 kHz [Kamysheva et al.].

Fig. 33. Loss tangent of KDP as a function of temperature at 9.2 GHz for fields along the a-axis, tan $\delta_a$, and along the c-axis, tan $\delta_c$ [Kaminow and Harding].



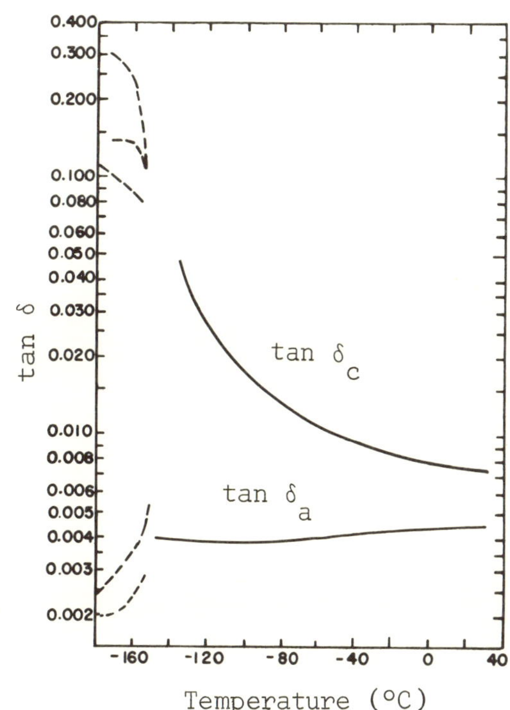

## Dielectric Properties - Electrical Resistivity

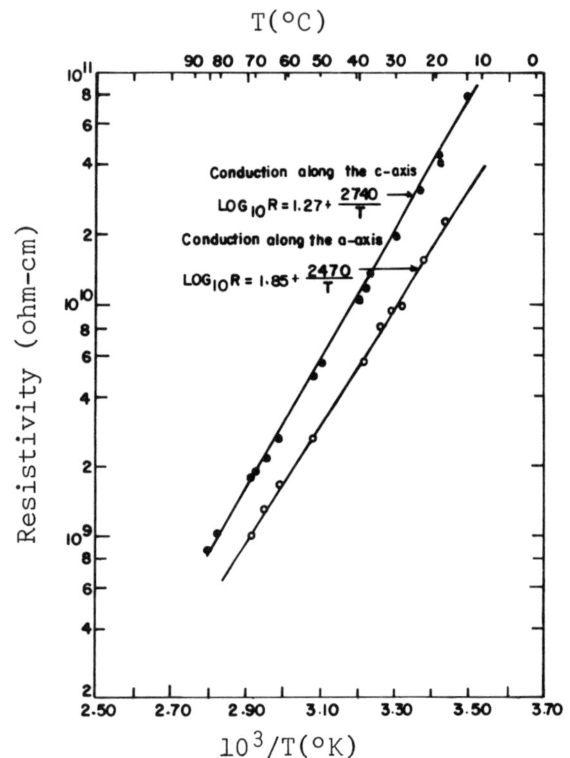


Fig. 34. Electrical resistivity of KDP as a function of reciprocal temperature [Mason].

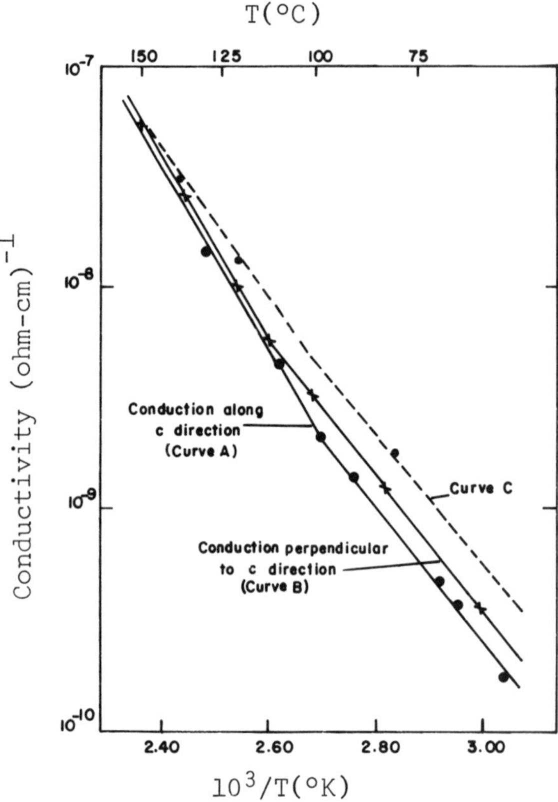


Fig. 35. Conductivity of KDP crystals in different directions. Curve B is obtained during heating and Curve C during cooling of the same sample during a run. [Harris and Vella].



Harris and Vella studied the electrical conductivity of KDP crystals and found it to be influenced by heat treatment effects (Figure 35). They concluded that heat treatment appears to affect the energy for dissociation into ion pairs. They also tested the thermal theory of dielectric breakdown and found it to apply to the breakdown of KDP crystals at 150°C. They measured a breakdown strength of 180 kV/cm for 0.2 mm thick samples.

## POTASSIUM DIHYDROGEN PHOSPHATE (KDP)

### Ferroelectric Properties

Potassium dihydrogen phosphate is an exceptionally interesting ferroelectric crystal due to the fact that the ferroelectric properties of this crystal are intimately related to the dynamics of the lattice. The ferroelectric behavior of $KH_2PO_4$ was first noted by Busch and Scherrer in 1935 and has been studied extensively since then, Jona and Shirane. The tetragonal form (at room temperature) of the crystal undergoes a phase change as the temperature is lowered to the Curie point ($T_C \simeq 123°K$). This phase transition is accompanied by the appearance of spontaneous polarization and shear strain which lowers the crystal symmetry to orthorhombic.

Potassium dihydrogen phosphate is representative of a class of order-disorder type ferroelectrics in which the hydrogen bond plays an important role in the ferroelectric transition. Känzig in 1957 and Jona and Shirane in 1962 have shown and discussed the importance of the establishment of long-range order of the hydrogen-bond network in the ferroelectric transition of KDP.

The characteristic feature of the crystal structure of KDP is the existence of four hydrogens in the tetragonal unit cell which form structurally equivalent O-H··O bonds connecting two neighboring $PO_4$ tetrahedra and oriented almost perpendicular to the c-axis. Reese and May have studied the order-disorder phenomena in $KH_2PO_4$ by calorimetry measurements. They noted from previous investigations that although the transition, which is accomplished by distortion from tetragonal to orthorhombic symmetry, changes the position of the potassium ions and phosphate groups sufficiently to account for the observed values of the spontaneous polarization, a key feature of the transition seems to be an ordering of the hydrogen-bond system. The hydrogens can take either of two positions in a double minimum potential connecting phosphate groups. In the paraelectric phase there seems to be no preference between these two positions, while in the ferroelectric state the hydrogen-bond system displays long-range order with regard to which of the two positions is occupied. Neutron diffraction experiments have shown that the hydrogen positions are symmetric between the bonding oxygens for temperatures above the Curie temperature ($T>123°K$), while in the ferroelectric phase ($T<123°K$), the hydrogens are found to be nearer one oxygen than the other and are found to change their position from one oxygen to the other as the direction of the dielectric polarization is reversed.

A remarkably large isotope effect, where, on complete deuteration, the Curie point increases from 123°K to 213°K, also is taken as evidence that the tunneling motion between the two minima of the potential along the hydrogen bond is a significant factor in the transition. That is, the isotope shift is a quantum effect caused by the lower tunneling probabilities of the deuterons. The theory of the fluctuating double minimum potential predicts two correlated phase transitions in KDP crystals. The lower transition point is identified as the Curie point and the higher one was suggested to be the melting or dissociation point, where some of the hydrogen bonds break. As shown in Figure 29, the higher transition has been observed at 180°C by Grinberg et al. Blinc et al., by means of thermogravimetric (TGA) and differential thermal analysis (DTA), have shown, however, that the high temperature phase transition occurs before the onset of thermal decomposition (213°C). These authors conclude that the transition is not connected with a static breaking up of the hydrogen bond network but rather with the onset of disordered hindered rotation of the $H_2PO_4$ groups around all three axes. Recently, detailed theories of the ferroelectric transition in KDP have been developed which include the tunneling motions of the protons and which reduce to a theory of the Slater type in the limit of zero tunneling. Reese and May discuss and analyze these various theories as a background to their calorimetric investigations which partly attempt to resolve the question of the transition in KDP.

Hill and Ichiki note that the hydrogen bond axis, determined by the two end oxygens, makes a very small angle, 0.5°, with the x,y crystal plane. If the equilibrium hydrogen position is along this axis, the proton ordering can contribute very little to the observed dielectric polarization, $5 \times 10^{-6} C/cm^2$, which appears parallel to the z axis. These investigators conclude, from their infrared absorption measurements (Figure 10), that the uncoupled hydrogen-bond vibrations which give rise to the absorption at 2670 and 2360 $cm^{-1}$ show no observable splitting of the ground or first vibrational state. This eliminates a double minimum potential with appreciable tunneling between minima as a model for this bond. In addition, they deduce that the hydrogen potential function is nearly parabolic with principal axes which are not along the O-O axis.

Recently, Kobayshi has elucidated the ferroelectric phase transition in $KH_2PO_4$ on the basis of the coupling between the proton tunneling mode and the optical mode vibration of the [K-$PO_4$] complexes along the c-axis. He states that KDP-type crystals can be considered as belonging to the "mixed" type ferroelectric class; that is to say, these crystals manifest both features of displacive-type and order-disorder type ferroelectrics.

## POTASSIUM DIHYDROGEN PHOSPHATE (KDP)

The existence of ferroelectric domains in KDP was demonstrated by x-ray measurements by de Quervain (in 1944) and Ubbelhode and Woodward (in 1945). Känzig predicted that the domain structure would have a layer-like structure with the domain walls containing the polar axis and one of the tetragonal axes. Känzig estimated a layer thickness of the order of $10^{-4}$ cm. Mitsui and Furuichi (in 1953) observed a gross domain pattern with a polarizing microscope and barely resolved a minute crack-like structure, thickness $\sim 10^{-3}$ cm, parallel to the tetragonal axis which they assumed was the basic domain pattern. Hill and Ichiki have reported on the observation by the aid of a helium-neon gas laser beam, of a thin, regular, layered domain pattern in KDP and some interesting optical properties which result from this structure.

Oettel has made an extensive and detailed study of ferroelectric domains and domain motion in KDP crystals. He found that the domains can be observed in these crystals without the aid of polarizers. The domain walls are visible presumably because of a slight discontinuity in the refractive index at the domain wall caused by strain. Preliminary calculations based on measurement of the threshold angle for total internal reflection indicate that the change in the refractive index at the domain wall is of the order of 0.6%. Domains are of the order of $2 \times 10^{-4}$ cm wide in 1 millimeter-thick plates and increase in width with plate thickness as $t^{\frac{1}{2}}$, however, a single plate may have domains of various widths ranging over a factor of 2 or 3. Saturation polarization is achieved when a crystal is completely clear of domain structure. Complete polarization reversal can be accomplished in less than 1/64 second in $KH_2PO_4$ crystals.

Many investigators have reported on the transition temperature ($T_c$) of KDP. The most quoted value in the literature is $T_c$ = 123°K; other values are detailed in Table 14.

Skalyo et al. report on the pressure dependence of the transition temperature by means of neutron diffraction measurements, to pressures of 10 kbar. The dependence is linear, $dT_c/dP = -2.40$°K/kbar.

Gladkii and Sidnenko measured polarization reversal at the phase change employing slowly changing electrical fields. They report $dT_c/dE = 1.25 \times 10^{-4}$°C-cm/V

TABLE 14. FERROELECTRIC TRANSITION TEMPERATURE OF KDP

| $T_c$(°K) | Method | Reference |
|---|---|---|
| 118. (clamped) | Dielectric (see Fig. 28) | Kaminow and Harding |
| 120-122 | Electrooptic | Sonin et al. |
| 121.127 | Dielectric | Craig |
| 121.97 | Cooling-Curve Apparatus | Stephenson et al. |
| 122.58 | Dielectric (see Fig. 27) | Reese and May |
| 122.88 | Specific Heat | Strukov et al. |
| 122.90 | Dielectric | Von Arx and Bantle |
| 123.2 | Dielectric | Cook |

Loiacono employed a DTA method and reports $T_c$ = 121.8°K on the heating cycle and 120.8°K on the cooling cycle.

Thermal Properties

| | | |
|---|---|---|
| Melting Point: | 252.6°C | Ballard et al. |
| Onset of Thermal Decomposition: | 213°C | Blinc et al. |
| Maximum Safe Operating Temperature: | 100°C | Union Carbide |

Kaminow has pointed out how a radiant thermal gradient due to r-f heating degrades the performance of KDP intensity modulators. For example, about 2 watts dissipated in a cylindrical KDP rod of any dimensions will reduce the degree of intensity modulation by one half. Loscoe and Mette also studied the effect of temperature gradients across KDP crystals in order to detect light deflection or misalignment of light in an optical communication system.

Thermal Properties - Thermal Expansion

Cook recently surveyed the literature on the thermal coefficient of expansion of KDP, as shown in Table 15.

TABLE 15. THERMAL COEFFICIENT OF EXPANSION OF KDP

| Temperature Range (°C) | Expansion Coefficient ($10^{-6}$/°C) | | Reference |
|---|---|---|---|
| | along a-axis | along c-axis | |
| -50 to +50 | 24.9 | 44.0 | Cook; see Fig. 36 |
| -40 to +90 | 26.6 | 44.6 | Mason; see Fig. 37 |
| 27 to 147 | 26.9 | 46.6 | Sirdeshmukh |
| -150 to +20 | 21.6 | 34.3 | Ubbelhode and Woodward |
| -150 to -25 | 22 | | Von Arx and Bantle |
| -150 to +20 | 20 | 42 | de Quervain |
| 20 | 26.5 | 44.4 | Haussühl |

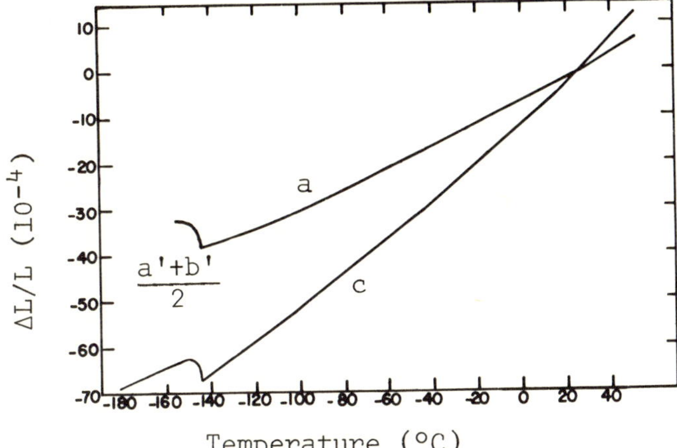


Fig. 36. Thermal expansion of KDP as a function of temperature [Cook].

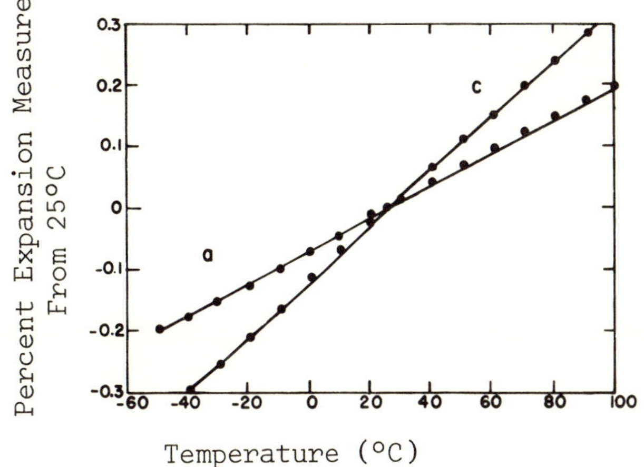


Fig. 37. Thermal expansion of KDP as a function of temperature [Mason].



Boiko and Fat report a decrease of 0.022 Å in the a-axis at 123.9°K, also re-reported in detail by Kobayashi et al. from 100 to 300°K.

## Thermal Properties - Thermal Conductivity

The ratio of the thermal conductivity of KDP single crystals with heat flow along the a-axis to the thermal conductivity with heat flow along the c-axis is 1.1 at 46°C and 1.4 at 74°C, McCarthy and Ballard.

TABLE 16. THERMAL CONDUCTIVITY OF KDP [McCarthy and Ballard].

| Orientation | T(°C) | Thermal Conductivity (mW/cm°K) |
|---|---|---|
| a-axis | 46 | 13.4 |
| | 74 | 17.6 |
| c-axis | 39 | 12.1 |
| | 77 | 13.0 |

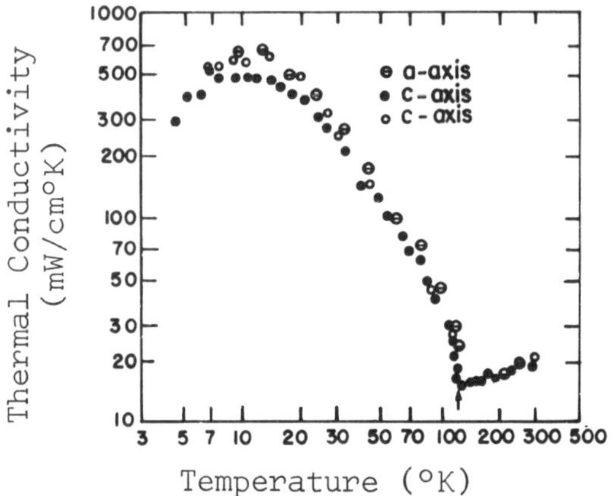


Fig. 38. Thermal conductivity of KDP as a function of temperature. The arrow indicates the Curie temperature of 122°K [Suemune].

## POTASSIUM DIHYDROGEN PHOSPHATE (KDP)

ARMSTRONG, J.A. et al. Interactions Between Light Waves in a Nonlinear Dielectric. PHYS. REV., v. 127, 1962. p. 1918-1939.

ASHKIN, A. et al. Observation of Continuous Optical Harmonic Generation with Gas Masers. PHYS. REV. LETTERS, v. 11, no. 1, 1963. p. 14-17.

BAIRD-ATOMIC, INC., Cambridge, Mass. Electro-Optic Light Modulators: KDP, ADP, KDDP. Publication No. RP-5, June 1967. 40 p.

MICHIGAN UNIV. WILLOW RUN LAB. Optical Materials for Infrared Instrumentation. by: BALLARD, S.S. et al. State-of-the-Art Report, Rept. no. 2389-11-S, Jan. 1959. Contract no. Nonr 1224-12. AD 217 367.

BANTLE, W. and C. CAFLISCH. The Piezoeffect in Ferro-Electric $KH_2PO_4$ Crystals. HELV. PHYS. ACTA, v. 16, 1943. p. 235-250.

BARKLA, H.M. and D.M. FINLAYSON. The Properties of $KH_2PO_4$ Below the Curie Point. PHIL. MAG., v. 44, 1953. p. 109-130.

BASS, M. et al. Optical Rectification. PHYS. REV. LETTERS, v. 9, no. 11, 1962. p. 446-448.

BASS, M. et al. Optical Rectification. PHYS. REV., v. 138, 1963. p. A534-A542.

BELYAEV, L.M. et al. Dielectric Constant of Crystals Having an Electro-Optical Effect. SOVIET PHYS. SOLID STATE, v. 6, no. 8, Feb. 1965. p. 2007-2008.

BJORKHOLM, J.E. and A.E. SIEGMAN. Accurate cw Measurements of Optical Second-Harmonic Generation in Ammonium Dihydrogen Phosphate and Calcite. PHYS. REV., v. 154, no. 3, Feb.1967. p. 851-860.

BLINC, R. et al. High Temperature Phase Transition in $KH_2PO_4$. J. CHEM. PHYS., v. 49, no. 11, Dec. 1968. p. 4966-5000.

BLOKH, O.G. Dispersion of $r_{63}$ for Crystals of ADP and KDP. SOVIET PHYS. CRYSTAL., v. 7, no. 1, 1962. p. 509-511.

BLOKH, O.G. and L.F. LUTSIV-SHUMSKII. The Longitudinal Electro-Optical Effect in ADP and KDP Crystals in the Near Ultraviolet and Infrared Regions. ACAD. OF SCI., USSR, BULL., PHYS. SER., v. 31, no. 7, 1967. p. 1162-1163.

BLOKH, O.G. and L.F. LUTSIV-SHUMSKII. The Spontaneous Electro-Optical Effect in ADP and KDP Crystals. ACAD. OF SCI., USSR, BULL., PHYS. SER., v. 31, no. 7, 1967. p. 1157-1161.

BLOKH, O.G. and L.F. LUTSIV-SHUMSKII. Photoelastic Effect in 45° X Cuts of $KH_2PO_4$ and $NH_4H_2PO_4$. SOVIET PHYSICS, SOLID STATE, v. 12, no. 1, July 1970. p. 256-257.

BLOCK, O.G. et al. Temperature and Wavelength Dependence of the $r_{41}$ Electro-Optical Coefficient of ADP and KDP Crystals. ACAD. OF SCI., USSR, BULL., PHYS. SER., v. 33, no. 2, 1969. p. 260-262.

BOIKO, A.A. and LIU TAN FAT. The Thermal Expansion of Potassium Dihydrogen Phosphate in the Vicinity of the Ferroelectric Phase Transition. SOVIET PHYS. CRYST., v. 14, no. 5, Mar. 1970. p. 712-715.

## POTASSIUM DIHYDROGEN PHOSPHATE (KDP)


BRODY, E.M. and H.Z. CUMMINS. Elasto-Optic Anomaly in $KH_2PO_4$. PHYS. REV. LETTERS, v. 23, no. 18, Nov. 1969. p. 1039-1041.

BURDANINA, N.A. et al. The Dielectric Properties of a gamma-Irradiated KDP Crystal. SOVIET PHYS. CRYSTAL., v. 15, no. 4, Jan. 1971. p. 721-723.

BUSCH, G. New Ferroelectrics (In Ger.). HELV. PHYS. ACTA, v. 11, 1938. p. 269-298.

BUSCH, G. and P. SCHERRER. A New Seignettoelectric Substance. NATURWISS., v. 23, 1935. p. 737.

BYSTROVA, T.G. and F.I. FEDOROV. Debye Temperatures of Tetragonal and Trigonal Crystals. SOVIET PHYS. CRYST., v. 12, no. 4, Jan. 1968. p. 493-498.

CALUCCI, E.J. Solid State Light Valve Study. INFORMATION DISPLAY, Mar.-Apr. 1965. p. 18-22.

CARPENTER, R.O. The Electro-Optic Effect in Uniaxial Crystals of the Dihydrogen Phosphate Type. III. Measurement of Coefficients. OPTICAL SOC. OF AMERICA, J., v. 40, no. 4, Apr. 1950. p. 225-229.

CARPENTER, R.O. Electro-Optic Sound-on-Film Modulator. ACOUSTICAL SOC. OF AMERICA, J., v. 25, no. 6, Nov. 1953. p. 1145-1148.

CLEVITE CORP. ELECTRONIC RES. DIV. Investigation of the Birefringence of Ferroelectric Materials, by: COOK, W.R. Jr. Final Rept., Nov. 1958-June 1960. Contract No. DA-44-009-ENG-3772. July 14, 1960. 29 p. AD 262-474L.

CLEVITE CORP. ELECTRONIC RES. DIV. Research on Birefringent Ferroelectric Materials, by: COOK, W.R. Jr. Final Rept., Apr. 1960-Mar. 1961. Contract No. DA-44-009-RNG-4434. May 1961. AD 262 652.

CLEVITE CORP. The Structure of Various Ferroelectrics: The Question of Ferroelectricity, by: COOK, W.R. Jr. Tech. Report. AFAL-TR-64-265, Sept. 1964. 59 p. AD 608-324.

COOK, W.R. Jr. Thermal Expansion of Crystals with $KH_2PO_4$ Structure. J. OF APPL. PHYS., v. 38, no. 4, Mar. 1967. p. 1637-1642.

CRAIG, P.P. Critical Phenomena in Ferroelectrics. PHYS. LETTERS, v. 20, no. 2, Feb. 1966. p. 140-142.

DIXON, R.W. Photoelastic Properties of Selected Materials and Their Relevance for Applications to Acoustic Light Modulators and Scanners. J. OF APPLIED PHYS., v. 38, no. 13, Dec. 1967. p. 5149-5153.

DOWLEY,. M.W. Parametric Fluorescence in ADP and KDP Excited by a 2573 Å CW Pump. OPTO-ELECTRONICS, v. 1, no. 4, Nov. 1969. p. 179-181.

FOUSKOVA, A. et al. Dielectric Measurements and Optical Observations of Domains in Potassium Dihydrogen Phosphate. ACAD. DES SCI., C.R., v. 262, no. 13, Mar. 1966. p. 907-910.

GARLAND, C.W. and D.B. NOVOTNY. Ultrasonic Velocity and Attenuation in $KH_2PO_4$. PHYS. REV., v. 177, no. 2, Jan. 1969. p. 971-975.

GERBER, E.A. High Frequency Vibrations of Plates Made from Isometric and Tetragonal Crystals. IRE PROC., v. 38, 1950. p. 1073-1078.

GLADKII, V.V. and E.V. SIDNENKO. Double Dielectric Hysteresis Loop of $KD_2PO_4$ Crystals. SOVIET PHYS. SOLID STATE, v. 13, no. 10, Apr. 1972. p. 2592-3.

GRÜNBERG, J. et al. Isotope Effect in the High Temperature Phase Transition of $KH_2PO_4$. SOLID STATE COMM., v. 5, no. 11, Nov. 1967. p. 863-865.

GRÜNBERG, J. et al. High Temperature Phase Transitions and Metastability in KDP-Type Crystals. PHYSICA STATUS SOLIDI, b, v. 49, no. 2, 1972. p. 857-869.

HARRIS, L.B. and G.J. VELLA. Conductivity of Single Crystals of Potassium Dihydrogen Phosphate. J. OF APPLIED PHYS., v. 37, no. 11, Oct. 1966. p. 4294.

HAUSSÜHL, S. Elastic and Thermoelastic Properties of $KH_2PO_4$, $KH_2AsO_4$, $NH_4H_2PO_4$ and $RbH_2PO_4$. Z. FUER KRYSTALLOGRAPHIE, v. 120, 1964. p. 401-414.

HEARMON, R.F.S. The Elastic Constants of Piezoelectric Crystals. BRITISH J. OF APPLIED PHYS., v. 3, Apr. 1952. p. 120-124.

HILL, R.M. and S.K. ICHIKI. Optical Behavior of Domains in $KH_2PO_4$. PHYS. REV., v. 135, no. 6A, Sept. 1964. p. A1640-A1642.

HILL, R.M. and S.K. ICHIKI. Infrared Absorption by Hydrogen Bonds in Single Crystal $KH_2PO_4$, $KDDPO_4$ and $KH_2AsO_4$. J. OF CHEM. PHYS., v. 48, no. 2, Jan. 1968. p. 838-842.

JERPHAGNON, J. and S.K. KURTZ. Optical Nonlinear Susceptibilities: Accurate Relative Values for Quartz, Ammonium Dihydrogen Phosphate and Potassium Dihydrogen Phosphate. PHYS. REV., B, Ser. 3, v. 1, no. 4, Feb. 1970. p. 1739-1742.

JONA, F. and G. SHIRANE. FERROELECTRIC CRYSTALS, 1962. N.Y., Macmillan Co.

KAMINOW, I.P. Microwave Modulation of the Electro-Optic Effect in $KH_2PO_4$. PHYS. REV. LETTERS, v. 6, 1961. p. 528-530.

KAMINOW, I.P. Strain Effects in Electrooptic Light Modulators. APPLIED OPTICS, v. 3, no. 4, Apr. 1964. p. 511-515.

KAMINOW, I.P. and G.O. HARDING. Complex Dielectric Constant of $KH_2PO_4$ at 9.2 GHz. PHYS. REV., v. 129, no. 4, Feb.1963. p. 1562-1566.

KAMYSHEVA, L.N. et al. Nonlinear Properties of KDP Crystals in Strong Fields. ACAD. OF SCI., USSR, BULL., PHYS. SER., v. 31, no. 7, July 1967. p. 1202-1206.

KÄNZIG, W. Ferroelectrics and Antiferro-electrics. SOLID STATE PHYS., v. 4,1957. N.Y., Academic Press. p. 1-197.

KOBAYASHI, K.K. Dynamical Theory of the Phase Transition in $KH_2PO_4$. PHYS. SOC. JAPAN, J., v. 24, no. 3, Mar. 1968. p. 497-508.

KOBAYASHI, J. et al. X-Ray Study on Thermal Expansion of Ferroelectric $KH_2PO_4$. PHYSICA STATUS SOLIDI, a, v. 3, no. 1, Sept. 1970. p. 63-69.

LANDOLT-BÖRNSTEIN. NUMERICAL DATA AND FUNCTIONAL RELATIONSHIPS IN SCIENCE AND TECHNOLOGY, New Series, v. 1, Elastic Piezoelectric, Piezooptic and Electrooptic Constants of Crystals. Group III. Crystal and Solid State Physics, ed. by: HELLWEGE, K.-H. and A.M. HELLWEGE. Berlin, Ger. Springer Verlag, 1966.

LOIACONO, G.M. A DTA Study of the Ferroelectric Transition in $KH_2PO_4$-Type Crystals. MAT. RES. BULL., v. 5, no. 9, Sept. 1970. p. 775-782.

LOSCOE, C. and H. METTE. Optical Misalignment Due to Temperature Gradients in Electrooptic Modulator Crystals. APPLIED OPTICS, v. 5, 1966. p. 93-96.

MAKER, P.D. et al. Effects of Dispersion and Focusing on the Production of Optical Harmonics. PHYS. REV. LETTERS, v. 8, no. 1, 1962. p. 21-22.

MASON, W.P. The Elastic, Piezoelectric and Dielectric Constants of KDP and ADP. PHYS. REV., v. 69, no. 5-6, Mar. 1946. p. 173-194.

MASON, W.P. PIEZOELECTRIC CRYSTALS AND THEIR APPLICATION TO ULTRASONICS. N.Y., Van Nostrand, 1950.

MASON, W.P. PHYSICAL ACOUSTICS, v. 1, pt. A. N.Y., Academic Press, 1964. p. 169-270.

McCARTHY, K.A. and S.S. BALLARD. New Data on the Thermal Conductivity of Optical Crystals. OPTICAL SOC. OF AMERICA, J., v. 41, 1951. p. 1062-1063.

MILLER, R.C. et al. Quantitative Studies of Optical Harmonic Generation in CdS, $BaTiO_3$ and $KH_2PO_4$-Type Crystals. PHYS. REV. LETTERS, v. 11, no. 4, 1963. p. 146-149.

HARVARD UNIV., CROFT LAB. Modulation of the Linear Electro-Optic Effect at Microwave Frequencies, by: MYERS, R.A. Contract No. NR-372-012, Tech. Rept. No. 433, Nov. 1963. AD 600 461.

UNIV. OF WASHINGTON, SEATTLE, WASHINGTON. Ferroelectric Domains and Domain Motion in $KH_2PO_4$, $KD_2PO_4$, by: OETTEL, R.E. M.S. in Electrical Engineering, 1965. 44 p.

OTT, J.H. and T.R. SLIKER. Linear Electro-Optic Effects in $KH_2PO_4$ and its Isomorphs. OPTICAL SOC. OF AMERICA, J., v. 54, no. 12, Dec. 1964. p. 1442-1444.

PEREVERZEVA, L.P. et al. Dielectric Anomalies in Crystals of $KH_2PO_4$, $KD_2PO_4$ and $RbH_2PO_4$ at High Temperatures. SOVIET PHYSICS, SOLID STATE, v. 13, no. 11, May 1972. p. 2690-2692.

PERFILOVA, V.E. and A.D. SONIN. The Quadratic Electro-Optical Effect in KDP Group Crystals. ACAD. OF SCI. USSR, BULL. PHYS. SER., v. 31, no. 7, July 1967. p. 1154-7.

PESHIKOV, E.V. Structural Sensitivity of the Ferroelectric Phase Transition and Dielectric Properties of Potassium Dihydrogen Phosphate Crystals Irradiated with Fast Neutrons. SOVIET PHYS. CRYSTALL., v. 16, no. 5, Mar. 1972. p. 820-823.

PHILLIPS, R.A. Temperature Variation of the Index of Refraction of ADP, KDP and Deuterated KDP. OPTICAL SOC. OF AMERICA, J., v. 56, May 1966. p. 629-632.

PISAREVSKII, Yu. V. et al. The Electro-Optical Properties of $NH_4H_2PO_4$, $KH_2PO_4$, $N_4(CH_2)_6$. SOVIET PHYS. SOLID STATE, v. 7, 1965. p. 530-531.

PRICE, W.J. and H.B. HUNTINGTON. Acoustical Properties of Anisotropic Materials. ACOUSTICAL SOC. OF AMERICA, J., v. 22, 1950. p. 32-37.

PURSEY, H. et al. Measurement of the Pockels Effect in KDP at GHz. J. OF APPLIED PHYS., v. 18, Mar. 1967. p. 285-292.

PYLE, J.R. Laser Modulation Using Linear Electro-Optic Crystals. Tech. Note PAD 125, Dec. 1966. 35 p. NASA N67-27128.

REESE, W. and L.F. MAY. Critical Phenomena in Order-Disorder Ferroelectrics. I. Calorimetric Studies of $KH_2PO_4$. PHYS. REV., v. 162, no. 2, Oct. 10, 1967. p. 510-518.

REESE, W. and L.F. MAY. Studies of Phase Transition in Order-Disorder Ferroelectrics. II. Calorimetric Investigations of $KH_2PO_4$. PHYS. REV., v. 167, no. 2, Mar. 1968. p. 504-510.

REITZ, E.A. et al. (Autonetics) Electro-Optic Projector Study. Tech. Rept. No. RADC-TR-66-394, Dec. 1966. AD 646 232.

ROBINSON, F.N.H. Nonlinear Optical Coefficients. BELL SYSTEM TECHNICAL J., v. 46, May/June 1967. p. 913-956.

ROSNER, R.D. et al. Clamped Electro-Optic Coefficients of KDP and Quartz. APPLIED OPTICS, v. 6, Apr. 1967. p. 778.

SHAMBUROV, V.A. and I.V. KUCHEROVA. Variation in Anomalous Birefringence in Crystals of $KH_2PO_4$. SOVIET PHYS. CRYSTAL., v. 10, no. 5, Mar. 1966; p. 558-560.

SIRDESHMUKH, D.B. and V.T. DESHPANDE. X-Ray Determination of the Lattic Parameters of Potassium Dihydrogen Phosphate at Elevated Temperatures. ACTA CRYST., v. 22, no. 3, 1967. p. 438-439.

SKALYO, J. Jr. et al. The Pressure Dependence of the Transition Temperature in KDP and ADP. J. PHYS. CHEM. SOLIDS, v. 30, no. 8, Aug. 1969. p. 2045-2051.

CLEVITE CORP. Reference Data on Linear Electro-Optic Effects. Engineering Memo., 64-10, by: SLIKER, T.R. May 15, 1964. 9 pp.

SLIKER, T.R. and S.R. BURLAGE. Some Dielectric and Optical Properties of $KD_2PO_4$. J. OF APPLIED PHYS., v. 34, no. 7, July 1963. p. 1837-1840.

SONIN, A.D. et al. Electro-Optic Properties of Potassium Dihydrogen Phosphate and Deuterated Potassium Dihydrogen Phosphate Crystals in the Region of Their Phase Transitions. SOVIET PHYS. SOLID STATE, v. 8, no. 11, May 1967. p. 2758-2759.

STEPHENSON, C.C. et al. Transition Temperatures in Some Dihydrogen and Dideutero Phosphates and Arsenates and Their Solid Solutions. J. OF CHEM. PHYS., v. 21, no. 6, June 1953. p. 1110.

STERZER, F. et al. Cuprous Chloride Light Modulators. OPTICAL SOC. OF AMERICA, J., v. 54, no. 1, Jan. 1964. p. 62-68.

STRUKOV, B.A. et al. Comparative Investigation of the Specific Heat of $KH_2PO_4$ Single Crystals. PHYS. STATUS SOLIDI, v. 27, 1968. p. 741-749.

SUEMUNE, Y. Thermal Conductivity of the $KH_2PO_4$ Type. PHYS. SOC. OF JAPAN, J., v. 21, no. 4, Apr. 1966. p. 802.

TOKUMARU, Y. and R. ABE. Effect of gamma-Irradiation on the Dielectric Constants of $KH_2PO_4$. JAPAN. J. APPL. PHYS. v. 9, no. 12, Dec. 1970. p. 1548-1549.

## POTASSIUM DIHYDROGEN PHOSPHATE (KDP)

UNION CARBIDE CORP., Linde/Electronics Div., San Diego, California. Linde Crystals. 4 pp.

VALPEY CORP., Holliston, Mass. Infrared Optics. 4 pp.

VAN DER ZIEL, J.P. and N. BLOEMBERGEN. Temperature Dependence of Optical Harmonic Generation in $KH_2PO_4$ Ferroelectrics. PHYS. REV., v. 135, 1964. p. 1662-1669.

VASILEVSKAYA, A.S. The Electro-Optical Properties of Crystals of KDP Type. SOVIET PHYS. CRYSTALL., v. 11, no. 5, Mar. 1967. p. 644-647.

VISHNEVSKII, V.N. and I.V. STEFANSKII. Temperature Dependence of the Dispersion of the Refractivity of ADP and KDP Single Crystals. OPT. AND SPECTRO., v. 20, no. 2, Feb. 1966. p. 195-196.

VON ARX, A. and W. BANTLE. The Inverse Piezoeffect in Ferroelectric $KH_2PO_4$ Crystals. HELV. PHYS. ACTA, v. 17, 1944. p. 298-318.

MASS. INST. OF TECHNOL., Insulation Lab. TABLES OF DIELECTRIC MATERIALS. by: VON HIPPEL, A. v. 4, Jan. 1953.

WARD, J.F. and P.A. FRANKEN. Structure of Nonlinear Optical Phenomena in Potassium Dihydrogen Phosphate. PHYS. REV., v. 133, no. 1A, Jan. 1964. p. A183-A190.

WARD, J.F. and G.H.C. NEW. Optical Rectification in Ammonium Dihydrogen Phosphate, Potassium Dihydrogen Phosphate and Quartz. ROYAL SOC. PROC., v. 299A, no.1457, June 1967. p. 238-263.

WEST, C.D. and A.S. MAKAS. Some Photoelastic Constants of Crystals: Ammonium Dihydrogen Phosphate and Potassium Dihydrogen Phosphate. AMERICAN MINERALOGIST, v. 35, 1950. p. 130.

YAMAZAKI, M. and T. OGAWA. Temperature Dependences of the Refractive Indices of $NH_4H_2PO_4$, $KH_2PO_4$ and Partially Deuterated $KH_2PO_4$. OPTICAL SOC. OF AMERICA, J., v. 56, no. 10, Oct. 1966. p. 1407-1408.

ZERNIKE, F. Jr. Refractive Indices of Ammonium Dihydrogen Phosphate and Potassium Dihydrogen Phosphate between 2000Å and 1.5 microns. OPTICAL SOC. OF AMERICA, J., v. 54, Oct. 1964. p. 1215-1220.

ZWICKER, B. Elastic Investigation of $NH_4H_2PO_4$ and $KH_2PO_4$. HELV. PHYS. ACTA, v. 19, 1946. p. 523-549.

ZWICKER, B. and P. SCHERRER. Electro-Optical Properties of the Ferroelectric Crystals $KH_2PO_4$ (In Ger.). HELV. PHYS. ACTA, v. 17, no. 5, Sept. 1944. p. 346-372.

# POTASSIUM TANTALATE NIOBATE (KTN)

## Introduction

Potassium tantalate niobate ($KTa_xNb_{1-x}O_3$) or KTN has been actively explored in the past few years for its very useful electrooptic properties just above the Curie temperature. It also has excellent acoustical parametric amplification characteristics and great promise for modulation, switching, and deflection of laser beams [Chen et al.].

While exploratory studies of the $KTaO_3$ - $KNbO_3$ solid solution system have covered the entire range of chemical compositions, the most useful composition for electrooptic purposes has been $KTa_{0.65}Nb_{0.35}O_3$. Data on other ratios are included herein for the sake of completeness and comparison purposes.

KTN can be made to be an insulating or semiconducting (n-type) crystal, interestingly enough, depending on the stoichiometry and crystal growing conditions [Wemple and Kurz]. Oxygen vacancies are believed to form donor states below the conduction band. Conductivity studies have shown this level to be relatively deep, (0.3 eV) leading to a spreading of the level through a large fraction of the Brillouin zone. DiDomenico and Wemple have focussed on the energy band structure of KTN and other ferroelectrics by an analysis of the optical properties and how the ferroelectric phase transition affects optical properties via its effect on the energy bands. A band gap value of 3.6 eV has been cited for KTN. Baer has compared the energy band gaps of KTN determined by various methods with his values obtained by means of interband Faraday rotation techniques.

Geusic et al. were apparently the first to discuss the applications of the electrooptic effect in KTN; the material exhibits useful electrooptic properties just above its Curie temperature when it is in the paraelectric state and has the perovskite crystal structure. KTN shows good infrared transmission characteristics to 5 microns, a low dielectric loss tangent and a driving voltage for modulation purposes lower than that required for KDP.

## Chemical and Physical Properties

According to Garn and Flaschen, $KNbO_3$ forms a continuous solid solution with $KTaO_3$. Reisman et al. found the $KTaO_3$ - $KNbO_3$ system to have an ideal solid solution behavior, as shown in Figure 1; their phase diagram differs somewhat with that given by Garn and Flaschen. Reisman and coworkers also found that these solutions followed Vegard's Law at 450°C. The solid solutions were prepared by fusing $Nb_2O_5$, $Ta_2O_5$ and $K_2CO_3$ until $CO_2$ evolution was completed. KTN is insoluble in water.

Hill et al. determined the subsolidus stability relations and showed the presence of an exosolution dome in the $KTaO_3$ - $KNbO_3$ system (Figure 2). Their study showed that the KTN composition $KTa_{0.65}Nb_{0.35}O_3$ is metastable at temperatures below 900 ± 20°C and thus it is not feasible to grow crystals of this composition hydrothermally. They found that hydrothermally-grown single crystals appeared to be related to the growth temperature rather than to the composition of the nutrient.

Reisman and Banks have determined that, within the limits of experimental error, the density variation is linear over the entire $KTaO_3$ - $KNbO_3$ composition range, indicating that the interaction is ideal and that the volume/unit cell ratio remains constant (Figure 3).

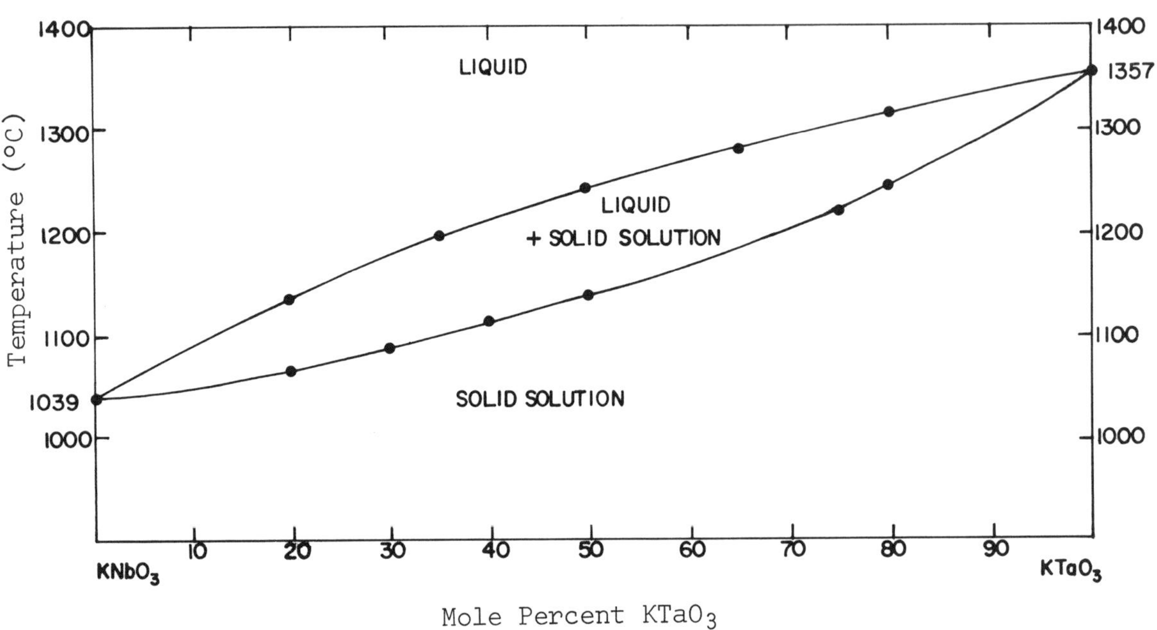


Fig. 1. Phase diagram for the system $KNbO_3$ - $KTaO_3$ [Reisman et al.].

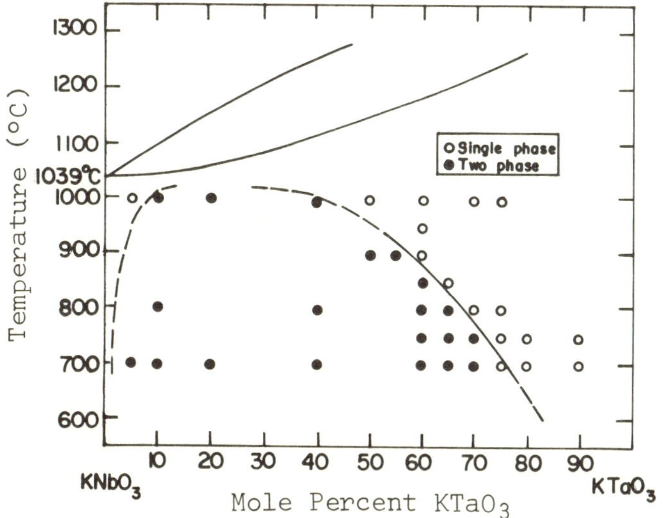


Fig. 2. Subsolidus phase relations in the system $KNbO_3$ - $KTaO_3$. Solidus and liquidus curves are from Figure 1 [Hill et al.].



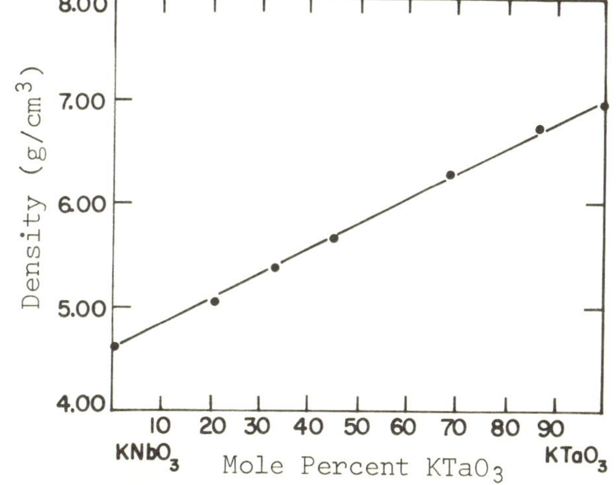


Fig. 3. Density phase diagram of the system $KNbO_3$ - $KTaO_3$ at 25°C [Reisman and Banks].



## Crystallography

The crystal growth of KTN has been studied extensively by a number of workers in various laboratories throughout the United States. Wilcox and Fullmer, of the Aerospace Corporation have reviewed and compared the various techniques of growing KTN, as shown in Table 1. These authors also reported on their results of growing these crystals by two techniques: (1) pulling in the <100> direction with a continuously increasing crystal diameter, and (2) growing [100] platelets while floating on the surface of the melt. They found the amount of tantalum in the crystals increased rapidly with increasing potassium in the melt. The rate of potassium oxide evaporation from the melt was significant.

TABLE 1. COMPARISON OF KTN GROWTH TECHNIQUES [Wilcox and Fullmer].

| Parameter | Modified Kyropoulous | <100> Czochralski | <111> Czochralski | Floating Crystal |
|---|---|---|---|---|
| Process | Cool down to obtain desired cross section; then pull | Pull immediately; neck down, then continuously enlarge diameter | Pull immediately; neck down, enlarge diameter, then hold constant | Drop very small seed on surface, lower temperature; lift out when large enough |
| $K_2O$ Composition in Melt | 50-52 mole % | 53-60 mole % | 53-60 mole % | 53-60 mole % |
| Seed Orientation | <100> | <100> | <111> | [100] |
| Heating Method | SiC resistance | RF induction | RF induction | RF induction |
| Temperature Gradient | Low | High | High | High |
| Crucible Size | 200 ml | 50 ml | 50 ml | 50 ml |
| Growth Rate | Pulled at 0.26 mm/hr | Pulled at 0.9 mm/hr | Pulled at 1 mm/hr | Vert. $\sim$ 0.7 mm/hr Horiz. $\sim$ 3 mm/hr |
| Anneal | Probably | No | No | No |
| Max. Diameter on Edge Length | $\sim$ 14 mm ? | > 12 mm | Constant diam: $\sim$ 5 mm Increasing diam: $\sim$ 10 mm | $\sim$ 15 mm |
| Max. Length or Thickness | ? | > 26 | --- | $\sim$ 3 mm |
| Additives | 0.1 mole % $SnO_2$ | None | None | None |
| Color | Colorless, sometimes blue or yellow | Colorless, sometimes pale blue near seed | Colorless, sometimes pale yellow | Colorless |
| Appearance in Crossed Polarizers | Large striations | | Microscopic figures, some gross strain curves | Few lines or marks |
| Concentration Variations | Very small | Not determined | Some | Very small |
| Thermal Strain | ? | Large cross sections crack | Some | Very small |

## POTASSIUM TANTALATE NIOBATE (KTN)

Homogeneity of composition in KTN is virtually impossible to achieve. KTN crystals are also characterized by striations perpendicular to the growth direction. These striae are caused by the existence of crystal layers or laminae having slightly different indices of refraction. Two different types of laminae seemed to occur: (a) very fine evenly spaced, and (b) coarse laminae. Many investigators have conducted varied experiments in the area of striae elimination. Striations have been attributed to variations of the Ta:Nb ratio and the cause of these variations to thermal fluctuations during growth.

Bonner and coworkers found that when using spectroscopically pure materials as components of the melt, and when the $K_2O$ content of the melt is approximately 50 mole percent, insulating crystals are obtained. When the $K_2O$ content of the melt is in the neighborhood of 62 mole percent, blue conducting crystals are obtained with the modified Kyropoulos technique. In another paper, they reported that it is quite important to use high purity $Ta_2O_5$ and $Nb_2O_5$ in the growth of KTN. Reduction can occur if divalent cations are present as a result of rapid crystal growth or the lack of free circulation of oxygen over the surface of the melt. Tin oxide additions are helpful to the crystal growing and prevention of discoloration. These authors also observed considerable variation in the Curie temperatures if the crucible was or was not rotated, as indicated in Figure 4.

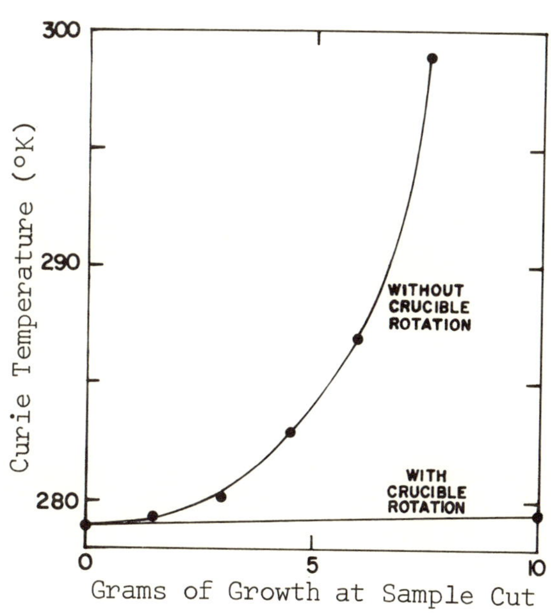


Fig. 4. Typical variation of Curie temperature with growth of potassium tantalate-niobate [Bonner et al.]

## POTASSIUM TANTALATE NIOBATE (KTN)

Fukuda and coworkers made a study of growing KTN crystals in which the $K_2O$ content varied. Colorless crystals were obtained with relatively small amounts of excess $K_2CO_3$ (52 to 55 mole %); whereas a relatively large amount of excess $K_2CO_3$ (above 60 mole %) produces colored crystals ranging from dark blue to colorless. About 0.1% (mole) addition of tin oxide in the melt with a large excess of $K_2O$ yielded brown crystals. According to these investigators, both the blue and brown color of the crystals seems to be caused by low valency state of cations ($Ta^{4+}$, $Nb^{4+}$, $Sn^{2+}$) in the crystals. They also found the electrical resistivity to vary with $K_2O$ and $SnO_2$ contents.

Denton, in a patent disclosure, reported on the growth of single crystals of KTN prepared from a mixture of $K_2CO_3$ (300 gm), $Ta_2O_5$ (304 gm), $Nb_2O_5$ (448 gm), and $SnO_2$ (0.7 gm). Growth was on an oriented seed lifted at 1/3 inch/day while being rotated at 40 rpm. The molten materials contained in a SiC were held at 1250°C with an oxygen-containing atmosphere to prevent reduction of Nb or Ta.

Hill and coworkers studied the crystal growth (using the hydrothermal method) of KTN at high temperatures and offered an explanation of the persistence of lamellar structures as caused by the presence of an exsolution dome in the subsolidus region of the phase equilibrium diagram. This indicates that at room temperature a single phase material is metastable; the equilibrium condition being the coexistence of two solid solution phases. Marshall and Laudise have also studied the hydrothermal growth method for KTN.

Gentile and Andres, at the Hughes Research Laboratories, have grown KTN crystals (of varying composition) for electrooptic modulator devices. The crystals, grown at constant temperature, utilizing a temperature gradient in the melt, resulted in a large yield, high optical quality and homogeneity with sizes up to 4 x 4 x 3 cm. Their Curie temperatures were found to range from -52°C to +16°C and the crystals were generally square in cross section, having clearly defined simple (100) faces.

Doyle and coworkers at Philco-Ford Corporation, have reported on their extensive attempts to grow large, high quality KTN crystals. Various melt/composition problems were encountered and corrected in the course of their experiments. Four major highlights in their program were:

a) startup and refinement of the rf Czochralski growth apparatus.

b) development of growth parameters for KTN 90:10 crystals which had well defined {100} faces and striae normal to these faces.

c) synthesis of small, high quality, optical gradient crystals.

d) refinement of growth techniques (seed rotation rate, crucible rotation rate, controlled depression of melt temperature, and controlled depression of the after heater temperature) to produce large, high quality KTN 80:20 crystals with well defined {100} faces and uniform striae normal to these faces.

Fay and coworkers, at Union Carbide Corporation, have reported on their extensive KTN crystal growing experiments to develop and improve electrooptic crystals for laser technology applications. A great deal of their efforts were devoted to crystal growth and laboratory evaluation methods to produce crystals of consistent quality.

Potassium tantalate niobate is a solid solution of two perovskite oxides, $KTaO_3$ and $KNbO_3$, which have very nearly the same unit cell size in their cubic phase, but quite different Curie temperatures [Chen et al.]:

| | | |
|---|---|---|
| $KTaO_3$ | $a_o = 3.989$ Å | $T_c \sim 4 \pm 2°K$ |
| $KNbO_3$ | $a_o = 4.021$ Å | $T_c = 698°K$ |

Triebwasser and Reisman and Banks have studied the various crystalline phases and their transition temperatures in the $KTa_xNb_{1-x}O_3$ system (Figure 5). For x = 0.65 the Curie temperature of KTN is approximately 10°C. Ferroelectric KTN belongs to the symmetry class 4mm. Above the Curie temperature, KTN has the ideal cubic perovskite structure with a lattice constant of $a_o = 3.988$ Å [DiDomenico and Wemple.]

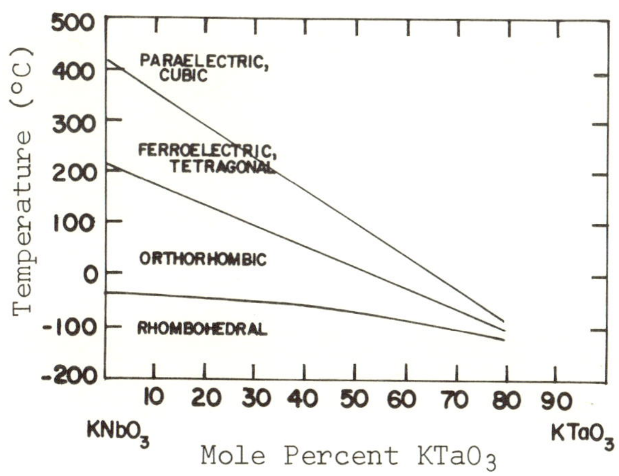


Fig. 5. Transition temperatures as a function of composition in the $KTaO_3$ - $KNbO_3$ system [Triebwasser]

Optical Properties - Refractive Index

Above its Curie point, KTN is optically isotropic in view of its cubic structure.

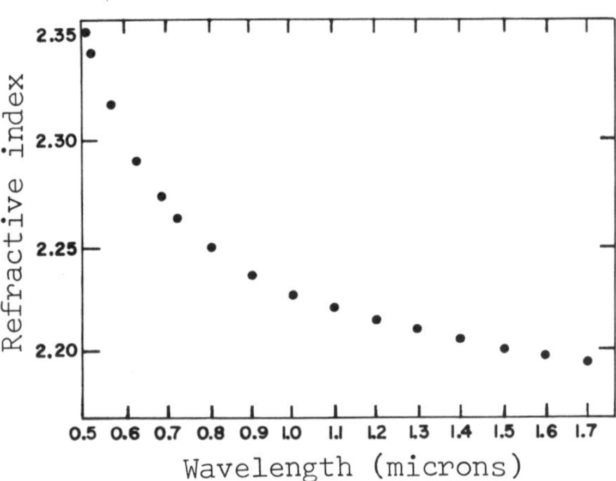


Fig. 6. Refractive index of $KTaO_3$ - $KNbO_3$ (composition 65:35) as a function of wavelength at room temperature [Chen et al.].

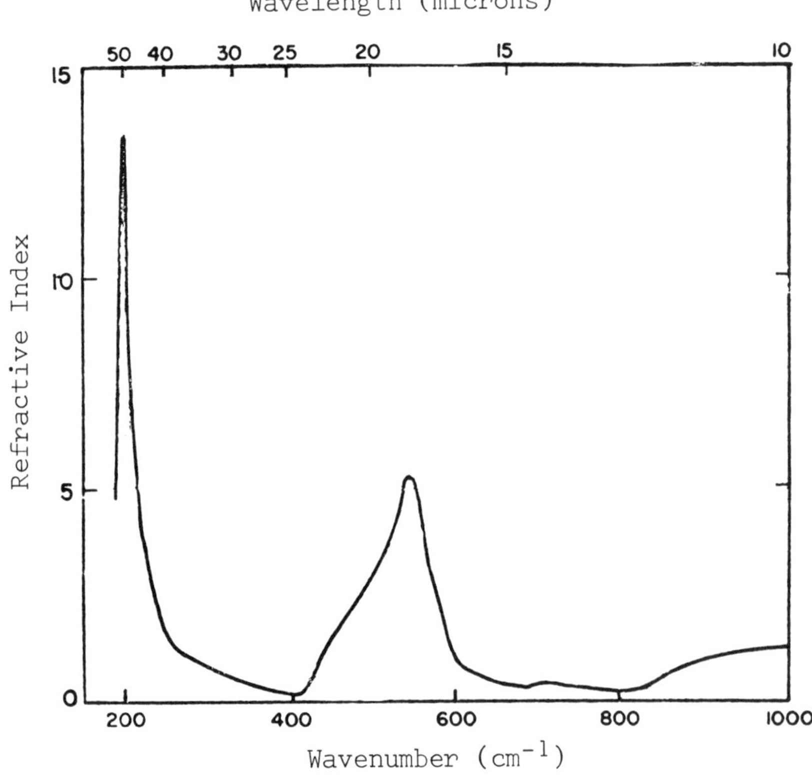


Fig. 7. Refractive index of $KTaO_3$ - $KNbO_3$ (composition 64:36) as a function of wave number at room temperature [Smakula].

## POTASSIUM TANTALATE NIOBATE (KTN)

### Optical Properties - Birefringence

KTN generally exhibits a static birefringence due to residual strain combined with a high strain optic coefficient. Smakula and Claspy studied various heat treatment cycles in an attempt to remove the strain, but found that heating to within a few degrees of the melting temperature for 10 hours or more, followed by a 40-hour cooling cycle, failed to remove the strain. They also noted that since the strain optic coefficient is so high, extreme care must be experienced in cutting and polishing to avoid introducing strain.

Eden described Texas Instruments Inc. experiments to sweep out the strain birefringence occurring in KTN crystals by applying bias electric fields. Early results at fields up to 5000 volts/cm appeared favorable. Most of the color fringes could be swept to the edges of the crystal, leaving only light tints in the center. Eventually, the crystal shattered after several runs at various field values, none of which exceeded 3000 volts/cm (he noted that in work at other laboratories, crystals of this type withstood much higher electric fields, e.g., up to 14 kV/cm at Bell Labs.). The cause of failure is not completely understood.

Haas and Johannes studied the birefringence of two KTN crystals with Curie points of 28 and 88°C, which, according to the data of Figure 5, corresponds to compositions of x = 0.62 and x = 0.52 ($KTa_xNb_{1-x}O_3$), respectively. The natural birefringence of the crystals at 5460 Å and 22°C was negative:

| $T_c$ | $n_o$ | $n_e$ | $n_o-n_e$ |
|---|---|---|---|
| 28°C | 2.300 | 2.312 | -0.012 |
| 88°C | 2.269 | 2.318 | -0.049 |

The birefringence of the crystal with $T_c$ = 88°C is shown in Figure 8. The refractive index $n_e$ = 2.318 remained practically constant in the temperature range investigated; the changes in the birefringence were due to changes in $n_o$.

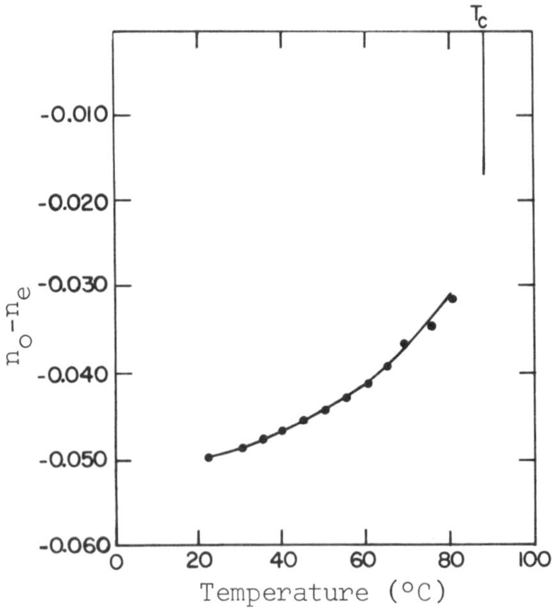

Fig. 8. Birefringence of a KTN crystal ($T_c$ = 88°C) as a function of temperature [Haas and Johannes]

## Optical Properties - Transmission

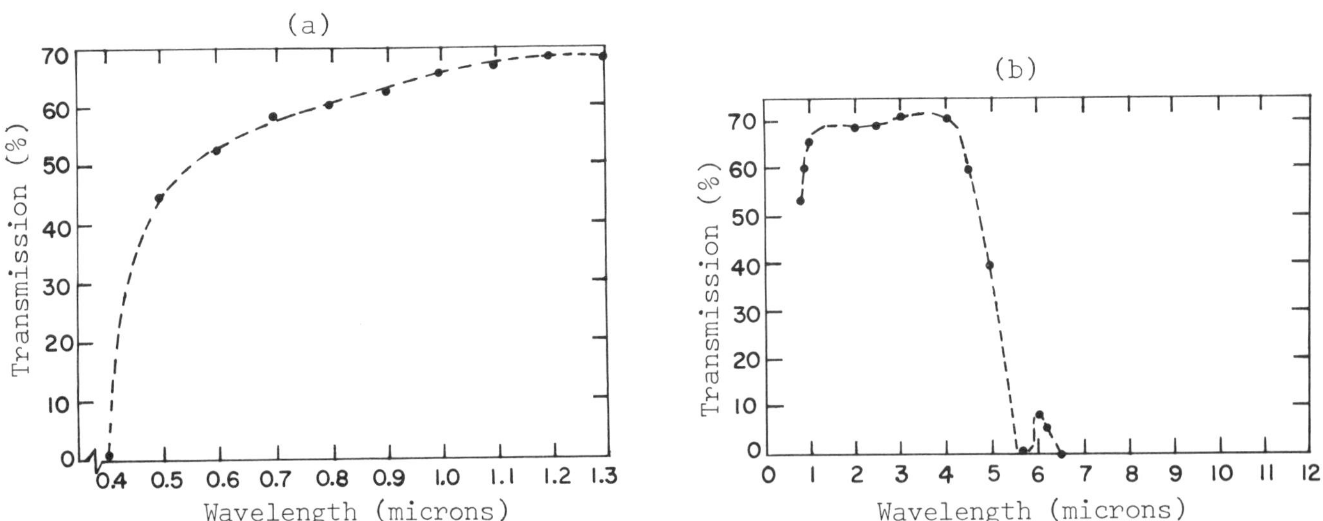


Fig. 9. Transmission of KTN as a function of wavelength. (a) 0.4 to 1.3 microns. (b) 1 to 12 microns [Hirschmann]

## Electrooptic Properties

Potassium tantalate niobate exhibits a strong linear electrooptic (Pockels) effect below the Curie temperature, which in turn has been shown above to vary with chemical composition. Above the Curie temperature, KTN exhibits a large quadratic electrooptic (Kerr) effect. The quadratic effect is very large in the vicinity of the Curie point and is found to be isotropic, Fay.

Because ferroelectric KTN belongs to the symmetry class 4mm, only three independent, non-zero electrooptic coefficients describe the linear effect:

$$r_{13} = r_{23}$$

$$r_{33}$$

$$r_{42} = r_{51}$$

Haas and Johannes have reported the following linear coefficients at 22°C and 5460 Å, with $n_e = 2.312$, $n_o = 2.300$ and $T_c = 28°C$:

$$r_{33}^{T} = 14 \times 10^{-10} \text{ m/V}$$

$$r_{13}^{T} \sim 10^{-10} \text{ m/V}.$$

The temperature dependence of the linear electrooptic coefficients is indicated in Figures 10-12. Similar results have been obtained by Anistratov for measurements at 20 to 72°C of $n_3^3 r_{33} - n_1^3 r_{13}$ which represents a change in birefringence for light parallel to the a-axis and a field along the c-axis. The crystals had a $T_c$=54°C.

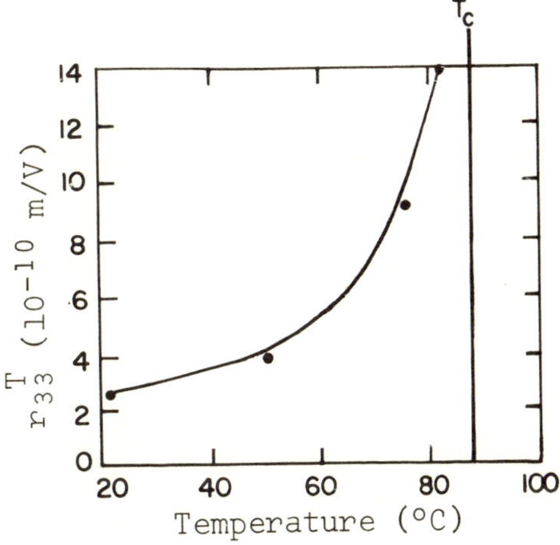


Fig. 10. Electrooptic coefficient $r_{33}^{T}$ for KTN ($T_c$ = 88°C) as a function of temperature [Haas and Johannes]

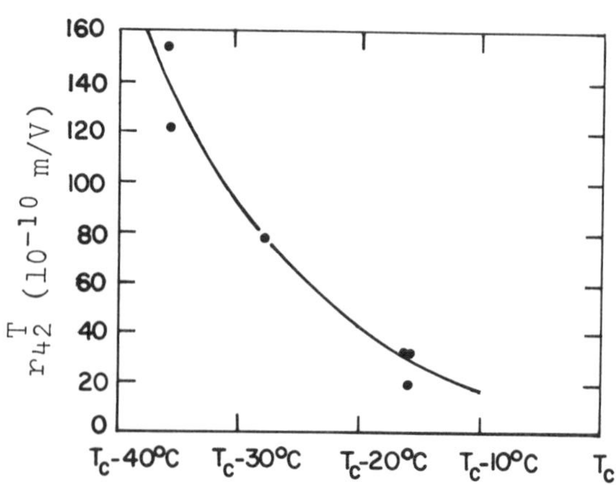


Fig. 11. Electrooptic coefficient $r_{42}^T$ as a function of temperature at 6328 Å for three separate samples of KTN with Curie points at 60, 52 and 42°C [Van Raalte]

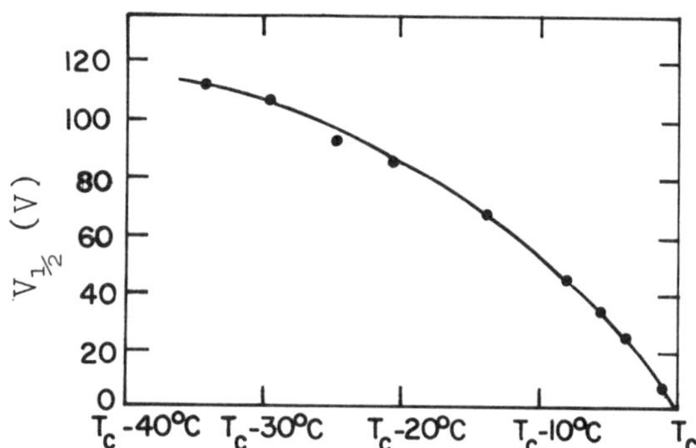


Fig. 12. Half-wave voltage (normalized for $\ell/d = 1$) as a function of temperature at 6328 Å for KTN sample with a Curie point of 60°C

$$V_{1/2} = \lambda\ (d/\ell) \big/ (n_e^{\,3}\, r_{33} - n_o^{\,3}\, r_{13})$$

$\ell$ = optical path length = 2.65 mm

d = electrode spacing = 3.35 mm

[ Van Raalte ]

Because paraelectric KTN belongs to the symmetry class Pm3m, three independent, non-zero electrooptic coefficients describe the quadratic effect: $g_{11}$, $g_{12}$ and $g_{44}$. Geusic et al. and Smakula and Claspy have reported quadratic coefficients in $KTa_{0.65}Nb_{0.35}O_3$ (Table 2). The former investigators found that the dc and 100 MHz values of $(g_{11} - g_{12})$ were identical; the latter authors found that a variation in Curie temperature from sample to sample resulted in no measurable difference in electrooptic coefficients. The wavelength dependence of $(g_{11} - g_{12})$ is indicated in Figure 13.

TABLE 2. QUADRATIC ELECTROOPTIC COEFFICIENTS FOR KTN.

| λ (microns) | $T_c$ (°C) | Electrooptic Coefficients ($m^4/C^2$) | | | | Reference |
|---|---|---|---|---|---|---|
| | | $g_{11} - g_{12}$ | $g_{11}$ | $g_{12}$ | $g_{44}$ | |
| 0.6328 | 10 | + 0.174 | +0.136 | -0.038 | +0.147 | Geusic et al. |
| 0.6328 | 0-26 | 0.156 | | | 0.141 | Smakula and Claspy |
| 1.15 | | 0.128 | | | 0.133 | |
| 3.39 | | 0.102 | | | 0.108 | |

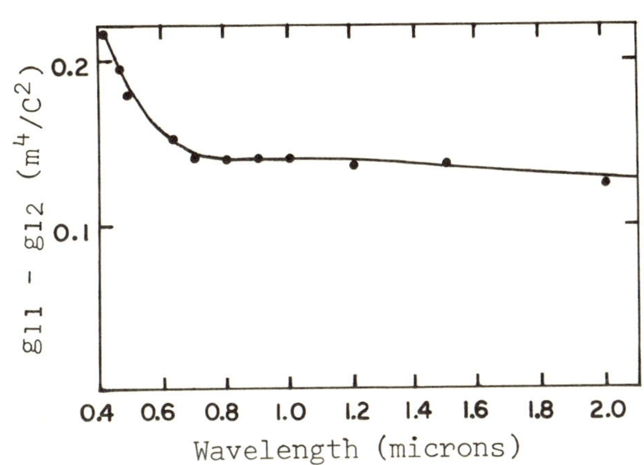


Fig. 13. Quadratic electrooptic effect ($g_{11} - g_{12}$) in KTN as a function of wavelength [Geusic et al.]



Photoelastic Properties

Piezooptic Coefficients ($KTa_{0.65}Nb_{0.35}O_3$):

$p_{11} - p_{12} \simeq 0.5$ Wemple and DiDomenico

Elastic Stiffness Constants:

$c_{11} = 2.64 \times 10^{11}$ N/m$^2$ Chen et al.

## Dielectric Properties - Dielectric Constant

The temperature dependence of the dielectric constant of KTN is shown in Figure 14. In the $KTaO_3$ - $KNbO_3$ system, the dielectric constant and the resulting Curie peaks are strongly dependent upon chemical composition. As indicated in Figure 15, the Curie point moves to lower temperatures and the peak dielectric constant increases with increasing $KTaO_3$ concentration. Thermal hysteresis in the heating and cooling curves of the dielectric constant has been reported by Smakula as well as Triebwasser. The latter author has indicated that the hysteresis generally decreases with increasing $KTaO_3$ concentration.

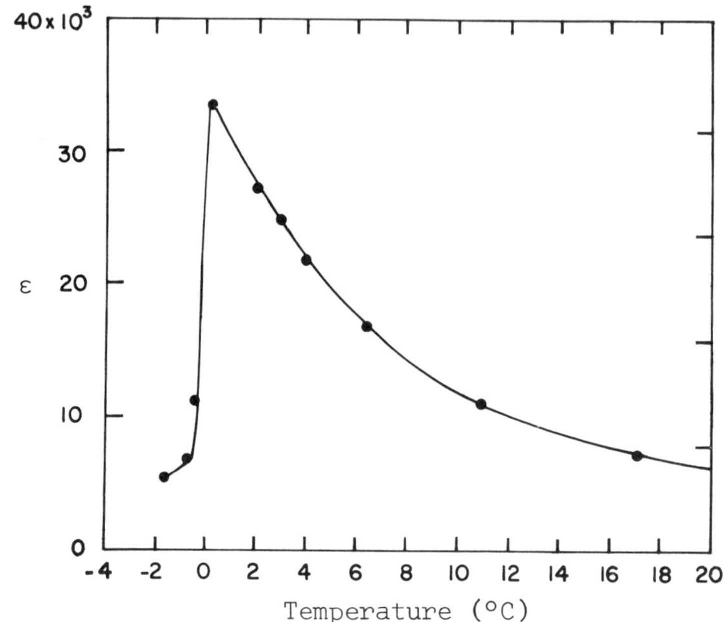


Fig. 14. Dielectric constant of $KTa_{0.66}Nb_{0.34}O_3$ as a function of temperature measured at 10 kHz with a field of 4V/cm. The data obey a Curie-Weiss law

$$\varepsilon = C/(T - T_c)$$

$$C = 1.45 \times 10^5 \ ^\circ K$$

$$T_c = 271.0^\circ K$$

Chen et al.

Fig. 15. Dielectric constant of KTN as a function of temperature measured at 10 kHz with a field of ∿ 5V/cm. The indicated percentages are in moles of $KTaO_3$ [Triebwasser].



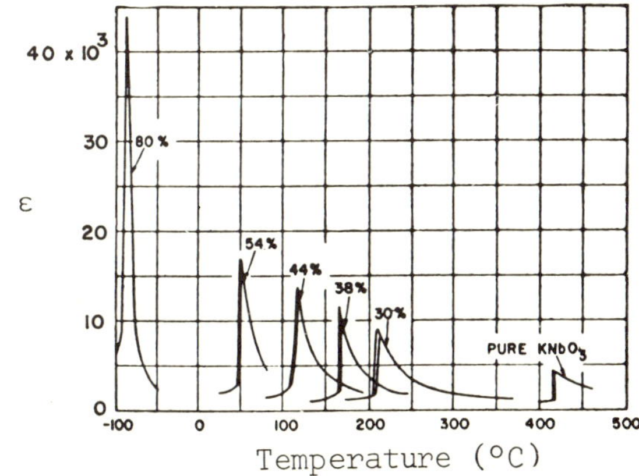


Typical results for the dependence of the dielectric constant of KTN on frequency, field, and direction of measurement are shown in Figures 16-18, respectively.

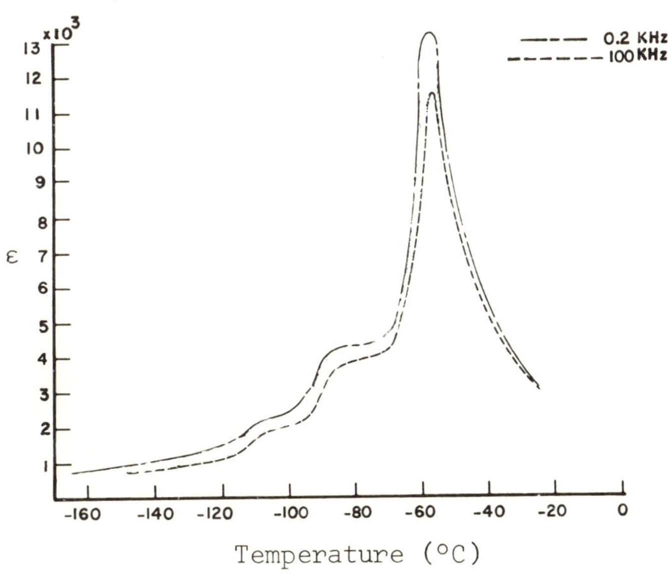


Fig. 16. Dielectric constant of $KTa_{0.80}Nb_{0.20}O_3$ as a function of temperature measured at two frequencies [Doyle and co-workers]

Fig. 17. Dielectric constant of $KTa_{0.80}Nb_{0.20}O_3$ as a function of applied field measured at 1 kHz and -56.5°C
[Doyle and coworkers]

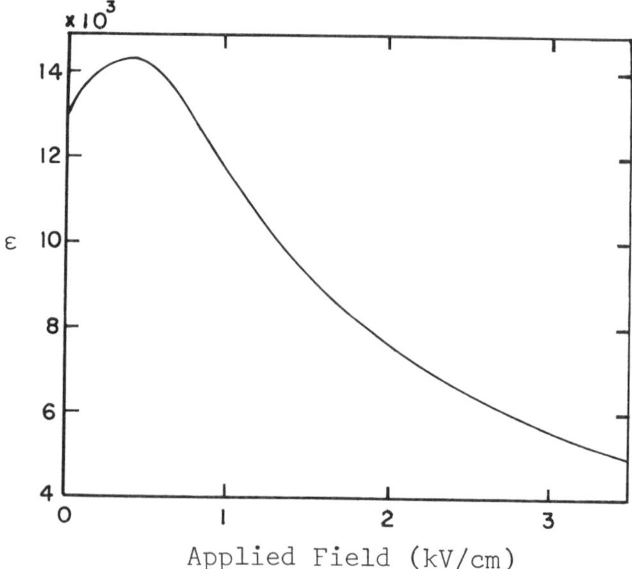


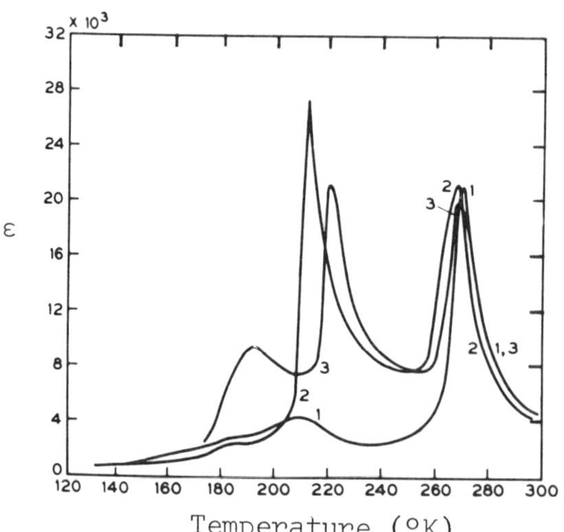


Fig. 18. Dielectric constant of $KTa_{0.63}Nb_{0.37}O_3:Sn_{0.001}$ as a function of temperature measured at 100 MHz and in three directions; direction 1 is in the growth direction, while 2 and 3 are normal to it
[Bonner et al.]

Dielectric Properties - Loss Tangent

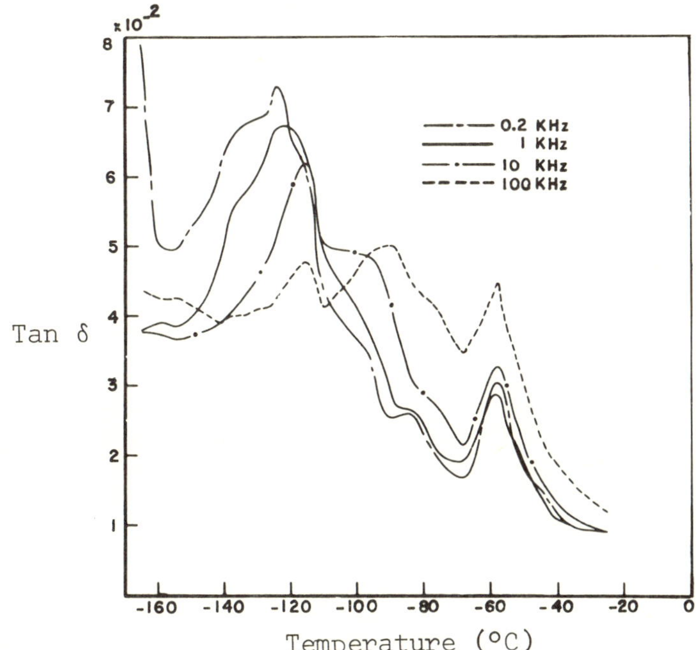


Fig. 19. Loss tangent of $KTa_{0.80}Nb_{0.20}O_3$ as a function of temperature and measured at three frequencies. Corresponding dielectric constant data are shown in Figure 16
[Doyle and coworkers]



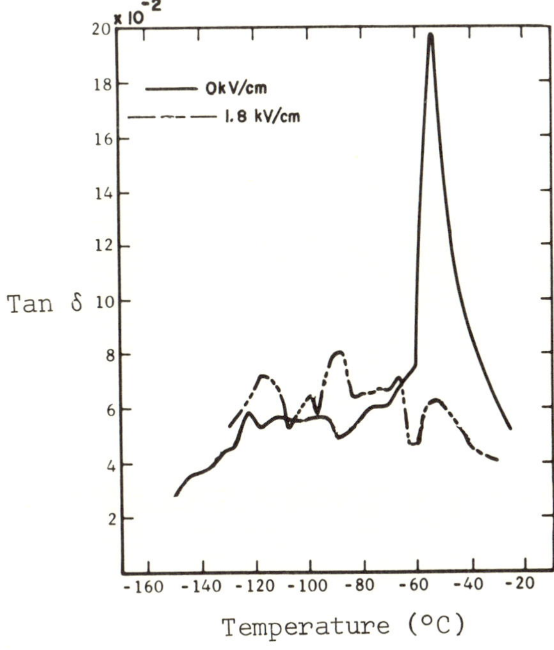


Fig. 20. Loss tangent of a small piece of KTN sample used to obtain data shown in Figure 19, as a function of temperature and measured at two values of bias field
[Doyle and coworkers]

## Dielectric Properties - Electrical Resistivity

Ideally, KTN should be a perfect dielectric insulator. Fay and Alford found that the conductivity is small but not negligible; the dc conductivity of several KTN crystals was found to range from $6 \times 10^{-15}$ to $3 \times 10^{-10}$ $(\text{ohm-m})^{-1}$. In addition to the steady state d-c conductivity, there is a relatively long term transient current. These workers could not identify the charge carriers involved in either the d-c or transient conduction processes.

Gentile and Andres, at Hughes Research Laboratories grew a wide range of KTN crystals at constant temperature, utilizing a temperature gradient in the melt. The resistivity of the samples ranged from $10^8$ to $10^{11}$ (ohm-cm), with no relationship to crystal composition.

Fukuda and coworkers found that the $K_2O$ content, as well as the $SnO_2$ dopant content, of the KTN melt affected the electrical resistivity. Their results are shown in Figure 21.

In these experiments, the resistivity was measured parallel to the grown axis <100> and the following conclusions drawn:
(a) Colorless, high resistivity ($\sim 10^{11\sim13}$ ohm-cm) crystals are obtained when relatively small amount of excess $K_2CO_3$ (52 $\sim$ 55 mole %) is used, whereas relatively large amount of excess $K_2CO_3$ (above 60 mole %) produces lower resistivity ($\sim 10^{8\sim10}$ ohm-cm) crystals, which are colored in variety of blue shades ranging from dark blue to colorless. (b) Sn-doped crystals show no remarkable increase in resistivity, in comparison with undoped ones. (c) About 0.1 mole % additions of $SnO_2$ in the melt with large excess concentration of $K_2O$ yields brown crystals.

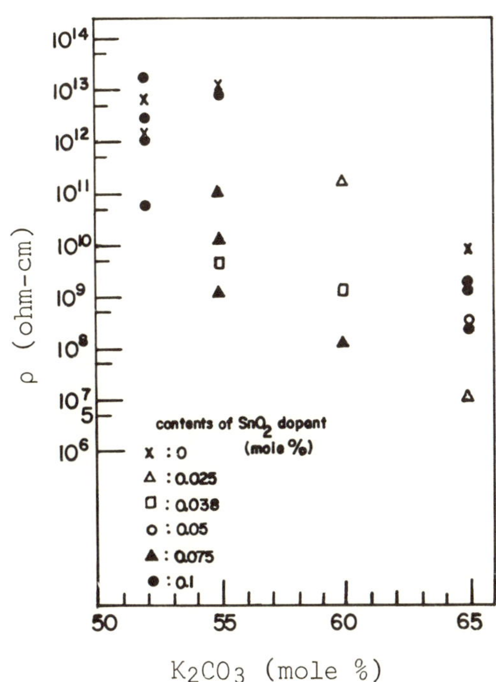


Fig. 21. Electrical resistivity at room temperature as a function of $K_2CO_3$ contents of starting materials [Fukuda et al.]

## Ferroelectric Properties

Various experimenters have shown that solid solutions of $KTaO_3$ and $KNbO_3$ have the perovskite structure and are ferroelectric; Triebwasser, for example, has shown (see Figure 5) that in the mixed crystal $KTa_xNb_{1-x}O_3$ the Curie temperature depends on the value of x. Bonner and coworkers conducted experiments which show that compositional inhomogeneities (ratios of $KTaO_3/KNbO_3$) occur during the growth of KTN crystals giving rise to fluctuations in the Curie temperature throughout the crystal. Curie temperature variations occurring during growth were reported in 1966 as ranging from -180°C to +450°C.

DiDomenico and Wemple have reported that 0.65:0.35 ratios in KTN have a first order phase cubic-tetragonal transition; crystals having a Nb:Ta ratio less than approximately 0.4 become second order. They have also discussed the crystallographic theories with respect to the transformation of high quality semiconducting crystals of KTN into single domain ferroelectric crystals below their Curie point. A first order phase transition is evidenced in the data presented by Chen et al., shown in Figure 22, in which the polarization-electric field curves agree with the Devonshire expansion of the free energy. Wemple et al. also report on the 65/35 material that under pressures up to 35 kbars, $dT_c/dP$ = 5.9°C/kbar.

POTASSIUM TANTALATE NIOBATE (KTN)

Ferroelectric KTN, according to Van Raalte must be poled to obtain single domain crystals. Before poling, the crystals appear cloudy, owing to a large number of domain boundaries. He was of the opinion that multiple domains are caused by the small extent of the tetragonal distortion of the lattice below the Curie point (x-ray data indicate a tetragonal distortion of less than 1/6% within 35°C of the Curie point). The crystals were poled by heating to 20°C above the Curie point while immersed in silicone oil (Dow Corning 705) with subsequent slow cooling under application of a dc field of 3-5 kV/cm. After poling, the samples were clear, although, after several minutes, some domain-boundary growth was often evident near the edges. As briefly noted above, this phenomenon, especially evident near the Curie point, is attributed to the small extent of the tetragonal distortion; these few domains could be eliminated by re-application of the poling field at room temperature or by application of a sinusoidal field, clamped above ground. This investigator also found that domain growth was less evident after multiple polings of the same crystal. Annealing and etching of the crystal to remove strains was not effective in retarding domain growth. Poling with the c-axis parallel and perpendicular to the lamellar pattern (striations) was equally possible.

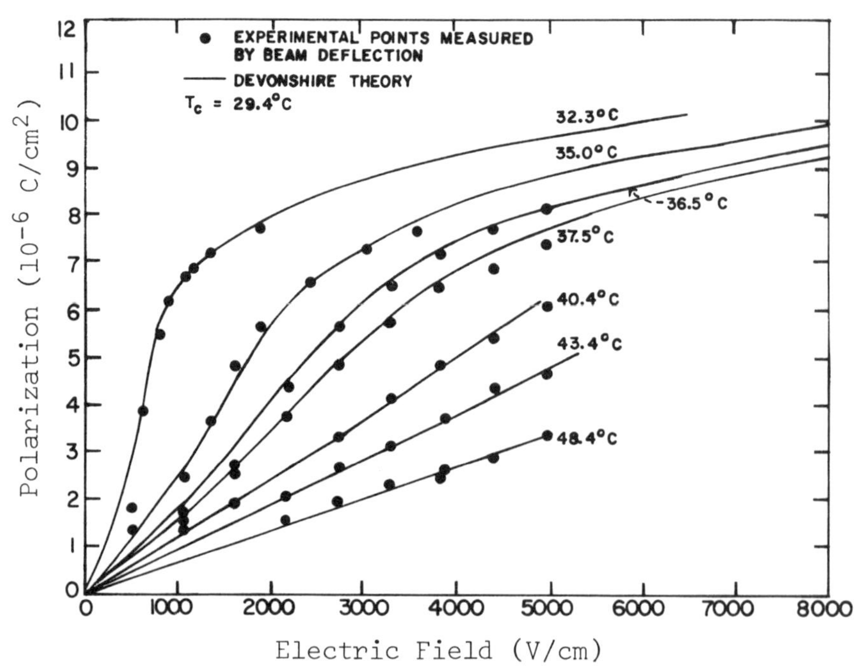


Fig. 22. Saturation effects in the induced polarization of KTN as a function of temperature [Chen et al.]

# POTASSIUM TANTALATE NIOBATE (KTN)

ANISTRATOV, A.T. Morphic Electro-Optic Effects in $KTa_xNb_{1-x}O_3$ Crystals. ACAD. OF SCI., USSR, BULL. PHYS., SER., v. 33, no. 7, 1969. p. 1014-1017.

BAER, W.S. Interband Faraday Rotation in Some Perovskite Oxides and Rutile. J. OF PHYS. AND CHEM. OF SOLIDS, v. 28, no. 4, Apr. 1967. p. 677-687.

BONNER, W.A. et al. Growth of Potassium Tantalate-Niobate Single Crystals for Optical Applications. AMERICAN CERAM. SOC., BULL., v. 44, 1965. p. 9-11.

BONNER, W.A. et al. Growth of $K(Ta,Nb)O_3$ Single Crystals for Optical and Semiconductor Studies. In CRYSTAL GROWTH, by: PIESER, H.S. International Conference on Crystal Growth, Boston, 1966. Oxford, N.Y., 1967. p. 437-440.

CHEN, F.S. et al. The Use of Perovskite Paraelectrics in Beam Deflectors and Light Modulators. IEEE PROC., Oct. 1964. p. 1258-1259.

CHEN, F.S. et al. Light Modulation and Beam Deflection with Potassium Tantalate-Niobate Crystals. J. OF APPLIED PHYS., v. 37, Jan. 1966. p. 388-398.

BELL TELEPHONE LABS., INC. Elastic Wave Devices Utilizing Mixed Crystals of Potassium Tantalate-Potassium Niobate, by: DENTON, R.T. Dec. 1966. 5 p.

DEMUROV, D.G. X-Ray Analysis and Dielectric Properties of Solid Solutions Based on the Ferroelectric $KTaO_3$. SOVIET PHYS.-CRYST., v. 16, no. 2, Sept.-Oct., 1971. p. 297-300.

DIDOMENICO, M. Jr. and S.H. WEMPLE. Paraelectric-Ferroelectric Phase Boundaries in Semiconducting Perovskite-Type Crystals. PHYS. REV., v. 155, no. 2, Mar. 10, 1967. p. 539-545.

DIDOMENICO, D. Jr. and S.H. WEMPLE. Optical Properties of Perovskite Oxides in their Paraelectric and Ferroelectric Phases. PHYS. REV., v. 166, Feb. 1968. p. 565-576.

PHILCO-FORD CORP. AERONUTRONIC DIV. NEWPORT BEACH, CALIF. Electronic Control of Laser Beams, by: DOYLE, W.M. et al. Final Tech. Rept. U-4378, Contract no. N00019-67-C-0263. May 1968. 73 p. AD 832 563.

PHILCO-FORD CORP. AERONUTRONIC DIV. NEWPORT BEACH, CALIF. Electronic Control of Laser Beams, by: DOYLE, W.M. et al. First Quarterly Tech. Rept. U-4439, Contract No. N00019-68-C-0469. July 1968. 17 p. AD 838-305.

PHILCO-FORD CORP. AERONUTRONIC DIV. NEWPORT BEACH, CALIF. Electronic Control of Laser Beams, by: DOYLE, W.M. et al. Final Rept. U-4566, Contract No. N00019-68-C-0469. Jan. 1969. 44 p. AD 847-930.

DUGAN, A.F. and W.M. DOYLE. Field-Dependent Dielectric Properties of $KTa_xNb_{1-x}O_3$. IEEE TRANS. ON ELECTRON DEVICES, v. ED-16, no. 6, June 1969. p. 522-525.

TEXAS INSTRUMENTS, INC. Solid State Techniques for Modulation and Demodulation of Optical Waves, by: EDEN, D.D. Final Tech. Rept. ECOM-03250-F. Sept. 1966. 283 p. AD 489 390.

FAY, H. Characterization of Potassium Tantalate-Niobate Crystals by Eelctro-Optic Measurements. MAT. RES. BULL., v. 2, no. 7, July 1967. p. 679-688.

UNION CARBIDE CORP. Single Crystal Perovskites as Electrooptically Active Materials, by: FAY, H. and W.J. ALFORD. Rept. No. AFML-TR-67-417. Dec. 1967. 157 p.

UNION CARBIDE CORP. ELECTRONICS DIV. Single Crystal Perovskites as Electro-optically Active Materials, by: FAY, H. Tech. Rept. AFML-TR-69-157. Contract No. AF 33(615)-5410. Apr. 1969. 64 p.

UNION CARBIDE CORP. Single Crystal Perovskites as Electrooptically Active Materials, by: FAY, H. et al. Prog. Rept. No. 1. Oct. 1966. 26 p. AD 650 410.

UNION CARBIDE CORP. Single Crystal Perovskites as Electrooptically Active Materials, by: FAY, H. et al. Prog. Rept. No. 2, Jan. 1967. 13 p. AD 650 411.

FUKUDA, T. et al. Growth and Eelctrical Properties of $KTa_xNb_{1-x}O_3$. PHYS. SOC. OF JAPAN, J., v. 24, no. 2, 1968. p. 430.

GARN, P.D. and S.S. FLASCHEN. Analytical Applications of a Differential Thermal Analysis Apparatus. ANALYTICAL CHEM., v. 29, no. 2, Feb. 1957. p. 271-275.

GENTILE, A.L. and F.H. ANDRES. A Constant Temperature Method for the Growth of KTN Single Crystals. MAT. RES. BULL., v. 2, 1967. p. 853-859.

GEUSIC, J.E. et al. Electro-Optic Properties of Some $ABO_3$ Perovskites in the Paraelectric Phase. APPLIED PHYS. LETTERS, v. 4, Apr. 1964. p. 141-143.

HAAS, W. and R. JOHANNES. Linear Electrooptic Effect in Potassium Tantalate Niobate Crystals. APPLIED OPTICS, v. 6, Nov. 1967. p. 2007-2009.

HILL, V.G. and R.I. HARKER. Development of Hydrothermal Method for Growing Large, High Purity Single Crystals of Beryllium Oxide, and Inititate Research on Growing Single Crystals of KTN and Chrysoberyl. Tem-Press Res. Rept. No. AFML-TR-66-343. Oct. 1966. AD 803 070.

HILL, V.G. et al. Subsolidus Stability Relations in the System $KTaO_3$ - $KNbO_3$. AMERICAN CERAM. SOC., J., v. 51, no. 12, Dec. 1968. p. 723-724.

HIRSCHMANN, E. Electro-Optic and Magneto-Optic Modulators. NASA Tech. Note D-2775. June 1967. 10 p.

MARSHALL, D.J. and R.A. LAUDISE. The Hydrothermal Phase Diagrams $K_2O$-$Nb_2O_5$ and $K_2O$-$Ta_2O_5$ and the Growth of Single Crystals of $K(Ta,Nb)O_3$. In CRYSTAL GROWTH, by: PREISER, H.S. International Conference on Crystal Growth, Boston, 1966. Oxford, N.Y., 1967. p. 557-561.

REISMAN, A. and E. BANKS. Reactions of the Group VB Pentoxides. VIII. Thermal, Density and X-Ray Studies of the Systems $KNbO_3$ - $NaNbO_3$ and $KTaO_3$ - $KNbO_3$. AMERICAN CHEM. SOC., J., v. 80, 1958. p. 1877-1882.

POTASSIUM TANTALATE NIOBATE (KTN)

REISMAN, A. et al. Phase Diagram of the System $KNbO_3$ - $KTaO_3$ by the Methods of Differential Thermal and Resistance Analysis. AMERICAN CHEM. SOC., J., v. 77, 1955. p. 4228-4228-4230.

REISMAN, A. et al. Reactions of the Group VB Pentoxides with Alkali Oxides and Carbonates. III. Thermal and X-Ray Phase Diagrams of the System $K_2O$ and $K_2CO_3$ with $Ta_2O_5$. AMERICAN CHEM. SOC., J., v. 78, 1956. p. 4514.

M.I.T. CRYSTAL PHYSICS LAB. A Study of the Physical Properties of High-Temperature Single Crystals, by: SMAKULA, A. Rept. No. AFCRL-67-0645. Sept. 1967. 124 p. AD 663 734.

SMAKULA, P.H. and P.C. CLASPY. The Electro-Optic Effect in $LiNbO_3$ and KTN. AIME METALL. SOC., TRANS., v. 239, no. 3, Mar. 1967. p. 421-424.

TRIEBWASSER, L. Study of Ferroelectric Transitions of Solid Solution Single Crystals of $KNbO_3$-$KTaO_3$. PHYS. REV., v. 114, 1959. p. 63.

VAN RAALTE, J.A. Linear Electro-Optic Effect in Ferroelectric KTN. OPTICAL SOC. OF AMERICA, J., v. 57, no. 5, May 1967. p. 671-674.

WEMPLE, S.H. and M. DIDOMENICO. Oxygen-Octahedra Ferroelectrics. II. Electro-Optical and Nonlinear-Optical Device Applications. J. OF APPLIED PHYS., v. 40, no. 2, Feb. 1969. p. 735-752.

WEMPLE, S.H. and S.K. KURTZ. Optical and Electrical Properties of Semiconducting $KTa_{0.65}Nb_{0.35}O_3$ (KTN). AMERICAN PHYS. SOC., BULL., v. 11, 1966. p. 401.

WEMPLE, S.H. et al. Electron Scattering in Perovskite-Oxide Ferroelectric Semiconductors. PHYS. REV., v. 180, no. 2, Apr. 10, 1969. p. 547-556.

AEROSPACE CORP. Some Factors Affecting the Growth and Properties of $KTaO_3$-$KNbO_3$ Mixed Crystals, by: WILCOX, W.R. and L.D. FULLMER. Rept. No. SSD-TR-65-68. June 1965. 36 p. AD 466 551.

## Introduction

Proustite (silver thioarsenite) is well known as a mineral, but natural specimens are small or are of poor optical quality. Large high-quality single crystals of synthetic proustite have recently become available for the first time. It has been actively investigated by a number of organizations (Hughes Research Laboratories, Bell Telephone Laboratories, and the Royal Radar Establishment in England) for electrooptic and nonlinear optic applications.

The material is a ternary compound composed of 65.4 wgt.% silver, 15.2 wgt.% arsenic and 19.4 wgt.% sulfur, having the chemical formula $Ag_3AsS_3$. It is brittle and fractures easily. The nonlinear optical coefficients of proustite are thirty to fifty times those of KDP and, since it is transparent from 0.6 to 13 microns, its potential application as a $CO_2$ laser modulator is worth considering. Proustite is interesting in that it also shows semiconducting and photoconductive behavior in addition to its good nonlinear optical properties.

## Chemical and Physical Properties

| | | | |
|---|---|---|---|
| Chemical Formula | $Ag_3AsS_3$ | | |
| Density | 5.49 g/cm$^3$ | | [Donnay] |
| Hardness | 2 (Mohs) | brittle | [Dana] |
| Color | scarlet | transparent | [Dana] |

Wehmeier, F.H. et al. have studied the $Ag_2S$-$As_2S_3$ system by means of the differential thermal analysis (DTA). Two stable phases, proustite ($Ag_3AsS_3$, 25 mol.% $As_2S_3$) and smithite ($AgAsS_2$) have been identified. In the quenched melt, xanthoconite ($Ag_3AsS_3$) is identified as well as a high sulfur-content Proustite. One other silver arsenosulfide has been studied by Matsumoto and Nowack, trechmannite ($AgAsS_2$) which has the same composition as smithite. The silver-arsenic-sulfur system has also been studied at 575°C to 920°C by Roland who found only liquids and binary solids in this temperature range.

## PROUSTITE

### Crystallography

Crystal growth studies on proustite have been conducted by a number of investigators. Gentile and Stafsudd have grown single crystals (2.4 cm diam.) by the Czochralski technique and using an indium rod for the nucleation of the crystal by the cold-finger technique. Best quality material (e.g., material without cellular structure) was obtained using growth rates of approximately 2mm/hr. Hulme and co-workers have also grown proustite crystals of good optical quality from the melt in sealed quartz tubes (approx. 1 cm diam.). In 1947, Peacock at the University of Toronto reported growing transparent crystals (up to 0.5 mm) by a hydrothermal process. Bardsley and Jones have reported on the Stockbarger, Bridgman and Czochralski growth techniques and the resulting quality of the crystals. Problems arising in the growth process are cellular structures and nonstoichiometry; the Stockbarger technique is shown to be the most useful method for production of large crystals.

| | |
|---|---|
| Crystal Symmetry | trigonal |
| Space Group | R3c |
| Point Group | 3m or $C_{3v}$ |

Lattice Constants

| $a_o$ (Å) | $c_o$ (Å) | | |
|---|---|---|---|
| 10.74 | 8.64 | | [Hulme et al.] |
| 10.77 | 8.67 | | [Butsko et al.] |
| 10.78 | 8.682 | | [Landolt-Börnstein] |
| 10.80 | 8.69 | $\alpha$= 103°31'<br>$a_{rh}$= 6.88 Å<br>Z= 6 | [Donnay] |

### Optical Properties - Refractive Index

TABLE 1. ABSOLUTE REFRACTIVE INDICES OF PROUSTITE AT 20°C [Hulme et al.]

| Wavelength (microns) | $n_e$ | $n_o$ |
|---|---|---|
| 0.5876 | 2.7896 | - |
| 0.6328 | 2.7391 | 3.0190 |
| 0.6678 | 2.7094 | 2.9804 |
| 1.014 | 2.5901 | 2.8264 |
| 1.129 | 2.5756 | 2.8067 |
| 1.367 | 2.5570 | 2.7833 |
| 1.530 | 2.5485 | 2.7728 |
| 1.709 | 2.5423 | 2.7654 |
| 2.50 | 2.5282 | 2.7478 |
| 3.56 | 2.5213 | 2.7379 |
| 4.62 | 2.5178 | 2.7318 |

PROUSTITE

Proustite is uniaxial negative, the birefringence, $n_o - n_e = 0.28$ at 6328 Å decreases to 0 at 22.7 microns [Hobden]. Refractive index measurements by Hobden at 0.58 to 10.6 microns and 20°C yield the following two dispersion equations:

$$n_o^2 = 9.220 + \frac{0.4454}{\lambda^2 - 0.1264} - \frac{1733}{1000 - \lambda^2}$$

$$n_e^2 = 7.007 + \frac{0.3230}{\lambda^2 - 0.1192} - \frac{660}{1000 - \lambda^2}$$

The temperature variation for both indices is: $dn/dT \simeq +1.5 \times 10^{-4}/°C$ in the visible and less in the infrared.

Although no large-scale systematic study has been made of laser damage in proustite, Bardsley et al. have reported the following observations:

1. 50 mW radiation at 6328 Å focused onto a proustite sample can cause burning; the room temperature absorption coefficient is $\sim 1\ cm^{-1}$ at this wavelength (cf Figure 2) and absorption would rise rapidly with temperature.
2. Proustite does not display refractive index inhomogeneities when a 50 pps ruby laser giving a mean power of 500 mW is focused onto the crystal by a 5 cm lens - a situation which would quickly produce observable damage in lithium niobate.

## Optical Properties - Transmission

As indicated in Figures 1 and 2, proustite transmits from 0.6 to 13 microns. Ernst and Witteman, using a $CO_2$ laser at 9.2 microns, obtained an absorption coefficient of $0.29\ cm^{-1}$ on a single crystal.

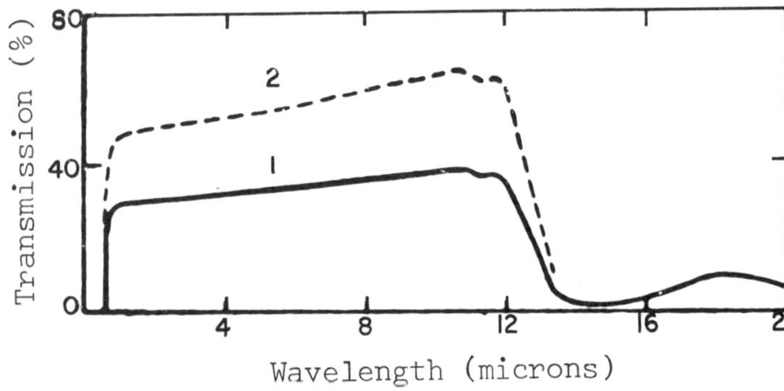


Fig. 1. Transmission of a proustite crystal 1.72 mm thick as a function of wavelength. (1) without including reflection; (2) with reflection losses included [Guseva et al.].

The absorption edge in single crystals of $Ag_3AsS_3$, is reported by Dovgii et al. at four temperatures as shown in Figure 2.

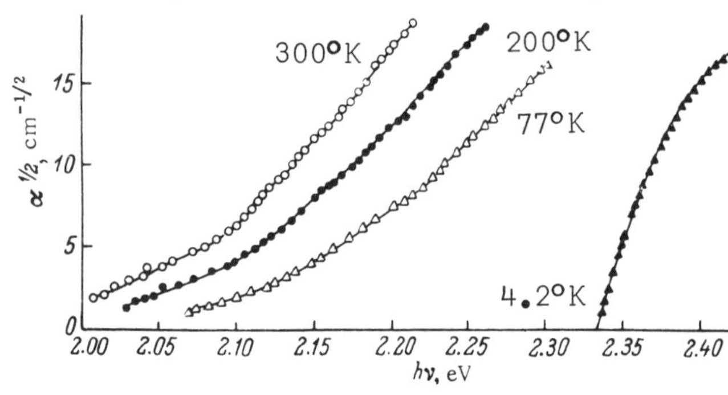


Fig. 2. Optical Absorption as a function of photon energy in single crystal $Ag_3AsS_3$ at 4 temperatures. E || c-axis.

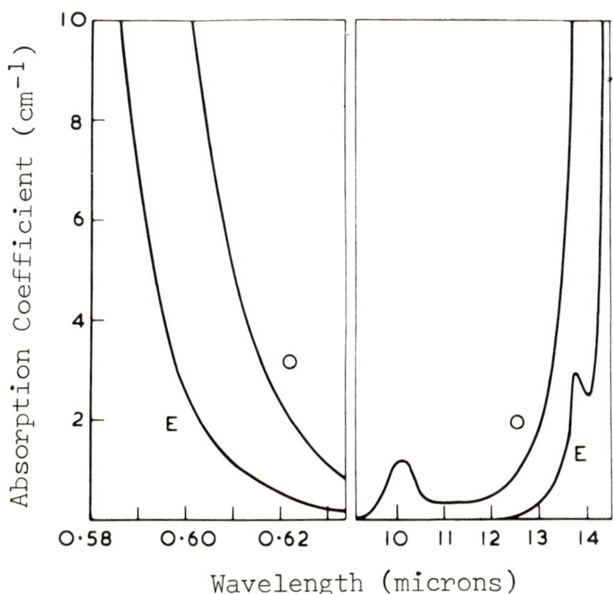


Fig. 3. Absorption coefficient for ordinary and extraordinary waves of an x-cut plate of proustite as a function of wavelength at 20°C. There are no absorption bands stronger than 0.1 cm$^{-1}$ from 0.63 to 9 microns [Hulme et al.].



## Optical Properties - Nonlinear Optical Behavior

The nonlinear optical coefficients in proustite, relative to $d_{36}^{2\omega}$ in KDP are [Hulme et al.]:

$$d_{22}^{2\omega} / d_{36}^{2\omega} \text{ (KDP)} = 50$$

$$d_{31}^{2\omega} / d_{36}^{2\omega} \text{ (KDP)} = 30$$

at 1.152 microns

Proustite shows considerable promise as a nonlinear optical material; it is readily grown, it has an exceptionally wide transmission band including part of the photomultiplier region, it allows any three-wave collinear process in that band to be phase-matched, and it has large nonlinear coefficients. Warner has demonstrated the suitability of using synthetic proustite for up-converting 10.6 micron radiation to the visible where it can be efficiently detected using photomultiplier tubes. Radiation from a $CO_2$ laser is detected by generating the sum frequency under phase-matched conditions in proustite at room temperature pumped with ruby laser radiation.

## Electrooptic Properties

The frequency response of the electrooptic effect in proustite is illustrated in

Figure 4. Trevelyan points out that it is not possible to make measurements using dc potentials because silver transportation occurs and the resistivity rapidly decreases; localized heating takes place causing the crystal to shatter. Measurements were possible to as low at 10 Hz.

Proustite, having trigonal 3m symmetry, posseses four independent linear electro-optic coefficients:

$$r_{22} = -r_{12} = -r_{61}$$

$$r_{51} = r_{42}$$

$$r_{13} = r_{23}$$

$$r_{33}$$

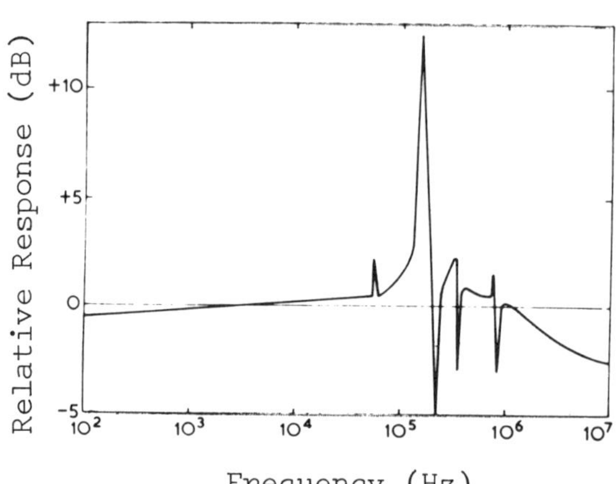




Fig. 4. Frequency response of the electrooptic effect in proustite [Trevelyan].

TABLE 2. ELECTROOPTIC COEFFICIENTS OF PROUSTITE AT 6328 Å.

| Coefficient | Value ($10^{-12}$ m/V) | Half-wave Voltage (kV) | Reference |
|---|---|---|---|
| $r_{13}^T$ | 2.78 | | Trevelyan |
| $r_{22}^T$ | 3.21 | | |
| $r_{33}^T$ | 1.57 | | |
| $n_e^3 r_{33}^T - n_o^3 r_{13}^T$ | 33.4 | | |
| $r_{22}^S$ | 1.07 ± 0.04 | 10.9 | Warner |
| $n_o^3 r_{13}^S - n_e^3 r_{33}^S$ | 140 ± 20 | 9.0 | |

## Photoelastic Properties - Elastic Constants

Trevelyan has determined the effective elastic constant of proustite using the relation

$$c = \rho v^2,$$

where $\rho$ is the density and v is the longitudinal acoustic wave velocity.

| direction | v ($10^5$ cm/sec) | c ($10^{10}$ N/m$^2$) |
|---|---|---|
| [0001] | 2.55 ± 0.15 | 3.58 |
| [00$\bar{1}$0] | 3.1 ± 0.2 | 5.28 |

The crystal appears to have potential application as an acoustically driven optical modulator or beam deflector.

## Dielectric Properties

Trevelyan has reported the following dielectric properties of proustite at 20 MHz:

$\varepsilon_{11}^{S} = 14.5$ $\varepsilon_{33}^{S} = 18.0$

$\varepsilon_{11}^{T} = 16.5$ $\varepsilon_{33}^{T} = 20.0$

$\tan \partial_{11} = 20 \times 10^{-3}$ $\tan \partial_{11} = 5 \times 10^{-3}$

The electric and dielectric properties of proustite have been studied in detail by investigators at the Royal Radar Establishment [Davies et al., Bardsley et al.]. They find that at room temperature, conduction in proustite is predominantly by transport of silver ions (Ag+) with transport number 1.02 ± 0.02. The commonly observed resistivity of proustite of $10^5$ ohm-cm is consistent with ionic conductivity. Photoconductivity measurements show that electronic conduction can also occur. Microscopic observations show that silver can grow from the cathode as dendrites during charge passage. These may penetrate the interior of the crystal and lead to cracks; a common observation with proustite is disintegration of the crystal on passing direct current.

These workers report a dielectric constant, measured at 300°K and in the range 1-50 MHz, of 21 ± 1 in all directions. Below 1 MHz, however, ionic conduction effects appear and the apparent dielectric constant can increase rapidly to as high as 1000 at 2 kHz. Measurements were made on single crystals of high purity.

## Thermal Properties

Melting Point

| | | |
|---|---|---|
| Solidified Melt | 472 ± 3°C | [Wehmeier et al.] |
| Single Crystal | 488 ± 3°C | |

Pyroelectric Coefficient

$p_3$ 80 x $10^{-11}$ Coul/cm$^2$°C at 20°C [Soref]

AMMANN, E.O. and J.M. YARBOROUGH. Optical Parametric Oscillation in Proustite. APPL. PHYS. LETTERS, v. 17, no. 6, Sept. 15, 1970. p. 233-235.

BARDSLEY, W. and O. JONES. On the Crystal Growth of Optical Quality Proustite and Pyrargyrite. J. OF CRYSTAL GROWTH, v. 3/4, 1968. p. 268-271.

BARDSLEY, W. et al. Synthetic Proustite; a Survey of Properties and Uses. OPTO-ELECTRONICS, v. 1, 1969. p. 29-31.

BUTSKO, M.I. et al. Investigation of Some Properties of Proustite ($Ag_3AsS_3$). Translation into English from Ukr. Fiz. Zh. (Kiev), v. 12, no. 12, 1967. p. 2052-2054. NASA N68-30659. 3pp.

DANA, J.D. and E.S. DANA. The System of Mineralogy. 7th Ed., v. 1, Elements, Sulfides, Sulfosalts, Oxides. Edited by: C. PALACHE, H. BERMAN and C. FRONDEL, Harvard Univ. Pub. John Wiley and Sons, Inc., N.Y., London, 1944. 808pp.

DAVIES, P.H. et al. The Electrical Properties of Synthetic Crystals of Proustite ($Ag_3AsS_3$). J. OF PHYS., D, (BRITISH J. OF APPL. PHYS.), Ser. 2, v. 2, 1969. p. 165-170.

DONNAY, J.D.H. (Ed.) Crystal Data. Determinative Tables. 2nd Ed. American Crystallographic Association. April 1963. ACA Monograph no. 5.

DOVGII, Ya.O. et al. Optical Properties of $Ag_3AsS_3$Single Crystals. SOVIET PHYS.-SOLID STATE, v. 13, no. 4, Oct. 1971. p. 995-996.

ERNST, G.J. and W.J. WITTEMAN. Second-Harmonic Generation in Proustite with a CW $CO_2$ Laser. IEEE J. OF QUANTUM ELECT., v. 8, no. 3, Mar. 1972. p. 382-383.

GENTILE, A.L. and O.M. STAFSUDD. Czochralski-Grown Proustite and Related Compounds. J. OF CRYSTAL GROWTH, v. 3/4, 1968. p. 272-274.

GUSEVA, L.M. et al. Proustite and Pyrargyrite as Optical Materials for the Infrared Region. OPT. I SPEKTROSKOPIIA, v. 24, Feb. 1968. p. 298-300. Trans. in: OPT. AND SPECTRO., v. 24, Feb. 1968. p. 156-157.

HANNA, D.C. et al. Reliable Operation of a Proustite Parametric Oscillator. APPL. PHYS. LETTERS, v. 20, no. 1, Jan. 1972. p. 34-36.

HOBDEN, M.V. The Dispersion of the Refractive Indices of Proustite ($Ag_3AsS_3$). OPTO-ELECTRONICS LETTERS., v. 1, no. 3, Aug. 1969. p. 159.

HULME, K.F. et al. Synthetic Proustite ($Ag_3AsS_3$): A New Crystal for Optical Mixing. APPLIED PHYS. LETTERS, v. 10, no. 4, Feb. 15, 1967. p. 133-135.

MATSUMOTO, T. and W. NOWACKI. The Crystal Structure of Trechmannite, $AgAsS_2$. Z. FUER KRISTALLOGRAPHIE, v. 129, no. 4, May 1969. p. 163-177.

ROLAND, G.W. The System Ag-As-S: Phase Relations between 920 and 575°C. METALL. TRANS., v. 1, no. 7, July 1960. p. 1811-1814.

SOREF, R.A. Interrelation of Pyroelectric and Nonlinear Optical Coefficients in Ferroelectric Crystals. IEEE J. OF QUANTUM ELECTRONICS, v. 5, no. 2, Feb. 1969. p. 126-129.

TREVELYAN, B. The Electro-Optic Effect in Proustite. OPTO-ELECTRONICS, v. 1, 1969. p. 9-12.

TREVELYAN, B. The Acoustic Properties of Proustite. OPTO-ELECTRONICS, v. 1, 1969. p. 61-62.

WARNER, J. Photomultiplier Detection of 10.6 Microns Radiation Using Optical Up-Conversion in Proustite. APPLIED PHYS. LETTERS, v. 12, Mar. 1968. p. 222-224.

WARNER, J. The Electro-Optic Effect in Proustite ($Ag_3AsS_3$). J. OF PHYS., D, (BRITISH J. OF APPLIED PHYS.), v. 1, July 1968. p. 949-950.

WEHMEIER, F.H. et al. The System $Ag_2S$-$As_2S_3$ and the Growth of Crystals of Proustite, Smithite, and Pyrargyrite. MAT. RES. BULL., v. 3, no. 9, 1968. p. 767-778.

# APPENDIX

The publications described below represent a cross-section of useful reviews of electrooptic materials properties and electrooptic modulator principles, design techniques and applications.

BAIRD-ATOMIC, INC. Cambridge, Mass. Electro-Optic Light Modulators: KDP, ADP and KDDP, and Supplementary Electronics. Publication No. EP-5, June 1967. 40 p.

The first company to exploit commercially electrooptic crystals in the development and manufacture of the electro-optic light modulator (EOLM) presents in this publication their historical experience in electro-optic phenomena.

BILLINGS, B.H. The Electro-Optic Effect in Uniaxial Crystals of the Type $XH_2PO_4$. I. Theoretical. OPTICAL SOC. OF AMERICA, J., v. 39, no. 10, Oct. 1949. p. 797-801.

Phenomenological description of linear electrooptic effect.

COMPTON, R.D. The Promising World of Electro-Optical Modulators. ELECTRO-OPTICAL SYSTEMS DESIGN, Sept./Oct. 1969. p. 60-71.

In this recent review, the editor of Electro-Optical Systems Design presents an accurate picture of present-day modulator activity and the characteristics that should be considered in applying modulators.

COOK, W.R., JR. Ferroelectric, Piezoelectric, and Electrooptic Materials. In DIGEST OF LITERATURE ON DIELECTRICS, vols. 30-32, 1966-1968. Nat. Acad. of Sci., Wash., D.C.

For the past several years, William R. Cook of Clevite Corp. has contributed comprehensive, yearly bibliographic reviews to this DIGEST; electrooptic materials have been included since 1966.

UNIVERSITY OF MICHIGAN, Willow Run Laboratories. Light Modulation Techniques. By: CURRIE, G.D. Contract No. DA-28-043-AMC-00013(E). Rept. No. ECOM-00013-64. Nov. 1966. AD 800 136.

A review of methods (mechanical, chemical, absorption, electrical and magnetic) for modulating light.

DiDOMENICO, M., JR. and L.K. ANDERSON. Broadband Electro-Optic Traveling-Wave Light Modulators. BELL SYS. TECH., J., v. 42, Nov. 1963. p. 2621-2678.

Survey of eight electrooptic crystals for this application: $BaTiO_3$, $SrTiO_3$, $KTaO_3$, CuCl, ZnS, KDP and ADP.

UNION CARBIDE CORP., Electronics Division, Indianapolis, Indiana. Single Crystal Perovskites as Electrooptically Active Materials. By, FAY, H. and W.J. ALFORD. Contract No. AF 33(615)-5410. Rept. No. AFML-TR-67-417. Dec. 1967

As part of a program to discover the develop perovskite compounds, this annual report contains tabulated data on numerous compounds reported in the literature and details a computer program for generating hypothetical perovskite compounds.

HIRSCHMANN, E. Electro-Optic Light Modulators. NASA Tech. Note D-3678. Nov. 1966. 13 p.

Compares the performance of KDDP, GaAs and HMTA light modulators.

PHILCO-FORD CORP., Blue Bell, Penn. Electronic Control of Laser Beams. By: JOHANNES, R. et al. Contract No. NOw 65-0226-f. Rept. No. B009-F. Mar. 15, 1966. AD 480 998.

A literature search from May 1964 to March 1966 resulted in the collection of data and information on the properties of 36 electrooptic materials.

JONA, F. and G. SHIRANE. FERROELECTRIC CRYSTALS, 1962. N.Y., Macmillan.

KAMINOW, I.P. and E.H. TURNER. Electrooptic Light Modulators. IEEE PROC., v. 54, no. 10, Oct. 1966. p. 1374-1390.

All available data on a large number of linear and quadratic electrooptic materials are tabulated; design considerations and operating principles for lumped, traveling wave, Fabry-Perot, Zig-Zag, Phase Reversal and p-n Junction Modulator configuration are outlined.

KÄNZIG, W. Ferroelectrics and Antiferroelectrics. SOLID STATE PHYS., v. 4, 1957. N.Y., Academic Press. p. 1-197.

LAND, C.E. and P.D. THACHER. Ferroelectric Ceramic Electrooptic Materials and Devices. IEEE PROC., v. 57, no. 5, May 1969. p. 751-768.

A report that thin polished plates of hot-pressed, rhombohedral, lead zirconate titanate (PZT) ferroelectric ceramics possess several useful and unique electrooptic properties.

LANDOLT-BÖRNSTEIN. NUMERICAL DATA AND FUNCTIONAL RELATIONSHIPS IN SCIENCE AND TECHNOLOGY, New Series, v. 1, Elastic, Piezoelectric, Piezooptic and Electrooptic Constants of Crystals. Group III: Crystal and Solid State Physics, ed. by: Hellwege, K.-H. and A.M. Hellwege. Berlin, Ger.: Springer-Verlag, 1966.

LANDOLT-BÖRNSTEIN. NUMERICAL DATA AND FUNCTIONAL RELATIONSHIPS IN SCIENCE AND TECHNOLOGY, New Series, v. 3, Ferro- and Antiferroelectric Substances. Group III: Crystal and Solid State Physics, ed. by: Hellwege, K.-H. New York, Springer-Verlag, 1969.

This recent volume represents an excellent compilation of experimental data on nearly 450 ferroelectrics and their solid solutions. Properties covered include optical, electrooptic, piezooptic, dielectric, thermal, elastic, electromechanical and magnetic.

## APPENDIX

LAUDISE, R.A. The Search for Nonlinear Optical Materials for Laser Communications. BELL. LABS. RECORD, v. 46, no. 1, Jan. 1968. p. 3-7.

This interesting article discusses the achievements of materials research at Bell Laboratories for the development of efficient optical modulators and harmonic generators.

MASON, W.P. PIEZOELECTRIC CRYSTALS AND THEIR APPLICATION TO ULTRASONICS. N.Y., Van Nostrand, 1950.

HARVARD UNIV., CRUFT LAB. Modulation of the Linear Electro-Optic Effect at Microwave Frequencies, BY: MYERS, R.A. Contract No. NR-372-012, Rept. No. 433, Nov. 1963. AD 600 461.

A historical development of light modulation and the limitations of modulation using acoustic, semiconductor, magnetic and electric effects are given. The modulation of light at microwave frequencies using KDP and ADP is studied.

NELSON, D.F. The Modulation of Laser Light. SCIENTIFIC AMERICAN, v. 218, no. 6, June 1968. p. 17-23.

A popular review of several modulation techniques.

NYE, J.F. PHYSICAL PROPERTIES OF CRYSTALS, 1957. London, Oxford University Press.

Formulates physical properties of crystals systematically in tensor notation, providing a common mathematical basis and mathematical relationships between properties.

OLDHAM, W.G. and A. BAHRAMAN. Electrooptic Junction Modulators. IEEE J. OF QUANTUM ELECTRONICS, v. QE-3, no. 7, July 1967. p. 278-286.

The use of p-n junctions, p-i-n junctions, metal semiconductor junctions, and heterojunctions as phase modulators is considered.

PERKIN-ELMER CORP. Investigation of Techniques for Modulating and Scanning a Laser Beam to Form a Visual Display - Final Report. Contract No. AF 30(602)-3122. Jan. 1965. AD 612 725.

An appendix to this report lists 79 electrooptic materials in order of decreasing Curie temperature.

ROSS, M. Visible and Near-IR Laser Modulation. ELECTRO-TECHNOLOGY, v. 83, no. 2, Feb. 1969. p. 21-28.

Popular review of physical principles of modulation.

# APPENDIX

CLEVITE CORP. Reference Data on Linear Electro-Optic Effects. Engineering Memorandum 64-10, by: SLIKER, T.R. May 15, 1964. 9 p.

This review, frequently referred to in the literature, details the important properties of KDP, KDDP, ADP, KDA, ADA, CuCl, ZnS, ZnSe and ZnTe.

SPENCER, E.G. et al. Dielectric Materials for Electrooptic, Elastooptic, and Ultrasonic Device Applications. IEEE PROC., v. 55, no. 12, Dec. 1967. p. 2074-2108.

Relates properties of lithium niobate, lithium tantalate, calcium pyroniobate, strontium barium niobate and bismuth germanium oxide to device performance.

AUTONETICS, Anaheim, Calif. Electro-Optic Projection Study. By: STITES, R.S. et al. Contract No. AF 30(602)-3263. Apr. 1965. AD 617 087.

The electrooptic properties of 25 crystals are compiled in this report.

WALSH, T.E. Infrared Modulation Techniques. ELECTRO-TECHNOLOGY, v. 83, no. 2, Feb. 1969. p. 29-33.

A popular review of magnetooptic, acoustic, electrooptic and semiconductor modulation techniques.

ZHELUDEV, I.S. and O.G. VLOKH. The Electrooptic Effect in Crystals. SOVIET PHYS. CRYST., v. 3, no. 5, Dec. 1959. p. 647-660.

A phenomenological description of both the linear and quadratic electrooptic effects is given and the use of ADP in polarizing interference filters, high voltage measurements, sound-track recording, distance-measuring devices and high-speed optical shutters, is presented.